Safety Engineering
Fourth Edition

John W. Mroszczyk

Editor

AMERICAN SOCIETY OF SAFETY ENGINEERS ❖ ❖ Des Plaines, Illinois

Library of Congress Cataloging-in-Publication Data

Safety engineering / John Mroszczyk, editor. – Fourth edition.
    pages cm
 Includes bibliographical references and index.
 ISBN 978-1-885581-67-9 (alk. paper)
 1. Industrial safety. I. Mroszczyk, John.
 T55.S21512 2013
 620.8'6–dc23
                                   2012037983

Printed in the United States of America
19 18 17 16 15 14   7 6 5 4 3 2

Contents

Preface		v
Acknowledgments		vi
Introduction		viii

Part I	**Attitudes and Standards**	
Chapter 1	Approaches to Safety Richard T. Beohm	1
Chapter 2	Standards and Legislation Richard T. Beohm	13
Chapter 3	Recognition and Control of Hazards Richard W. Stickle	27

Part II	**The Human Element**	
Chapter 4	Work Systems and Ergonomics Daniel J. Ortiz	43
Chapter 5	Personal Protective Equipment Scott E. Brueck	83

Part III	**The Workplace Environment**	
Chapter 6	Environmental Controls Richard T. Beohm	111
Chapter 7	Walking and Standing Surfaces Keith Vidal and Kirk E. Mahan	131
Chapter 8	Egress and Life Safety Kevin L. Biando and Richard T. Beohm	159
Chapter 9	Fire Prevention and Suppression Kevin L. Biando and Richard T. Beohm	181
Chapter 10	Noise and Noise Control Richard T. Beohm and Scott E. Brueck	217
Chapter 11	Explosion Richard T. Beohm and Richard W. Stickle	247
Chapter 12	Radiation Glenn M. Sturchio	259
Chapter 13	Hazardous Materials Richard T. Beohm	271

Part IV   **Equipment Design**

Chapter 14   Mechanical Hazards
                      Steven W. Hays   289

Chapter 15   Electrical Hazards
                      Steven W. Hays and Richard T. Beohm   303

Chapter 16   Tools and Machine Controls
                      Dennis Cloutier and Robert N. Andres   329

Chapter 17   Principles of Risk Assesessment
                      and Machine Safeguarding
                      Robert N. Andres   359

Part V   **Applications**

Chapter 18   Material Handling and Storage
                      Kirk E. Mahan   391

Chapter 19   Wood and Metalworking Operations
                      Michael Lowish   405

Chapter 20   Prevention Through Design
                      John W. Mroszczyk   415

Chapter 21   Case Studies
                      John W. Mroszczyk   425

Appendices
Appendix A   **443**
Appendix B   **444**
Appendix C   **446**
Appendix D   **447**
Appendix E   **448**
Index   **451**

Preface

Like the subject itself, the new fourth edition of *Safety Engineering* encompasses many disciplines. Building on the vision of author Gilbert Marshall, the new edition remains focused on providing the best solutions to preventing workplace injuries. All of the chapters were written by experienced safety professionals with years of management experience in their respective subject areas. Professor Marshall's clear and concise presentation in the first and second editions has left its imprint, which is evident throughout the new fourth edition.

Safety Engineering, Fourth Edition, was edited by Dr. John Mroszczyk, who shares a vision embraced by many safety professionals, believing both: (1) that more attention must be given to safe design of all equipment, tools, consumer products, and building structures, and (2) in the critical role safety professionals fill by virtue of training and experience in recognizing unsafe conditions and acts in all industries. The skill set required of safety professionals is evident in the scope of *Safety Engineering*. The new edition has been extensively updated and revised. A few examples include new information on PPE for exposure to nanoparticle cryogenic liquids, arc flash and electrical hazards, and ionizing radiation. A new chapter, "Prevention through Design," provides key questions for hazard analysis as the first step in design development. All references have been reviewed, and most chapter questions for college instruction are new. Also for the first time, ASSE will make available the set of answers to the end-of-chapter questions to instructors who adopt the book.

A practical, solutions-driven reference, *Safety Engineering*, Fourth Edition, brings the peer-reviewed knowledge and experience of safety professionals to address the most frequently encountered hazards through proven principles and best practices. And those preparing for professional practice will find the concise presentation of occupational safety and health principles a valuable foundation for career development.

Acknowledgments

Many changes have occurred since Gilbert Marshall, Professor Emeritus, University of Vermont, revised the second edition of *Safety Engineering* in 1994. Changes have occurred in world culture, technology, and the way risk and safety are perceived. Safety is as important—if not more so—today as it was in 1982, when Professor Marshall's first edition was published. Interest in the safety-engineering field and its importance in reducing injuries and fatalities have grown since then. There have been national initiatives, a global emphasis on safety, a new focus on human error, and the case for the "business of safety." There is increased focus on *Design for Safety:* the incorporation of safety-engineering principles into projects and products in the early design stages in order to eliminate and reduce injuries and fatalities.

In spite of our rapidly changing society, the basic scientific and engineering principles of safety still apply. Hazard identification, risk assessment, and hazard elimination or risk reduction still require a thorough analysis, good judgment, and creativity.

Each chapter has been reviewed by recognized subject-matter experts and revised as deemed appropriate to make the fourth edition relevant to the safety issues of today and the next decade. A new chapter on "Prevention Through Design" has been added.

The American Society of Safety Engineers is deeply grateful to Professor Marshall for giving the Society the rights to this book and for making this fourth edition of *Safety Engineering* possible. Professor Marshall's book has served the safety profession for the past 30 years and will continue to serve the profession well into the future. We remain indebted to him for his contribution. It is an excellent professional reference for practicing engineers and safety professionals. It also serves as a college text for engineering and safety students, providing a solid foundation in the fundamentals of safety engineering.

The Society would like to extend its sincere appreciation to the following contributors to this new edition who willingly gave their time and energy to make this fourth edition possible. Thank you!

The American Society of Safety Engineers would like to extend particular recognition to **John W. Mroszczyk**, Ph.D., PE, CSP, who, as editor of the fourth edition of *Safety Engineering,* assembled the team of contributing authors and approved all first- and second-draft revisions. He also revised his chapter and wrote a new chapter, "Prevention through Design." Dr. Mroszczyk is a former ASSE Engineering Practice Specialty Administrator and a contributing author and section coordinator for both the first and the second (2012) editions of *The Safety Professionals Handbook,* published by ASSE. His article "Prevention of Fall Fatalities and Injuries," which was published in ASSE's Construction

Practice Specialty Newsletter, *Blueprints,* received the Best Technical Article Award for 2011. Dr. Mroszczyk is a consulting engineer with Northeast Consulting Engineers, Danvers, MA. He received a Ph.D. in Applied Mechanics and a Masters of Science in Mechanical Engineering from MIT.

Robert N. Andres, CPE, CMfgE, BCFE, CSP, Safety and Noise Consultant, Environmental and Safety Associates, LLC, Syracuse, NY and Naples, FL

Richard T. Beohm, PE, ARM, CSP, Consulting Fire Protection and Safety Engineer, Thomaston, GA

Kevin L. Biando, PE, The Biando Group, LLC, Norcross, GA

Scott E. Brueck, MS, CIH, Senior Industrial Hygienist, National Institute for Occupational Safety and Health (NIOSH)

Dennis Cloutier, Consultant, Cloutier Consulting Services, LLC, Lawrenceburg, TN

Steven W. Hays, CSP, Safety and Health Manager, Georgia-Pacific, LLC

Michael D. Lowish, CSP, Manager, Safety and Health, Georgia-Pacific Corporation

Kirk E. Mahan, PE, CSP, Director-Safety, Georgia-Pacific, Atlanta, Georgia

Daniel J. Ortiz, MPH, CSP, Division Chief, Human Systems Integration Division, Director, OSH Programs Office, Georgia Tech Research Institute, Atlanta, GA

Richard W. Stickle, PE, CSP, Principal Safety Engineer, ManTech International Corp, Fairfax, VA

Glenn M. Sturchio, Ph.D., CHP, Radiation Safety Officer, Mayo Clinic, Rochester, MN

Keith Vidal, MA, PE, CXLT, Consulting Engineer, Vidal Engineering LC, St. Louis, MO

Introduction

Safety engineering is a multidisciplinary field where scientific and engineering principles and methods are applied to assess and eliminate hazards. Safety engineers must be knowledgeable in the physical, chemical, biological, behavioral, and environmental sciences, human factors, mathematics, business, training, hazard-identification techniques, risk assessment, and educational techniques. They must also be knowledgeable in various operations (construction, manufacturing, transportation), design, and the basic engineering disciplines (mechanical, electrical, civil, chemical). Interest in safety engineering has grown since the last edition—national safety initiatives, a global emphasis on safety, a new focus on human error, and the case for the "business of safety."

Every day, we hear of serious injuries and fatalities involving buildings, vehicles, machines, equipment (ranging from tools to sports), consumer products (including toys), and even amusement rides, just to name a few. Many of these incidents could have been prevented with safety-engineering interventions. The following are just five real-life examples:

- A US Bureau of Mines study showed that engineers design equipment with little consideration about how the equipment is to be used by the operator, the sequence of use, or which functions are most important or most frequently used. (The typical engineer does not consider human factors and does not recognize the creation of an equipment *system* involving human beings.)

- A roofer died from injuries sustained after falling 30 feet through a skylight fixture. A skylight guardrail or cover would have prevented this fatality.

- A Thanksgiving Eve 2006 chemical-plant explosion woke the town of Danvers, Massachusetts. Approximately 24 homes and 6 businesses were demolished, 70 homes were damaged, and 300 residents within a half-mile radius of the plant were evacuated. A valve left open by an employee filled the building with a vapor. When the thermostat went on, the spark ignited the vapor. An investigation concluded that adequate safeguards would have prevented the explosion.

- A worker in a metal-container company lost a finger in a machine. The machine was inadequately guarded.

- A worker in a glass plant lost a finger when it became caught in a glass-former press while performing maintenance. Workers had not been trained in lockout/tagout procedures for isolating the energy sources of machines to prevent their accidental operation.

Safety engineers know that the most effective way to prevent injuries and fatalities in the workplace is to address hazards in the design phase rather than attempting to manage them after the fact. While this design philosophy ("Design for Safety," aka "Safety in Design") has been around since the early 1900s, interest has increased over the past decade. In the United Kingdom, the codified Construction Design and Management (CDM) Regulations define the role of design professionals in construction-worker safety and health. The 2002–2012 National OHS Strategy, endorsed by Australia's Workplace Relations Ministers Council, includes the elimination of hazards in the design stage.

In the United States, the OSHA Alliance Construction Roundtable is applying the Design for Safety model to construction. *Design for Construction Safety* (DfCS) goes against the traditional approach to a construction project, where safety is managed after the project is underway. Under DfCS, the designer's role is extended to include construction-site safety, constructability, and maintenance. Another initiative, *Prevention through Design*, has been put forward by NIOSH, focusing on research, education, practice, and policy. The goal of *Prevention through Design* (PtD) is the prevention and control of occupational injuries, illnesses, and fatalities across eight industry sectors by addressing hazards early in the design process.

A new look at human error is emerging. All too often, injuries and fatalities are blamed on a convenient alibi—human error. Humans have well-known traits and behaviors that can result in injuries and fatalities, depending on the environment and the situation. Additional hazards can arise from the machine-human-operator, product-human-user, or environment-human interface. By considering human behavior in the initial hazard-identification process, safety engineers can attempt to eliminate human-behavior-related hazards or reduce the risk through engineering design.

The oft-mentioned *business* of safety is also getting more and more attention. Lost productivity, workers' compensation costs, and wage payments from workplace injuries and illnesses cost companies billions of dollars each year. The median days away from work for goods-producing industries is nine days, while many work cases are a month or more. Clearly, the application of safety engineering has the potential to save companies (and the overall economy) billions of dollars.

The safety-engineering profession has the knowledge and skill to reduce injuries, illnesses, and fatalities resulting from work- and non-work-related activities. Alignment and harmonization of Design for Safety efforts within the United States and globally will reduce occupational injuries and fatalities worldwide. Management and owners should come to realize that investing in safety early on will have payback down the road in reduced maintenance, workers compensation, and wage costs.

This new fourth edition of *Safety Engineering* provides an excellent professional reference for practicing safety engineers and safety professionals in performing the work resulting in these cost savings—and to the ultimate goal of reducing and eliminating occupational injuries. It also is an excellent text for safety and engineering students.

Dr. John W. Mroszczyk

Approaches to Safety

Richard T. Beohm, PE, CSP, ARM

Background

The desire to maintain the well-being of self, family, and friends is evident throughout the history of the human race. Even animals, with their survival instincts, have a strong sense of self-preservation that extends to offspring and sometimes to mates. Despite this natural tendency toward self-protection, however, people have also invariably demonstrated a willingness to take chances in exchange for possible gains. Ancient cave dwellers no doubt had to take risks to kill wild animals for food.

During the early period of industrial growth both in Europe and in the United States, the concept of "common law" that originated in medieval times continued to develop. Common law is law that grows from custom and precedent rather than from the enactments of a legislative body. In the matter of safety and health in the workplace, reference was made to three parties: the master (or the owner/employer), the servant (the employee), and the stranger (a visitor or anyone else on the premises, invited or not). It was generally felt that when a person accepted employment, he or she also accepted any risks involved in the job. The master should have the courtesy to point out the hazards of the job, but even that was not required: a servant should be smart enough to avoid danger.

There were many cases in which it was not clear whether an injured worker was in fact a servant (employee) or someone doing business with the employer. This question became more serious when workers' compensation laws were written. If a worker could be classified as an independent contractor, he or she was not considered an employee and was, therefore, not eligible to collect workers' compensation benefits. Other problems concerning the legal status of workers became apparent at this time. A person without fixed working hours was considered a "casual" worker rather than a "regular" employee and, therefore, could not collect insurance benefits. In 1924 a girl between 14 and 16 years old was working in a manufacturing plant in Vermont. She had not deposited with her employer an employment certificate required by a child-labor law and, therefore, was not entitled to workers' compensation benefits when she was injured.[1]

A stranger was often treated with greater concern than a servant. The rationale behind this was that a stranger on the premises presumably had no knowledge of the potential hazards within the plant, but an employee was there by his or her own choice and had accepted the risk of working there.

Under common law, workers were expected to assume some responsibility for the safety of their fellow workers, although they were seldom actually held responsible in case of an accident. Then, as now, workers were usually afraid of losing their own jobs if they intervened on behalf of their fellow workers.

Most of our present-day laws concerning labor, safety, and health have their roots in the common law of centuries past. What has changed a great deal over the course of time is the level of concern for the welfare of the worker and his or her family. Today, a court decision involving compensation for injuries suffered in an accident will usually favor the worker or injured party rather than the employer or property owner. In a 1978 case in Vermont, for example, a man skiing at a ski resort got off the well-traveled trail and hit some underbrush, according to the court testimony. He was paralyzed as a result of the accident. The court awarded him $1 million.[2]

Building codes and other safety codes adopted by states and cities place responsibility on the owner of the building or property, although the occupant or lessee obviously has some control over their condition. Most leases and rental contracts contain clauses stating that the lessee must exercise reasonable care in maintaining the condition of that property. The owner may bring the lessee to court to enforce this clause. However, the owner is invariably named as the liable party by state laws and city ordinances.

Similarly, in the workplace, the employer is always the one responsible for observing the laws and standards of occupational safety and health. Some of these standards are performance standards; that is, they relate to the way a task is done, as distinguished from specification standards that relate to physical conditions. Performance standards are satisfied or violated by the worker. Management must provide proper training and supervision in order to control the worker's performance, but that, of course, does not guarantee that the worker will observe the standards. Nevertheless, the employer is responsible for the actions of the employees who are on the job. If an inspector from the Occupation Safety and Health Administration (OSHA) finds an employee violating a standard enforced by that agency, the employer may be fined. The employer may in turn impose sanctions upon the employee at fault, but the best way to avoid fines is to make the observation of safety and health standards a condition of employment.

Terminology and Definitions

The terms "safe" and "safety" are part of our everyday vocabulary, but an exact and complete definition of them is difficult to set forth. One may say that something is safe if it is free of hazards or free of recognized hazards. Such a definition may suffice for casual conversation, but technically it is not adequate. Nothing is really free of hazards, and a hazard may be present without being recognized. An object may be considered foolproof, meaning that there is no way to misuse it, but, again, nothing is really foolproof.

The real issue seems to be how nearly foolproof or free of hazards something must be before we are willing to call it safe. It must be kept in mind that a condition we might call "safe" in one situation might not be safe at all in another.

Dr. William W. Lowrance is given credit for a widely recognized definition of safety. In a presentation at the American Chemical Society national meeting in March, 1977, he stated: "For most purposes I find it useful to define safety as a judgement of the acceptability of risk, and risk, in turn, as a measure of the probability and severity of harm to human health. A thing is safe if its attendant risks are judged to be acceptable." But this definition is not unambiguous either, and Lowrance himself raises questions about it. What is meant by "acceptable risk"? To whom is the risk posed? By whom is it judged acceptable? A condition that is acceptable to the employer may not be acceptable to the employee, and vice versa. On many occasions employees have elected to work in hazardous ways or conditions even when specifically told not to by a supervisor. And on many occasions employers have instructed employees to work in dangerous manners or environments against the will of the employees.

Differences of opinion about what is safe may be the result of selfish interests, but they may also be honest and sincere. Such differences of opinion have often been the starting point of an effort to establish a standard. A standard, to be discussed further in Chapter 2, is a statement of a condition or level of performance that is acceptable to all concerned and that is then used to evaluate conditions and performance.

Another point of controversy in Lowrance's definition is the measurement of safety in terms of human health, to the exclusion of property. Safety is often evaluated this way because human life and well-being are generally considered more valuable than material possessions. But unsafe conditions or performances also result in accidental damage to property, which, it must be remembered, does have some value. Elsewhere, Lowrance defines risk as a "measure of the probability and severity of adverse effects."[3] This definition would include property damage as well as human injury.

Other terms and definitions have been discussed at great length. Lawrence B. Gratt, Chairman of the Society of Risk Analysis, Committee for Definitions, presented a paper titled "The Definition of Risk and Associated Terminology for Risk Analysis" at a special session of that committee in 1987. The terms "risk analysis," "risk assessment," and "risk management" are often used and were discussed at that meeting, but they have slightly different meanings to different people. There are no simple definitions. Gratt's committee finally arrived at an acceptable definition of risk: the probability of an adverse effect to human health, property, and the environment, and the severity thereof.

Risk management, of which risk control and safety play an important part, consists of the following: identify, assess, prioritize, reduce or mitigate, and prevent all levels of risk in the EH&S system. The ISO 31000 Standard provides principles and guidelines on the implementation of risk management. Risk management is a key element in OSHA's 18001 and ANSI Z10 Management Systems.

Another term used here is "accident." An accident is commonly defined as an unplanned and undesirable event that interrupts a planned activity and that may or may not result in injury or property damage. It should be remembered, however, that an accident always has the potential to cause injury or property damage. By this definition, unintentionally dropping a wrench is an accident. If the wrench hit only the floor, it would cause no injury and probably no property damage. If while on the way to the floor the wrench hit a delicate instrument on a bench and then hit the worker's toe, it could cause both injury and property damage.

During the last few decades, an increasing number of court cases involving work-related injuries and illnesses has caused a rethinking of what constitutes an accident. Because of the wording of workers' compensation laws and other insurance and liability laws, the courts have broadened the meaning of the word "accident" to cover illnesses that result from exposure to a hazardous environment. An accident then becomes any unplanned and unexpected event that causes injury or illness. This definition may be suitable for court litigation, but it does not satisfy the needs of a safety engineer.

Safety engineering is the application of basic scientific and technical principles to the mitigation or reduction of loss of life, property, and the environment by recognized hazards. It requires a broad knowledge of mathematics, chemistry, physics, and other basic sciences and a familiarity with one or more of the recognized branches of engineering.

Safety engineers are usually qualified by obtaining at least a four-year technical or engineering degree, as well as professional certification (CSP) or professional registration (PE). In safety engineering, it is useful to distinguish between an accident and an exposure; they are treated differently. It is good to be aware of the legal interpretation, but throughout this book, "accident" will be used in terms of the earlier definition.

The terms "injury" and "illness" also need to be distinguished. In safety engineering, an injury is a bodily impairment resulting from an accident. The impairment is immediate; it occurred at a fixed time and place. An illness, on the other hand, is a bodily impairment resulting from exposure to a harmful substance or environment. It does not occur immediately and is not evident until sometime after the exposure. Generally the exposure occurs over a period of time and has a cumulative effect.

There are situations, of course, that do not fall neatly into either of these classifications. How quickly is immediate? A mosquito bite may happen in a second but not be noticed for a few hours. Is that an injury or an illness? Despite their lack of preciseness, however, these definitions are generally adequate for our purposes and will be used accordingly.

Perhaps the major point to be made in this rather academic discussion of terms is that it is far more important to reduce accidents than merely to reduce injuries. In the words of a current saying, "Safety is no accident." If we can eliminate accidents and hazardous exposures, we can eliminate injuries and illnesses; the reverse is not necessarily true. This important concept needs to be emphasized: *the ultimate goal of all efforts in safety engineering should be to reduce accidents and harmful exposures.*

Cost of Accidents

Just how important is it to reduce accidents? Is the cost of accidents so great that industry should expend the time, money, and effort required to create safer work environments? A few statistics will help to establish a better understanding of the cost of work-related accidents.

There are two major sources of statistics on accidents and their costs in the United States. The data they present do not always coincide because they are obtained from varying sources and may be tabulated in slightly different ways. One source is the Bureau of Labor Statistics (BLS), an agency within the U.S. Department of Labor. The other is the National Safety Council (NSC), a private nonprofit organization. The data presented here are from the NSC report, *Injury Facts*, 2010 edition, and pertain to 2008 on-the-job accidents. Table 1-1 shows the cost of disabling unintentional injuries. The NSC adopted the BLS's Census of Fatal Occupational Injuries (CFOI), which was implemented nationwide in 1992.

The point to be made here is that enormous losses are sustained because of accidents, and industry could afford to spend an amount at least equal to these losses in an effort to reduce accidents. These figures, of course, take into account only the tangible, dollar-value aspects of accidents, but such factors as reduced efficiency and productivity, as well as human suffering, should be considered as well. It is extremely difficult to make a meaningful evaluation of many of these factors.

Much has been done in recent years to analyze accidents and to extract this needed data. But these techniques require a great deal of time, and that means added cost. In general, management has chosen to ignore everything but readily available hard data, and, sadly, some choose to ignore even that when it is available. Far too often the excuse given for failing to attempt to reduce accidents is that it costs too much. An advertisement for a prominent machine-tool manufacturer contained the statement: "If you need a new machine tool, you are already paying for it." A paraphrase of this statement might be: "If you need to make a greater effort to reduce accidents, you are already paying for it."

An important aspect for the safety professional to consider is the "business of safety" and how safety can pay in the long run when it is positively integrated into corporate management. An effective safety management system is critical to the success of any organization. It benefits the company's bottom line and improves its overall safety culture. The business of safety is an emerging and important concept in today's world of doing business. EH&S should be an integral part of a company's business operation to function in a safe and responsible manner in today's global markets. It is a new paradigm to reduce fatalities and injuries.

Table 1-1. Cost of disabling unintentional injuries

According to the National Safety Council's 2010 edition of *Injury Facts*, preliminary data showed the number of disabling unintentional injuries reached 3.2 million in 2008, while 4,303 fatalities occurred. That translated to:

110 million lost workdays—70 million from injuries.

An estimated total cost of $183 billion. Of that amount, wage and productivity losses were estimated at $88.4 billion, medical costs at $83.3 billion, and administrative costs at $37.7 billion. In previous years, the number of days lost for permanent disabling injuries was 40 million.

The total per worker for loss of productivity and injuries was estimated to be $1,250.

The cost per disabling injury was $48,000 and the cost per death was estimated at $1.3 million.

The 2,000 companies that participate in OSHA's VPP program showed a 50 percent reduction in injuries, which reflects in reduced workers compensation premiums. IBM, G.E., General Motors and the American Foods Group are good examples of creating a sustainable safety culture into their organization.

Safety professionals need to show management that besides injury cost, there are also indirect costs such as loss of productivity, retraining and legal costs, as well as the effect on morale. For every dollar spent on injury-related direct costs, $3–5 are spent on indirect costs. Liberty Mutual Insurance reports that, in 2009, injuries and illnesses in the workplace cost $50.1 billion in workers' compensation.

Safety professionals need to promote a positive culture of safety to corporate management and develop a safety leadership program to educate workers and give them the skills they need to promote working conditions among fellow workers. They should consider the ANSI/AIHA Z10 standard in integrating EH&S culture into the operations management systems.

Cost-effectiveness studies and cost/benefit analysis are largely tools of management, and they apply to accident prevention as much as to any other area of management. The cost-effectiveness of accident prevention is the subject of several books and articles, a few of which are listed in the bibliography at the end of this chapter. Often an engineer can persuade management to accept a safety proposal by proving through a cost/benefit analysis that it is worthwhile.

Insurance carriers and brokers, especially those with industrial clients, usually have a property and casualty loss-control department with representatives available to help their clients analyze accidents and find ways to reduce or prevent them. Many companies now include in their organization a Safety, Health, and Environmental department or individual. This is an excellent field for engineers to enter. Safety, health, environmental, and fire protection activities are closely aligned to those of risk management within the company or agency.

Attitudes Toward Safety

The approach to safety taken by top management in a company determines to a large degree the approach that will be taken by subordinates. In reviewing several case studies, C. Donald Schott noted that all the success stories had certain factors in common, namely:

1. All managers were willing to do something to eliminate the given problem, even though proposals did not always seem feasible to some.
2. Management aggressively attacked the problem.
3. Management was willing to restudy the manufacturing process totally and not be bound by tradition.
4. The least expensive idea was tried first.[4]

A study made for the National Institute of Occupational Safety and Health (NIOSH) compared the safety practices of industrial companies with exceptionally good safety records with the practices in average companies.

The conclusions correlate well with those of Schott. Among the successful group, the report emphasized, "safety in each instance was a real priority in corporate policy and action." Elements of success were summarized as follows:

1. A strong management commitment to safety expressed not only through stated policy and adequate financial support, but also through active involvement in the program's implementation and demonstrated concern for worker well-being;

2. Efficient hazard identification, engineering control, job-safety training, and safety evaluation programs designed to anticipate and manage hazards, not just count and investigate accidents (after the fact);

3. An effective employee communication, feedback, and involvement program designed to motivate management and employees to deal with one another and safety problems in positive, humanistic ways; and

4. A safety program that is integrated into the larger management system and is designed to deal with safety as an intrinsic part of plant operations.[5]

Managers have exhibited a wide range of attitudes toward safety. At one extreme are those who resist safety measures and defy those who try to enforce them. The owner of a small printing company was asked what safety training he provided for his workers. He replied, "When I hire a man I tell him, 'If you lose your fingers, I can't use you anymore.'" Some managers, particularly in small companies, take a stand of noncompliance regarding OSHA standards, sometimes simply because they don't like being told what they must do or must not do.

Other managers express indifference to safety, claiming to leave it up to the workers. This is probably the worst approach a supervisor can take because it gives a false impression to the workers. The message is, "You don't have to worry about safety. We've been doing it this way for years." This attitude belongs in the same category as the notion that accidents always happen to someone else!

Many managers have learned that an emphasis on safety at the highest level is worthwhile. Several of the largest U.S. corporations have had effective safety programs for a long time. It is very interesting to note that every company with an exceptionally good safety record is a company in which everyone, beginning with top management, has become safety-conscious.

International Business Machines Corporation issues a safety and health manual to all its managers in which the chairman of the board states: "IBM's commitment to safety is every bit as important as our commitment to excellence, to outstanding service, to respect for the individual. There can be no compromise with safety. Together, we must make IBM a safe place to work all the time, for everyone." Such a message from the chairman of the board of directors adds particular significance to any safety program.

One might expect that production workers would be more concerned about accident prevention than any other personnel in a plant. However, this does not seem to be the case, probably because of ignorance of the hazards, for the most part, and of the magnitude of potential accidents. Sometimes a person who is aware of a hazard still takes a chance, whether to experience a thrill, to show off and impress someone, to retaliate for a reprimand, to attract attention, or to earn extra pay. Psychologists tell us that each of us will experience some of these feelings at one time or another. Most people occasionally do something potentially unsafe—ski jumping, car racing—just for the thrill of it. Young people often take needless chances in the face of danger. They sometimes feel immortal, not having experienced some of life's tragedies. Moderation and caution come with age as lessons are learned. Senior citizens sometimes become accident-prone due to their diminished capabilities and senses, attributable to their aging as well as to life experiences and habits that were acquired during a simpler time.

A few case studies will also illustrate this willingness to take chances with safety. A young woman working on an incentive plan was operating a small power press to assemble several metal parts. The workplace was quite well designed and enabled her to use both hands conveniently on the two-hand control of the press. She found, however, that by taping down one control she could use her free hand to prepare a part for the next cycle and, thereby, gain a few seconds. She was repeatedly instructed to obey the safety rules, but she persisted until threatened with the loss of her job.

In another case, a worker reported to the nurses' first aid station with an earache shortly after beginning work. He said he had considered staying home but had finally decided to come in anyway. A nurse examined the ear, applied some medication with a cotton swab, and sent him back to work. His job that morning involved cleaning the inside of the compressing chamber of a hydraulic baler. The baler ram was in "hold," but for some reason it began drifting slowly upward. The worker's chin was caught by the moving ram, and he was raised up with it until his head was crushed against the top of the chamber. He then dropped to the bottom of the baler pit, where he was found dead.

In this case the elements are not so clear, but many questions are raised. What was the worker's attitude that morning? It was suggested that he was depressed or despondent and that he wanted attention. Why didn't he notice the moving ram? On the other hand, why wasn't someone assigned to work with him in that confined space? Why wasn't the power turned off while a worker was inside the chamber? Was it the man's attitude, or unsafe practices sanctioned by management, that caused this accident?

The attitude of labor union leaders is important in industrial safety, too. When unions were first being organized in the 1800s, the creation of safe working conditions was a major issue. Samuel Gompers and John L. Lewis made safety and the welfare of the workers their primary concerns.

In 1978–79, under the leadership of George Meany, the AFL-CIO established the Workers' Institute for Safety and Health as a research and education organization within the labor movement. The union's major emphasis has always been on wages and fringe benefits, but this organization facilitates the promotion of better safety and health conditions as well. In an interview with Michael Flannery of the *Chicago Sun Times*, Mr. Meany quipped, "What the h— good are wages if you are going to lose your health in carrying out your occupation?"

Not all union leaders have taken this approach. One union official who was the safety committee chairman of his local union expressed concern that since wage scales are often based, at least in part, on the hazards inherent in the job, the guarding protection required by OSHA safety requirements could result in a reduction in wages.

A bitter labor dispute in the automobile industry in the mid-1970s brought attention to some rather poor working conditions. The union was able to get quite a lot of publicity and newspaper coverage on the issue. But when the new contract was signed, all the gains were in wages and fringe benefits; the demands relating to working conditions had been dropped. Apparently, worker safety was not as important as monetary gains, at least to the union representatives.

Another element in workers' attitudes toward safety is their off-the-job activities. Managers should take an interest in this for two reasons. First, if a worker is hurt off the job, he or she is just as disabled as he or she would be if the accident had occurred on the job. Second, attitudes developed off the job carry over to work, and vice versa. Similarly, worries and concerns about personal or family matters can affect a worker's ability to concentrate on the job.

As part of their safety program, many companies provide their employees with copies of *Family Safety and Health* magazine, a quarterly publication of the National Safety Council. One steel company reported that within two years of starting a family safety program, they could count 12 lives saved as a direct result of the program. Some companies and governmental entities offer their employees various kinds of wellness programs to promote health (both physical and mental), fitness, and safety, both at home and on the job. The results of an effective program should yield more productive employees as well as fewer workers' compensation and health claims and days off due to sickness and injury.

Whenever the managers of a company decide to initiate a new program, they have to take human nature into account. One of the most salient features of human behavior is a resistance to change. One would think that a change for the benefit of the worker, such as an improvement in working conditions, would immediately be accepted, but this isn't always the case. People are especially hesitant to accept change if it's someone else's idea and they had no part in developing it. For a safety program to work, the ideas and changes must have not only the endorsement of management but also the acceptance and participation of everyone in the workplace.

The world is now experiencing a fast track of change, some of it good and some bad. Some of the reasons for this are the increase in technology, instant worldwide information and communication, a drastic increase in world population with migrations across borders, natural disasters due to climate change, and an increase in man-made environmental hazards affecting our water, air, and soil.

Engineers and architects are now incorporating the sustainable building design concept to increase the efficiency of natural resources while reducing the impact on human health and the environment throughout the building's life cycle. New buildings are attempting to achieve the "green" USGBC Leadership in Energy and Environmental Design (LEED) Certification, a recognized standard in measuring building sustainability, to improve human health and welfare.

The September 11, 2001, terror attack on the NYC twin towers and subsequent attacks and threats of attacks have had a profound influence on the culture of safety and security around the world. It has led to the importance of emergency response planning and upgrades to the existing National Warning System, such as the Integrated Public Alert and Warning System (IPAWS). FEMA and Homeland Security now play a larger role since the flooding of New Orleans, wildfires across America, and the possibility of weapons of mass destruction.

The economic and financial recession of 2008 is a major concern that impacts safety. Many companies are going "lean," and safety and quality are taking a backseat to profits. In March 2010, 9.7 percent of U.S. workers were unemployed. The Bureau of Labor Statistics reports that of the 300 million people in the United States today, 63 million are over 65 years old and 10 million are still working. Engineers and architects must consider an aging population in their designs. The Americans with Disabilities Act (ADA) requires conformance to such standards as ADAAG/ANSI A 117.1 in their designs.

Another factor is the immigration or migration of workers across borders to seek work and a better life. This has a profound affect on safety in the United States due to the ethnic work culture and language barrier. The NSC reports that in 2008, 316 Hispanic construction workers were killed on the job. The incident rate for workers killed on the job in 2008 was 9 deaths per 100,000 workers, about the same as for non-Hispanic whites.

As the world becomes smaller and changes constantly at an even faster pace, there is a movement called "cultural change" in doing things differently in an effort to move away from the old model and toward a more business-driven model that embraces flexibility, self-determination, and empowerment at the front-line staff. The "culture of safety" is also driven by many factors, including politics, global and environmental concerns, climate change, economic recession and job loss, codes and standards, litigation, and insurance premiums. Safety culture is a subcomponent of corporate culture and is affected by other operational processes and systems. It is an important factor for corporations as it affects virtually all other elements of an organization, including production, quality, job satisfaction, and expenses.

Safety culture change can be created to be positive, proactive, and participative. Sustainability drives corporate social responsibility (CSR) changing safety culture by engaging top management to blend business models with safety. It should be integrated into daily business processes and focused on leadership and involvement. CSR is a growing practice worldwide by companies that wish to have both sustainability and an environmentally responsible operation and reputation. The safety engineer/professional should be a part of the company's corporate culture as an active participant like other social and business movements such as "green," "lean," and "sustainability." It should be integrated in all endeavors from the conceptual phase to commissioning to maintenance and operation in order to prevent or mitigate losses and reduce risk.

Preventable disasters are still occurring, as chemical industrial explosions like the BP refinery in Texas in 2005; the BP oil platform Gulf explosion, fire and environmental catastrophe in 2010; and the 2008 sugar dust explosion in Georgia, to name a few.

Engineers are obligated to protect public health, safety, and welfare and are, in essence, safety engineers and professionals.

Engineering students require a broad technical education, which also should include such topics as workplace safety, safety codes and standards, and the principles of safety engineering. Chemical engineering undergraduates need to know hazards of adverse chemical reactions in order to prevent disastrous chemical explosions such as occurred at the BP refinery. Engineers and safety professionals need to continuously learn to keep current with new technology and best practices of their profession. PE boards and the Board of Certified Safety Professionals (BCSP), which credentials SHE practitioners in order to enhance the safety of people, property, and the environment, require a number of continuing education units each year for competency, which can be obtained through professional development and continuing education seminars, courses, writing articles, teaching, correspondence courses, and being active in trade and professional societies.

The editor of a safety and health publication once asked readers, "Why do accidents happen?" A great many responses were received. About 30 of these were selected as being representative. They were all from persons with experience in this area.

Nearly half of these letters listed attitudes as the major cause of accidents. Some went on to state ways in which attitudes toward safety, good or bad, are developed. All of these were factors which are controlled by management and include the following:

1. Training of employees
2. Communication between employees and management
3. Work environment
4. Performance appraisal
5. Housekeeping practice
6. Maintenance of tools and equipment

These factors and the attitudes they foster should be considered key factors in a safety program. Furthermore, it should be stressed that it all starts with the people in top management, who set the mood for everyone else in the company.

QUESTIONS

1. Name some of the driving forces that affect the culture of safety in today's world.
2. What is safety engineering, and how does it fit into the "culture of safety" concept?
3. Name several advantages of integrating the business of safety into a corporate culture.
4. Name several disastrous consequences of putting profit before safety.
5. What are three aspects of risk management, and how are they important to EH&S?
6. Name several means for the safety engineer/professional to maintain his/her competency.
7. List four attitudes that could contribute to or cause an accident.
8. Name some characteristics or types of people that could put them at risk for an accident.
9. How does an accident at home affect a person at work? How does it affect the employer?
10. What is the difference between an injury and an illness? Give several examples of each.
11. What is the ultimate goal of all safety engineering efforts?

12. How can safety engineers reduce or eliminate accidents and hazardous exposures?
13. Name two elements for successfully reducing on-the-job accidents.
14. Name at least two factors as a major cause of accidents.

NOTES

1. Wlock v. Fort Dummer Mills, 98 Vt. 449, 129 Atl. 311 (1924).
2. Sunday v. Stratton Corp., 136 Vt. 293 (1978).
3. William W. Lowrance. *Of Acceptable Risk* (Los Altos, CA: William Kaufmann, Inc., 1976): 94.
4. C. Donald Schott. "Can You Afford It?" *Professional Safety* (December, 1979): 33.
5. U.S. Department of Health, Education and Welfare, NIOSH. *Safety Program Practices in Record-Holding Plants*, Publication No. 79–136 (Washington, D.C.: Government Printing Office, 1979).

BIBLIOGRAPHY

Asfahl, Ray. 2009. *Industrial Safety and Health*. 6th ed. New York: Prentice-Hall.

Bonin, James J., and Donald E. Stevenson, 1989. *Risk Assessment in Setting National Priorities*. New York: Plenum Press.

Gloss, David S., and Miriam G. Wardle, 1984. *Introduction to Safety Engineering*. New York: Wiley.

Grimaldi, John V., and Rollin H. Simonds, 1988. *Safety Management*. 5th ed. Homewood, IL: Irwin.

Haight, Joel, ed., 2012. *The Safety Professionals Handbook*. 2nd ed. 2 vols. Des Plaines, IL: ASSE.

Hammer, Willie, 2000. *Occupational Safety Management and Engineering*. 5th ed. New York: Prentice-Hall.

Isenhagen, Susan, 1988. *Work Injury: Management and Prevention*. Rockville, MD: Aspen.

Kletz, T., 1988. *Learning From Accidents in Industry*. Stoneham, MA: Butterworth.

National Safety Council, *Safety and Health*. Skokie, IL.

2

Standards and Legislation

Richard T. Beohm, PE, CSP, ARM

Introduction

A large part of our daily routine is based on standards of one sort or another. A standard is an accepted or agreed-upon rule for the measurement of a quantity or quality. We use standards of time, distance, mass, weight, value, and so on, as well as standards of performance. You must achieve a certain grade point average to graduate from your school!

Many standards involve specific quantities—for example, 60 seconds in a minute, 1,000 millimeters in a meter. Others do not involve quantities but are just as specific—the colors in the U.S. flag are red, white, and blue. Still others may refer to quantitative entities without specifying exact quantities—a coffee break may be taken at midmorning; you must study math until you know it well. Some rules, such as this last one, hardly seem rigorous enough to be called standards at all. Almost anything can be called a standard if it prescribes a condition or level of performance and is accepted by those affected by that condition or performance.

Development of Safety and Health Standards

Over the years many groups have developed standards to protect the safety and health of themselves or others. As industry grew in Europe in the nineteenth century, a concern for the safety of workers increased, although very little was actually done to promote or enforce standards. The earliest laws awarding some benefits to injured workers were enacted around 1840, but it wasn't until the 1890s and early 1900s that a body of work-related law began to appear.

In the United States, as in Europe, the first laws dealt with compensation rather than accident prevention, but interest was developing in the establishment of codes and standards. Two tragic boiler explosions—one in 1854 in Hartford, Connecticut, and one in 1865 on a Mississippi River steamboat—led to the formation of a "club" of people concerned with this hazard. The Hartford

Steam Boiler Inspection and Insurance Company was formed in 1866. Its inspection of boilers was very successful and prompted many manufacturers of steam boilers to express an interest in establishing a uniform set of construction rules. This eventually led to the development of the "Rules for the Construction of Stationary Boilers and for Allowable Working Pressures," a set of standards that was adopted by the American Society of Mechanical Engineers in 1915. These boiler codes have been revised and are still used today.

The development of these codes is typical of the development of safety and health standards. The first step is a recognition of the need for guidelines for the design and operation of equipment. Frequently this recognition occurs only after many people have been injured or killed in a serious accident. Guidelines usually originate with individuals or a small group of people whose ideas are then adopted by a trade group, such as the American Welding Society or the American Boiler Manufacturer's Association. Some standards, like the boiler codes, have been developed by professional organizations. Some standards are developed for a specific situation and are not appropriate beyond that area; others have wider applications and may be adopted by several industries. As they become generally accepted, standards and codes are often adopted by organizations with a wide range of membership.

The American Society of Agricultural Engineers (ASAE) has a long list of standards used by manufacturers and operators of all types of agricultural equipment.

These standards are all consensus standards, but organizations like the ASAE often have two categories of codes, which they call "standards" and "recommendations." A company with membership in ASAE is expected to comply with the ASAE standards, but may follow the recommendations at the discretion of the management. Of course, there is no law enforcing the ASAE standards, but a company enhances its reputation when it claims to follow the industry standards.

To achieve greater credibility, availability, and acceptance for their standards, some groups seek to have their codes adopted by a national standards organization. In the area of safety and health there are two predominant national standards organizations, the American National Standards Institute (ANSI) and the National Fire Protection Association (NFPA). ANSI originated in 1918 when several major engineering societies formed an association for the purpose of formulating industrial standards. It was first known as the American Engineering Standards Committee. In 1928 it was reorganized as the American Standards Association (ASA). Later the name was changed to the United States of America Standards Institute (USASI), and finally to the American National Standards Institute. Occasionally the acronyms ASA and USASI are still found in literature. ANSI currently has about 1,400 standards.

ANSI does not originate or develop standards, but adopts standards from other organizations in a wide range of fields. ANSI standards must not only be proper but also of sufficient significance to merit adoption as national consensus standards. Since ANSI adopts standards from other organizations, a given standard may be published by two organizations under different numbers.

Unlike ANSI, NFPA has developed many of its own standards and has also limited its scope. All NFPA standards are, in one way or another, related to protection from fire, explosion, and related hazards. Among the best-known NFPA standards are the National Electrical Code (NEC) and the Life Safety Code (LSC). Currently there are 300 NFPA standards.

Another organization that is gaining importance is the International Organization for Standardization, or ISO, which consists of member organizations (not individuals) from all the major nations of the world. The representative member from the United States is ANSI. Since no international body exists to enforce them, ISO standards are, of course, consensus standards, but ISO recommendations have been adopted into law by many governments. As the exchange of information among countries increases, the significance of ISO grows.

When a standard is proposed to ISO by a member nation, it is discussed and evaluated. It may then be adopted as proposed, modified and then adopted, or rejected entirely, but in any case it maintains its status in the originating country. ANSI has proposed many standards to ISO, some of which have been adopted as proposed and many of which have been adopted after modification. ANSI may adopt any ISO standards for use in the United States or retain a different one.

The ISO and European Community Standards and Directives for safe products are now emerging as being essential for global-international trade and economics. European Community Standards and Directives are the CE mark certification, DIN 24980, EN (European Norm) 292, EN 1050, EN 954, BS 5304, and others.

A number of other federal statutes were enacted, such as the Architectural Barriers Act of 1968; the Individuals with Disabilities Education Act of 1975; the Fair Housing Act of 1968, amended in 1988; and the Americans with Disabilities Act of 1990, which incorporated ANSI A117.1 and the Uniform Federal Accessibility standards.

Terrorist attacks and shootings have led to a number of code upgrades and standards for increased safety and security. NFPA 72-2010 now addresses Mass Notification Systems (MNS) and Emergency Communication Systems (ECS), a part of MNS. The U.S. Department of Defense (DOD) now requires for their buildings antiterrorism codes such as UFC 4-010-01 and UFC 4-021-01 (MNS), which are available on the web.

It must be remembered that even at the national or international level, these standards are only recommendations to be followed voluntarily. They become mandatory only when adopted by a city, by a state, or by the federal government. Today there are about 94,000 U.S. standards, of which 41,500 are private-sector standards.

A growing number of standards are requiring risk assessment. ANSI standards that require risk assessment include Z244.1, Control of Hazardous Energy Lockout/Tagout and Alternative Methods; the B11 machine tool series, B155.1, Requirements for Packaging Machinery and Packaging-Related Converting Machinery; and R15.06, Industrial Robots and Robot Systems: Safety Requirements. In addition, ANSI/AIHA Z10, Occupational Health and Safety Management Systems, specifically calls for SH&E, including risk assessment, to be considered in designs. NFPA standards associated with combustible dust also incorporate risk assessments.

Mandatory Safety and Health Standards

Former Chief Justice of the United States Supreme Court Earl Warren said in an address in 1962, "Society would come to grief without ethics, which is unenforceable in the courts, and cannot be made part of law." Why, then, is it necessary to make laws about things that should be a matter of ethics? In the area of safety and health, is it necessary to construct and enforce standards and codes? It seems that although people in positions of responsibility should consider the welfare of others as a matter of conscience, they frequently fail to uphold standards of safety and health, either from ignorance or from avarice.

The fact that good safety and health practices have been violated by many people is evident from the statistics mentioned in Chapter 1. To compel observation of the standards and codes that have been developed, many city, state, and federal laws have been passed. Building codes have been enforced by cities and states for many years. Other standards commonly enforced by states, and sometimes by city ordinances as well, concern boilers, drinking water, elevators, ski lifts, mines, and highways. Of course, many violations of these laws continue to occur, but offenders can be punished. There can be no doubt that such state and local laws have helped improve the safety and health of our society.

On the national level, relatively little legislation was enacted until the 1960s, when such laws as the Federal Metal and Nonmetal Mine Safety Act, the Construction Safety Act, and the Coal Mine Health and Safety Act were passed. Prior to that, with the exception of the Walsh-Healey Act in 1936, legislation was meager. In 1893 the Federal Safety Appliance Act went into effect, requiring railroads to install available safety appliances. The Federal Bureau of Mines was created in 1910 but was not permitted to enforce safety standards until 1952.

In June of 1936, Congress enacted a law now known as the Walsh-Healey Act. This act applied to the manufacturers and suppliers of materials and equipment, exclusive of agricultural commodities, selling to any federal agency. One section of this act stipulates that "no part of such contract will be . . . manufactured or fabricated in any plants, factories, buildings, or surroundings, or under working conditions which are unsanitary or dangerous to the health and safety of employees. . . ." Although this legislation provided for the establishment of standards, enforcement was weak, and the number of workers protected was rather limited. Nevertheless, it did raise some concern among employers, and employees as well, and fostered discussion of standards, particularly those concerning limitations on noise.

The Walsh-Healey Act, as an example of legislation designed to improve working conditions, raises some interesting questions. As Chief Justice Warren noted, ethics cannot be enforced. Whether or not standards such as those adopted in the Walsh-Healey Act are matters of ethics may be argued, but some people react to them as if they are. They do not like being told what they must or must not do about such things. They sometimes regard the law as punitive. People have often spent more time and effort, and probably more money, in trying to find ways to avoid obeying the law than they would have in complying with it.

It has been noted that the person most influential in creating safe or unsafe working conditions is the same one who controls the company's money. This is not always so, of course, but cost is the excuse most often given for not improving unsafe working conditions. As we have seen, the cost of providing a safe and healthful workplace is more than recovered by improved productivity, lower maintenance costs, and much lower insurance costs. Engineers and designers should take an active role in convincing managers that this is true.

Occupational Safety and Health Act of 1970

During the 1960s there was more legislative activity concerning occupational safety and health than there had ever been before in this country. In addition to the passage of several bills controlling safety and health, an extensive study of occupational accidents, statistics, and standards was undertaken. This study culminated in the introduction in the U.S. Senate of the Williams-Steiger Occupational Safety and Health Act of 1970, also referred to as the Occupational Safety and Health Act, or OSHA. It differed from other such acts in that it applied to every business engaged in commerce that had employees. It established several agencies and committees, including an enforcement agency, advisory and review committees, and a group to study the workers' compensation laws and policies in the individual states.

The best-known agency is the enforcement agency, the Occupational Safety and Health Administration, also known as OSHA. For the sake of clarity, the administration will be referred to here as OSHA and the act as the OSH Act. The act charged the Secretary of Labor with promulgating and starting to enforce standards within two years. Because this was such a short time to get standards developed and ready to be enforced, OSHA turned to the ANSI and NFPA standards that already existed for a wide range of fields, as well as to earlier legislation. Practically all of the OSHA standards, then, were originally developed within industry as consensus standards and adopted by national standards organizations before OSHA adopted them and made them mandatory.

The stated purpose of the OSH Act is "to provide for the general welfare, to assure so far as possible every working man and woman in the Nation safe and healthful working conditions, and to preserve our human resources."

The general industry standards promulgated by OSHA are recorded in the Federal Register and listed in the Code of Federal Regulations as 29CFR1910, and the standards with which general industry must comply are often referred to as Part 1910. Each standard carries the number 1910 followed by a decimal code. For example, the general industry standards regarding noise are contained in 1910.95. Standards for the construction industry are numbered 1926. The entire 1910 set of general industry standards is also divided into subparts as shown in Table 2-1.

In the mid-1980s another standard, known as the OSHA Cancer Policy, was added. This is referred to as Part 1990 and has been attached to the OSHA 1910 standards. Most of the standards enforced by OSHA are referred to as specification standards, as distinguished from performance standards. However, the trend today is to write all standards in performance-based language as much as possible. Specification standards describe physical conditions, whereas performance standards describe how a job is to be done or what is to be accomplished. Even specification standards are often vague since they may describe a condition qualitatively without stating exactly how that quality is to be attained. For example, Section 1910.141(d)(1) states: "Washing facilities shall be maintained in a sanitary condition." This statement leaves open the question of just how clean the washroom must be and who decides if it is sanitary or not. On the other hand, some standards seem to be very prescriptive or specific; Section 1910.23(e)(1) states: "A standard railing shall consist of top rail, intermediate rail, and posts, and shall have a vertical height

Table 2-1. Subparts of 29CFR 1910

Subpart	Standards Covered
A,B,C	(General discussion of scope and applications)
D	Walking-working surfaces
E	Means of egress
F	Powered platforms
G	Occupational health and environmental control
H	Hazardous materials
I	Personal protective equipment
J	General environmental controls
K	Medical and first aid
L	Fire protection
M	Compressed gas and air equipment
N	Material handling and storage
O	Machinery and machine guarding
P	Hand and portable powered tools
Q	Welding, cutting, and brazing
R	Special industries
S	Electrical
T	Commercial diving operations
Z	Toxic and hazardous substances

of 42 inches nominal from upper surface of top rail to floor, platform, runway, or ramp level."

In 1978 over 900 standards were revoked. They had been found to be either picayune, obsolete, or insignificant. There are still many controversial standards, as well as many situations not covered by any standard. The probability that this would happen was recognized when the act was written and a statement was included to cover it. This "general duty clause" appears in the act but not among the standards and states: "(a) Each employer (1) shall furnish to each of his employees employment and a place of employment which are free from recognized hazards that are causing or are likely to cause death or serious physical harm to his employees; (2) shall comply with occupational safety and health standards promulgated under this Act." It also states: "(b) Each employee shall comply with occupational safety and health standards and all rules, regulations, and orders issued pursuant to this Act which are applicable to his own actions and conduct."

Revision of standards is an ongoing task. Research and experience render many standards out-of-date or insufficiently clear. In the area of health, especially, as more information becomes available through research, new knowledge must be incorporated into the standards. For example, the carcinogenic properties of vinyl chloride were not recognized until several years after the OSH Act was passed.

Revising an OSHA standard, however, is a very time-consuming task. After a revision is proposed, it has to be reviewed to assess or ascertain what its probable impact will be. By law, public hearings must be held to allow all interested parties to respond to the proposal. Sometimes there has been considerable disagreement. Not surprisingly, such a revision may take well over a year. Although over 900 standards were revoked in 1978, over 1,100 were originally being considered for revocation. Disagreements, primarily between labor unions and manufacturers, caused about 200 of the proposed 1,100 to be retained. There are several companies that publish and distribute information about changes in the OSHA standards and enforcement. The Government Printing Office also puts out such information. There is also a quarterly CD-ROM service.

States with their own occupational safety and health administrations generally adopt any changes made in the federal OSHA standards. In both cases, state and federal, changes are often effective at some future date.

Although it is not the intent of this book to cover the OSHA standards and policies thoroughly, engineers should be at least somewhat familiar with them. It is recommended that engineering students become acquainted with the scope of the 1910 and 1990 standards.

OSHA has a proposed ergonomics standard and is updating the Confined Space Standard 1910.146. A recent standard is ANSI/ASSE Z359, the Fall Protection Standard. There have been 54 major OSHA standards since 1971. OSHA subdivision Z addresses toxic and hazardous substances, including 13 carcinogens. Title 29 of CFR, Part 1990 classifies carcinogens.

OSHA standards are written in legal terminology, which makes some of them difficult to interpret. Industry has long requested a simplification of the terminology but so far without success. Many state plans have included a consultation service to help people in industry to understand the standards and also to help identify violations. Occasionally this consultation service can offer suggestions about how to correct a hazard—that is, a violation of a standard—but that is not its purpose. The federal OSHA also has a consultation service. It is funded by OSHA, but it is administered by private companies to avoid conflict of interest.

In selecting work sites to inspect, OSHA, both state and federal, has set the following priorities:

1. A site where imminent danger has been determined
2. A site at which there has been an accident involving a fatality or three or more serious injuries
3. A site at which there has been a valid employee complaint about a hazardous condition
4. Sites that have been targeted as "high hazard"

The inspector, called a compliance officer, goes to the selected site unannounced and presents his credentials. When admitted, the compliance officer tours the site, accompanied by the employer or a representative, and notes all the violations he or she finds. In some cases a representative of the employees also goes on the tour, especially if the plant is unionized.

At the end of the tour the officer reviews the violations with the employer. The officer then takes his or her findings back to the OSHA office and discusses them with the director and staff. Citations and fines, if any, are mailed to the employer, who is given a certain period of time to pay the fines, correct the violations, or appeal the citations.

Fines are levied according to the severity of the violation or, more to the point, according to the severity of the potential accident. There are six categories of citations:

1. Other-than-serious violations
2. Serious violations
3. Willful violations
4. Criminal/willful violations
5. Repeated violations
6. *De minimis* violations

Fines also vary depending on any evidence that the employer has taken steps to correct hazards or shows "good faith." Willful violations, or failure to make any improvements after a citation, can bring extremely heavy fines.

In many cases an employer may be aware of a violation but not know how to correct the hazard. OSHA does not normally provide any such guidance, although individual compliance officers frequently offer suggestions.

In some cases the standards specify in detail how a particular hazard is to be treated, but sometimes other equally effective methods are found. In fact, on occasion companies have discovered a way to achieve safe working conditions that is better than the way specified in the standards. The policy of OSHA has been to accept an alternative if it is judged by the compliance officer and staff to be at least as effective as the specified method.

OSHA compliance officers should have a good technical background, an understanding of a wide range of industrial operations, and the personality to work effectively with many people in a variety of circumstances. In addition, they must be knowledgeable of the thousands of standards that are to be enforced. Such persons were and are still hard to find. Until such persons were found and trained, compliance officers often had found picayune violations and overlooked serious ones.

This author, on one occasion, was working with the owner of a small plant in which there was a serious hazard and no apparent way of correcting it short of completely dismantling the building and reconstructing it with a different type of operation. One piece of equipment was handmade as part of the building and became a common way of designing that particular operation in many plants in the area. Fearing the consequences of an OSHA inspection, the author contacted the regional OSHA director and described the hazard without mentioning the company or even the industry. The director immediately recognized the situation and replied, "Oh yes, that's a common problem in the ____ industry. There's another plant in the northern part of the state where a consultant has been working for nearly a year and cannot find a way to correct it." The director was obviously aware of the dilemma faced by companies in this industry and avoided making inspections.

This situation puts OSHA directors and compliance officers in a difficult position. This director was aware of a serious hazard and did nothing about it. He knew that if the plant was inspected it would be cited for serious violations and possibly for imminent danger. But all the companies in this industry were small and would have gone bankrupt if they had been required to correct this hazard, thus throwing many people out of work. Was this director derelict in his duty? Is it more important to allow companies to remain in business and provide employment, even hazardous employment, than to create safe workplaces? Or should that particular kind of employment be banned anyway?

Mention was made earlier of special emphasis programs that identify target industries and target health hazards. Early in the administration of OSHA it was realized that certain industries had safety records far worse than the record of industry in general, and they were identified as target industries.

Top five deadliest industries in 2010 by total number of fatalities:

Construction	751
Transportation and warehousing	631
Agriculture, forestry, fishing, and hunting	596
Government	477
Manufacturing	320

Top five deadliest industries in 2010 by fatality rate:*

Agriculture, forestry, fishing, and hunting	26.8
Mining	19.8
Transportation and warehousing	13.1
Construction	9.5
Wholesale trade	4.8

*per 100,00 full-time employees

Companies involved in these industries were given higher priority when OSHA directors selected inspection sites. It is difficult, if not impossible, to compare the records of these industries today with those records of the 1970s. Frequency rates have been replaced with incidence rates, which measure injuries in a different way. Furthermore, the categories of industries have changed. For example, longshoring is no longer listed as an industry. Available data suggest that all of these industries, except meat products, have improved, while some other industries have poorer records today than before.

Other special emphasis programs have centered on health hazards such as asbestos, cotton dust, and lead. Foundries have also been targeted. These target companies all have a high inspection priority. Some other specific industries are construction, maritime, logging, and oil and gas extraction.

Since the 3rd edition of this book, there have been many proposed revisions to the OSHA regulations, including general industry, construction, shipyards, the hazard communication rule (GHS), hearing protection, nanoscale materials, underground storage tanks, fall protection, combustible dust, bloodborne pathogens, mining and confined space, and cranes, as well as new DOT, FDA, and EPA regulations and proposals.

OSHA is to modify the Hazard Communication Standard (1200) to make it consistent with the international Globally Harmonized System of Classification and Labeling Chemicals (GHS). OSHA expects to adopt the GHS in September of 2012. Employers will have three years to comply with the standards after its adoption. See Chapter 13 for details on the GHS.

OSHA has a proposed ergonomics standard and is updating the Construction Confined Space Standard. A recent standard is ANSI/ASSE Z359, the Fall Protection Standard. There have been 54 major OSHA standards since

1971. OSHA subdivision Z addresses toxic and hazardous substances, including 13 carcinogens. Title 29 of CFR, Part 1990 classifies carcinogens.

Effectiveness of Mandatory OSHA Standards

Today, after 40 years of OSHA enforcement, there is still a great deal of controversy about this law. Some condemn it; others defend it; but all agree that very few legislative acts have had as much impact on as many people.

Engineers, as well as managers, would do well to try to assess the effectiveness of OSHA. Records show a slight improvement in some areas but no significant change in others. Fatalities in 2008 were down 2.5 percent from 2007. Injuries are down in some industries but have risen in others. There are many ways of explaining these changes. Better record keeping is required now, so very likely some injuries are included that would have been missed before. The size of the total workforce has increased, of course, but the figures are based on million-employee-hours worked, not on the absolute number of injuries.

Since OSHA standards are aimed primarily at unsafe conditions, it is reasoned that no amount of enforcement of these standards would reduce unsafe acts. But there does seem to be a relationship between unsafe conditions and unsafe acts. Reasons for this will be discussed in Chapter 4. Certainly the existence of OSHA has increased awareness of hazards and promoted the ideal of the safe workplace. Whether the consciousness of safety makes workers practice safety is another matter. Unchanged or even poorer safety records in some industries indicate that not all workers have been so motivated.

There are many people who so resent being told what to do that they respond with defiance and even try to do the opposite. When this attitude exists among shop workers, it invariably existed first in management. When managers scoff at safety rules, workers tend to do the same. When managers take an interest in safety, on the other hand, workers generally respond with safe performance. Too often managers in industry have looked upon OSHA standards and other governmental regulations as punitive devices to be evaded. The attitude of management is so important that it needs to be emphasized over and over again. No company safety program can succeed unless everyone in management supports and participates in it. By the same reasoning, OSHA cannot be successful in fulfilling its objectives until managers are convinced that complying with safety and health standards is for their own benefit.

A multi-employer work site is any work site, permanent or temporary, where more than one employer (or his or her employees) work, but not necessarily at the same time. The most common multi-employer work sites are temporary ones, at which construction activities take place. Other examples include permanent work sites at which outside contractors perform activities at that work site, including, but not necessarily limited to, construction, environmental or janitorial services, repairs and deliveries (OSHA). The employer that has control over the site is responsible for the safety and health of their employees and the employees of others at the site. The employer that created the hazard could be liable and cited by OSHA. Construction sites usually have an onsite safety manager to control and mitigate hazards.

Employers of multi-employer work sites need to keep their employees informed of hazards posed by chemicals. This could be accomplished by a written plan and employee training. Employers bringing hazardous materials on-site should notify all workers of their presence and provide MSDSs. Hazard communication should also be in Spanish to inform Latino and Hispanic workers. Supervisors should be aware of cultural differences regarding safety and correct unsafe workers appropriately.

Companies can choose to participate in OSHA's Voluntary Protection Program (VPP) and, in so doing, establish a new relationship with OSHA and not be subject to programmed inspections.

OSHA's goal in the VPP is to encourage companies to develop their own work-site safety and health programs to create a safe workplace for all employees.

Certification of Products

Over the past few decades, as people became more safety-conscious, they also became more concerned about the products they use. Greater interest in product control led Congress to pass the Consumer Product Safety Act in 1972, which created the Consumer Product Safety Commission.

It has become increasingly important to assign responsibility correctly for the safe design, construction, and operation of products. Unfortunately the legitimate right to safe products has also enabled the less scrupulous to take advantage of consumer laws for their own monetary benefit. Many product liability claims have played an important role in improving product design and construction, but many others have been gross abuses of the right to sue for damages.

A few case studies will help illustrate several aspects of liability. A young man working on a farm was trying to operate a silo unloader. The device was not working properly, so he climbed up into the silo to see what was wrong. He found that because part of the silage was frozen and part was not, the device had dug itself into the softer material and gotten stuck. While the machine was still running, the man tried to pull it out of the hole. The drive wheels were powered by an exposed drive shaft which had a setscrew with a square head protruding from a collar on the shaft. The only place where the man could stand to pull the machine was quite close to this setscrew. Although the shaft was turning very slowly, the setscrew caught in the man's belt buckle and began winding his clothing around the shaft. The only control switch was located far beyond his reach and nobody was close enough to hear his calls. He was killed before the motor stalled.

Who was at fault for this accident? There were obviously several contributing factors. The man should not have entered the somewhat restrictive area in the presence of moving machinery without someone else at the control. He should not have attempted to pull on the machine while it was running. The designer and manufacturer of the silo unloader, on the other hand, should have recognized several problem areas and designed the machine accordingly. It should have been realized during the machine's design that it would get stuck in partly frozen silage, and provisions should have been made for retrieving it safely. The drive shaft should have been shielded and the head screw, in particular, should not have been left exposed. In fact, the manufacturer had used headless setscrews in other places on the same machine and could have done the same at this point. The case resulted in the award of a large sum of money to the family of the man, paid by the manufacturer.

In another case, a man working in a bakery was overseeing the operation of an automatic cupcake machine. Cake dough was formed into individual round shapes and carried along on a conveyor. At one point a wheel with a curved finger caught the dough, turned it over, and transferred it to another conveyor. The wheel was mounted in a box that was completely enclosed except for a hinged cover on the top. Access was needed only for maintenance and repair, and the worker had been instructed by his employer never to open the cover. One day, however, the dough got clogged on the finger of the wheel and would not pass through. Without stopping the machine, the worker opened the cover and reached in to pull out the dough. One of his fingers was mangled and eventually had to be amputated.

Again, the question of fault is raised. The worker later sued both his employer and the manufacturer of the machine. The manufacturer could have placed a limit switch on the cover, so that if the cover were lifted, the machine would stop. The box containing the wheel was mounted at about shoulder height and the control switch was located right beside it. If you were the designer or manufacturer of this machine, would you have anticipated this kind of activity and included a limit switch or some other safety device to prevent such an accident? How far does a manufacturer have to go to make a machine free of hazards?

An elderly woman was shopping in a store that had an escalator between the first and second floors. While she was on the second floor the escalator stopped and required repair. The store management placed a warning sign at the bottom but had not placed one at the top before the woman started walking down the escalator. She fell and was injured, and she subsequently sued the store. Her lawyer approached a consulting engineer and asked if he could find some OSHA standard that the store management had violated. Of course, OSHA standards protect only employees, not the general public. The lawyer admitted privately that the woman was quite unsteady on her feet and that he was looking for a way to place the blame on the store. The engineer refused to help the lawyer, but when the case went to court, the woman was nevertheless awarded $75,000 (Lofgren 1989). Fortunately, most lawyers refuse to employ such unscrupulous tactics.

The janitors of a university building were taking a coffee break in a supply room and were joking with one another. An electric circuit-breaker panel in the room had an opening that had not been covered as required by the electrical code. One of the janitors took a cigarette lighter from someone else and tossed it through the opening. After more joking, he attempted to retrieve it by bending a coat hanger into the shape of a hook and inserting it into the panel. When this didn't work, they finally called an electrician, who recovered the lighter and covered the opening. In spite of this foolish horseplay, there were no short circuits or injuries, but if there had been an injury, who would have been at fault? The university undoubtedly would have been responsible. The uncovered panel not only violated the electrical code, but also invited the unwise actions of the janitor (Lofgren 1989). This case is a good illustration of why even the simplest safety standards should be enforced.

Problems of liability have prompted manufacturers to try to protect themselves, as well as their customers, by having their products tested and certified by some recognized agent. Several testing laboratories have been in existence for many years, and the significance of their stamp of approval is increasing.

Industrial as well as consumer products and components often carry the label of either Underwriters Laboratories, Inc., or Factory Mutual Systems. The familiar UL and FM labels signify that the product has been tested and found to satisfy all the appropriate existing safety standards. It should be noted, however, that if an electric cord carries the UL label, that does not mean that the product to which the cord is attached has been tested and found safe. The Good Housekeeping Seal of Approval is another time-honored certification. A company gains credibility by the use of these seals and stamps of certification, but, of course, using them illegally carries a stiff penalty.

One of the agencies created by the OSH Act, the National Institute of Occupational Safety and Health (NIOSH), certifies respirators and certain monitoring equipment. But OSHA does not certify any kind of product. Several companies have been reprimanded because their advertisements stated or implied that their product was approved by OSHA. NIOSH now has a new Prevention through Design Initiative.

Many injuries and deaths could be prevented with a little forethought to safety. Designers of products must now consider all aspects of their product's use, including its reasonably foreseeable misuse, in the conceptual stages of design. Today's legal culture is intolerant of unacceptable and unreasonable risks. Designers must realize that unsafe acts and unsafe conditions are often symptoms rather than causes and that most accidents are foreseeable and predictable.

The safety engineer/professional is confronted with new and emerging technology and science such as nanotechnology and bioengineering as well as new clean sources of fuel, all of which will present new hazards to consider. Risk control for these hazards and their risks should become an integral part of the corporate culture of safety as well as the business of safety to eliminate or minimize risk, losses, and OSHA, EPA, and other government fines and penalties, as well as civil litigation. Every organization should have a compliance officer or department to handle and keep up with the many government regulations.

Despite advancements in technology, manufacturers are still failing to meet the recognized standards of industry such as OSHA, ANSI, and European Community Standards, as well as the Standards of Care set by such authorities as the National Safety Council, American Society of Safety Engineers, and others.

QUESTIONS

1. What are some of the international standards, and why are they important?
2. What are some of the U.S. codes and standards?
3. What is the Accessibility code, and why is it important today?
4. What is the "general duty clause" in the OSH Act, and why is it important to OSHA?
5. What are the benefits of workers' compensation?
6. Why did OSHA consider a change to its Hazard Communication Standard?
7. What is the purpose of NIOSH, and what is its connection to OSHA?

8. What is the leading cause of non-fatal accidental injuries requiring hospital attention?

9. How does OSHA encourage companies to comply with safety regulations?

10. What are several reasons for OSHA to inspect a work site?

11. Where did OSHA first obtain the standards it adopted?

12. Name three important NFPA standards.

13. What is the boiler code, its history, and its purpose?

14. How are OSHA fines levied?

15. What does OSHA consider to be a major contributing cause of accidents?

16. Name several organizations that certify products.

BIBLIOGRAPHY

Gloss, David S., 1988. *Handbook for Industry Standards of the OSH Act.* New York: Wiley.

Lofgren, Don J., 1987. *Dangerous Premises: An Insider's View of OSHA Enforcement.* Ithaca, NY: ILR Publishers Division of New York State School of Industrial & Labor Relations.

Noble, Charles, 1986. *Liberalism At Work: The Rise and Fall of OSHA.* Philadelphia: Temple University Press.

REFERENCES*

American National Standards Institute, Inc.
   1430 Broadway, New York, NY 10018

Architectural and Transportation Barriers Compliance Board (ATBCB)
   1331 F Street, NW, Suite 1000
   Washington, D.C. 20004-1111

National Fire Protection Association
   1 Batterymarch Park, Quincy, MA 02269-9101

National Safety Council, Injury Facts, Itasca, IL: NSC.

OSHA Compliance Manual
   Neenah, WI: J. J. Keller & Associates, Inc.

1910 OSHA Guide - 2012.
   Neenah, WI: J. J. Keller & Associates, Inc.

---

*Standards, reports, and data relevant to the topics discussed in Chapter 2, as well as to many other topics discussed in the book, are available from the sources listed.

3

Recognition and Control of Hazards

Richard W. Stickle, PE, CSP

Setting Goals

In Chapter 1, safety and acceptable risk were discussed in general terms. In order to design without hazards, a safety engineer, with the help of management, must be more specific about what is safe and what is not safe in a particular environment. Hazards must be identified exactly. Furthermore, the limits of acceptable risk must be fixed. Assessing the probability and severity of risks is the most difficult aspect of hazard control, and too often it is not done by any logical process. A lack of quantitative limits of acceptable risk does not mean that no limits exist; it merely means that management has not stated them.

It is not easy for a company to say that it will accept one fatality, 45 serious injuries, and so on, next year. But if that was the record last year and nothing is done to prevent those accidents from happening again, that becomes the acceptable risk. On the other hand, if that record is found to be unacceptable, something can be done about it. If a company really means that it will not accept a fatality, it can, and will, expend the effort and the money to assure that an accident with the potential to cause a fatality will not occur.

Does that sound a bit unreasonable? After all, won't accidents happen in spite of our efforts? Not at all! Accidents do not just happen; they are caused. There is no such thing as a freak accident or an accident without a cause. If safety engineers do their jobs thoroughly, the causes and potential causes of accidents will be found and recognized, and then limits of acceptable risk can be set. Those limits become the goals toward which the safety engineer works.

Today larger companies and organizations have risk-management departments. Risk management is the process of making and carrying out decisions that will minimize the adverse effects of accidental losses upon the organization. Risk-management decision making first involves identifying the loss exposures, analyzing those exposures, and then considering the feasibility of using alternative techniques such as risk control or risk financing to deal with them. The best technique is selected, implemented, and finally monitored to improve the risk-management program. Occupational health and safety departments either work closely with or are a part of risk management.

The safety engineer is a valuable asset in providing risk-management information, assessing exposures and losses, and formulating risk control techniques.

Many insurance carriers, brokers, and risk-management agencies utilize some type of computerized information management system to collect and sort data on their losses. Sometimes these "loss runs" can be of use to the safety engineer in analyzing a company's loss exposure.

Many casualty insurance companies apply an "experience modification rating" to adjust the average determined manual insurance rate to reflect the insured company's loss record. This encourages companies to establish effective loss control/safety programs to reduce the cost of their premiums.

There are other ways of assessing a company's safety and health performance. One of the most common methods, though not necessarily the best, is to compare the record of the number of accidents, or the number of lost days, or the number of reported injuries and illnesses, of a given company with that of other companies. This, of course, requires that all companies keep the same kinds of records.

Record Keeping

For a long time, industry and business, insurance companies, and government agencies have been keeping track of the safety performance of companies in a wide range of categories. Frequency rates and severity rates have been the guidelines for judging the performance of all companies. These are slowly being replaced by performance guidelines called incidence rates. Since all of those methods require records to be kept for a period of several years, the older ones are still being used to some extent.

Lost-time frequency rate is defined as the number of accidents causing lost days per million employee-hours worked. Mathematically it is expressed as:

$$F = \frac{N}{\frac{\text{(total hours worked)}}{1,000,000}} \quad \text{or} \quad F = \frac{1,000,000N}{H}$$

Lost-time frequency rate does not reflect the severity of accidents, except insofar as it counts only the accidents that result in days lost from work. Like any sort of ratio or index, it is meaningless until it is compared with other frequency rates. A company can compare its frequency rate with its own past frequency rates, with those of other companies in the same industry, or with those of all industries. Comparison indicates whether it is doing better or worse, and, to some extent, how much better or worse.

Severity rate is defined as the number of days, including weekends and holidays—that is, calendar days—lost per million employee-hours worked. Mathematically it is expressed as:

$$F = \frac{D}{\frac{\text{(total hours worked)}}{1,000,000}} \quad \text{or} \quad F = \frac{1,000,000D}{H}$$

Severity rate indicates severity of accidents only in terms of the number of days lost, but it is a somewhat better measure of severity than is the frequency rate.

In both these ratings, the day the accident happened is not counted as a lost day, nor are any partial days. If a worker had an accident on Monday morning and returned to work on Thursday afternoon, only Tuesday and Wednesday would be considered lost days. On the other hand, if an accident occurred on Friday afternoon and the injured worker was sent home or to a doctor or to a hospital but returned to work on Monday morning, Saturday and Sunday would be considered lost days.

If a worker contracted dermatitis and was sent home, any lost days would be counted. But if that worker was treated in the company dispensary and returned to the job the same day, the incident would not be counted at all.

When loss of a body member or loss of life result from an accident, the frequency rate is calculated as usual, but the severity rate is not. A table of scheduled charges is used for establishing lost days for loss of body members, as shown in Table 3-1. Please note that Table 3-1 is based on 1989 dollars. In addition, there is an automatic charge of 6,000 days for loss of life or permanent total disability.

A major problem with the frequency and severity rates was the fact that records were incomplete. Data are maintained primarily by two organizations, the National Safety Council and the Bureau of Labor Statistics, which is an agency within the Department of Labor. These organizations collect data from whatever sources are available to them. Until 1971 there was no requirement that accidents be reported unless workers' compensation was involved, and even then reporting was not complete.

When the Occupational Safety and Health Act was passed in 1970, this problem was recognized, along with inherent deficiencies of the frequency and severity rates. A new record-keeping system was therefore developed and included in the OSH Act. The new system requires more detailed information and widens the range of companies who must report.

The new system involves measurements called incidence rates and alternate incidence rates. The incidence rate is defined as the number of recordable cases per 100 employees per year. It is expressed mathematically as:

$$I = \frac{C}{\dfrac{(\text{total hours worked in 1 year})}{(100 \times 40 \times 50)}}$$

or

$$I = \frac{200,000\,C}{H}$$

Recordable cases are comprised of all work-related injuries requiring more than first aid, and all occupational illnesses, and include the following:

1. Deaths, regardless of the time between occupational injury or illness and death
2. All occupational illnesses

3. All occupational injuries resulting in any of the following:
   a. Lost workdays—either days away from work or days of restricted work activity
   b. Medical treatment other than first aid
   c. Loss of consciousness
   d. Restriction of work or motion
   e. Temporary or permanent transfer
   f. Termination of injured or ill employee

Table 3-1. Scheduled Charges For Traumatic or Surgical Loss of Member and Impairment

Fingers, Thumb, and Hand						
			Fingers			
Amputation involving all or part of bone*	*Hand*	*Thumb*	*Index*	*Middle*	*Ring*	*Little*
Distal phalange	—	300	100	75	60	50
Middle phalange	—	—	200	150	120	100
Proximal phalange	—	600	400	300	240	200
Metacarpal	—	900	600	500	450	400
Hand at wrist	3000	—	—	—	—	—

Toe, Foot, and Ankle			
Amputation involving all or part of bone*	*Foot*	*Great Toe*	*Each of Other Toes*
Distal phalange	—	150	35
Middle phalange	—	—	75
Proximal phalange	—	300	150
Metatarsal	—	600	350
Foot at ankle	2400	—	—

Arm	
Any point above elbow, including shoulder joint	4500
Any point above wrist and at or below elbow	3600

Leg	
Any point above knee	4500
Any point above ankle and at or below knee	3000

Impairment of Function	
One eye (loss of sight), whether or not there is sight in the other eye	1800
Both eyes (loss of sight) in one accident	6000
One ear (complete industrial loss of hearing), whether or not there is hearing in the other ear	600
Both ears (complete industrial loss of hearing) in one accident	3000
Unrepaired hernia	50

*If the bone is not involved, use actual days lost. The tuft of the distal bone of a finger or toe is considered bone if it shows in x-rays.

Source: This material is reproduced with permission from American National Standard for Uniform Recordkeeping, 216.3-1989, Copyright ©1989 by the American National Standards Institute. Copies of this standard may be purchased from the American National Standards Institute at 1430 Broadway, New York, N.Y. 10018.

Lost days, in this calculation, include only workdays, not weekends or holidays. The incidence rate does not count the amount of lost time, but only the number of accidents in recordable cases.

The time period for this calculation must be one year. It is assumed that normal working time is 40 hours per week, 50 weeks a year. Thus, for 100 employees, the number of hours worked is $100 \times 40 \times 50 = 200{,}000$ hours. If the actual number of hours worked is more or less, that total is divided by 200,000 to give a comparable rate.

The incidence rate includes many more cases than the old frequency rate but still fails to address severity. Therefore, an alternate incidence rate was developed and defined as the number of working days lost or charged per 100 employees per year. It is expressed mathematically as:

$$I' = \frac{D}{\dfrac{(\text{total hours worked in 1 year})}{(100 \times 40 \times 50)}}$$

or

$$I' = \frac{200{,}000D}{H}$$

Days are lost workdays, not calendar days as in the old severity rate. Loss of body members and loss of life are automatically charged according to the same schedule used for the old severity rate (Table 3-1). The following example will illustrate the use of these several rates.

The employees of a manufacturing company with an average payroll of 2,500 worked a total of 5,500,000 hours in a given year and experienced the following:

1 fatal heart attack (assumed to be work related)
2 hernias (38 workdays, 49 calendar days, lost)
3 sprained backs (8 workdays, 10 calendar days, lost)
10 burns (30 workdays, 35 calendar days, lost)
41 cuts and contusions (89 workdays, 95 calendar days, lost)
55 cases requiring only first aid

$$F = \frac{1{,}000{,}000 \times 57}{5{,}500{,}000} = \frac{57}{5.5} = 10.36$$

$$S = \frac{6189}{5.5} = 1125.3$$

$$I = \frac{200{,}000 \times 57}{5{,}500{,}000} = 2.07$$

$$I' = \frac{200{,}000 \times 6165}{5{,}500{,}000} = 224.18$$

Was this a good record or a poor one? One cannot really say until these figures are compared with similar data from other companies of comparable size and the same industry. The values of I and I' for all industries in 1978 were 2.56 and 47 respectively. Frequency and severity rates from an earlier year (1956) were 6.38 and 733 respectively for all industries.

If it could be shown that the heart attack was not work related, F and I would be reduced very slightly but S and I' would be reduced drastically.

$$F = \frac{56}{5.5} = 10.18$$

$$S = \frac{189}{5.5} = 34.36$$

$$I = \frac{200,000 \times 56}{5,500,000} = 2.04$$

$$I' = \frac{200,000 \times 165}{5,500,000} = 6.00$$

Such modifications make this company's record look much better. Whether or not this is a good record is still subject to interpretation, but such figures do provide some measure of performance.

To add further value to these figures, the data from all sources are divided into categories known as Standard Industrial Classifications (SICs). The performance of one company can then be compared with that of other companies in the same general classification. There are other factors that are known to contribute to differences in performance data, such as the size of the company and its geographical location, but this breakdown of data is not generally available.

One of the major problems with performance indices, and with safety and health records in general, is that accidents are not reported unless injuries or damage are quite significant, as Figure 3-1 illustrates. The most severe results are reported because they are obvious. The less severe are far more common but less obvious and often not reported. If the area under the curve in Figure 3-1 were labeled "cost," which it really is, it would be seen that the total cost of injuries and property damage not reported is probably greater than that which is reported.

A major goal of the new record-keeping system was to acquire more data that would be more meaningful. No matter how it is done, the paperwork requires the time of someone in each and every company. Quite understandably, small companies find this an unacceptable burden when added to all the other records and reports required by governmental agencies and programs. As a result, many smaller companies may not bother to record the less obvious accidents, and the total performance statistics are only slightly, if any, more comprehensive than before.

Once some idea is gained of what the overall safety performance is, attention should be given to more in-depth analyses of how and why accidents occur. Accidents will not be reduced until causes are known and corrected.

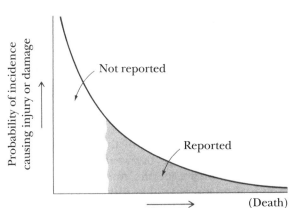

Figure 3-1. Probability of incidence versus severity.

Techniques for Analyzing Hazards

A hazard is the presence of a risk situation—a condition that is a prerequisite for a mishap. *Hazard analysis* is the technique used to systematically identify, evaluate, and resolve hazards. Over 100 techniques have been developed and utilized to analyze and find the hazards that exist in a wide variety of situations. The techniques have been developed as both management and engineering tools and are usually considered to be "System Safety Engineering." The best of these tools attempt to look at the whole program as a system which includes man, machine, and the environment which surrounds them.

System safety was initially developed during the 1940s and 1950s by the aerospace industry and the military to analyze large systems such as aircraft and intercontinental ballistic missiles (ICBMs). The basic tenet of system safety is proactive, and its goal is to identify and eliminate hazards before they can become accidents, or losses, throughout the life cycle of the project or activity.

During the intervening years, system safety has grown into other fields and been called by different names; however, the basic principles and many of the tools remain the same. Only a few of those tools will be discussed here.

Before using any of these techniques, a person should have a clear under-standing of what he or she intends to accomplish with it. If the data and infor-mation used as input are very general, the outcome will also be general. More specific input requires more time and effort, but produces more specific results.

Additionally, the earlier in the life cycle of the project that system safety methods are applied, the more economical they make the project, since this allows for safety to be "designed in" during the conceptual phase and, there-fore, eliminates the necessity of later changes when the cost for those changes would be higher.

The safety engineer will usually be looking for exact causes of a specific potential accident. One writer refers to this causal factor as the single-point failure[1]—that is, the one most basic thing that could fail that, by itself or with some other element, could cause a specific accident. It is this single-point failure that the safety engineer must eliminate.

Probability plays an important part in our daily lives. It involves weighing the chances, or likelihood, that something or other will take place, such as

"Sooner or later there's going to be a fatality here." The theory of probability provides a mathematical framework for such statements.

The percentage of "successful" outcomes of a repetitive operation is called the probability of the event. Probabilistic models deal explicitly with the uncertainty inherent in an event that leads to an accident. The concept of safety itself is one of uncertainty. As there is no such thing as absolute safety, human activity will always, and unavoidably, involve risks. The reliability of manufactured, or fabricated, components or systems is a major source of uncertainty. Real-world systems are complex, and there is no practical way to evaluate enough scenarios with a deterministic model.

There are many forms of probabilistic models:

- Statistical modeling involves the description of random phenomena by an appropriate probability distribution.
- A probability distribution is a mathematical function that defines the probability of the event.
- Binomial distributions are often encountered in systems that behave in accordance with the Bernoulli process that can govern the generation of statistical data.
- The normal distribution is widely used to describe biological phenomena, such as human height, and is very applicable to ergonomics and human factors engineering. It is sometimes called the Gaussian distribution, or "bell-shaped" curve.
- The Poisson distribution is widely used to describe random events having a large opportunity to occur but a small likelihood that any one of these opportunities will actually result in an occurrence, such as a fire or aircraft accident.

Relevant variables in a decision problem are usually too numerous or complex to be handled mathematically. They are handled by using computer simulations, which can predict a range of outcomes from probabilistic situations.[2]

Several of these "analytical" techniques include the probability of an event's happening, and a classification of the seriousness of a resulting accident if it does occur. This is usually the weakest part of the analysis. Unless records have been kept in which a specific event occurred several times, there is little upon which to base an estimate of its future probability. Often a particular event can be compared with another event to establish a probability. Sometimes manufacturers of equipment can provide reliability data, but these may be of questionable accuracy. Several modeling techniques, such as Monte Carlo, can also be used to establish probability.

System safety engineering is an element of systems engineering involving the application of scientific and engineering principles for the timely identification and control of hazards within the system. It uses various tools such as Hazard Analysis, Preliminary Hazard Analysis (PHA), Fault Hazard Analysis (FHA), Failure Mode and Effects Analysis (FMEA), Fault Tree Analysis (FTA), and others to accomplish its goal of identifying hazards.

The initial step in controlling or eliminating a hazard is the recognition that the hazard exists. No single method of analysis is adequate to evaluate an entire program; to properly evaluate a program a combination of analyses is utilized. Perhaps the most basic methodology of hazard identification is the

Hazard Analysis. Hazard analysis in the early developmental stage of a project's life cycle is important, since it is always easier to design out a hazard than design to control it at a later stage, and it provides the groundwork for the other types of analysis that may be done.

The hazard analysis is a systematic look at the whole project. A system safety engineer not only will bring his own experience to the analysis, but also will know how to use historical and statistical records to aid him in the search for hazard potential. These records may include such things as accident and failure reports from similar projects, and many projects will produce "lessons learned" reports to aid future projects. Additionally, the safety engineer should go beyond the historical data to determine areas of high accident potential that might not have an accident history or severe accident potentials with a low probability of occurrence.

The hazard analysis is usually done more than once during the life of a program. The initial or preliminary hazard analysis is generally the first analysis of a new program or a program that is being modified and is carried out during the conceptual portion of the program. Subsequent hazard analyses, often called fault or system hazard analyses, differ from the preliminary hazard analysis mostly in the depth of the analysis and the amount of information that is available later in the program.

As the design of the project matures, it is possible to make analyses in greater depth. The analyses can now reflect with greater accuracy how a system will operate, any problems that may be encountered during its operation, the failures that may occur, and any hazardous conditions that may exist. There are various types of analyses that can be used, but there is no one analysis that will satisfy all the requirements for analysis.

Failure Mode and Effects Analysis

The first analysis method to be described is the failure mode and effect analysis (FMEA). It was developed originally to study reliability, but it can be used for safety analysis also. In fact, reliability of both equipment and people are the elements of safety. The FMEA is what is known as a bottom-up analysis, since the FMEA starts with the components of the system and analyzes those failures and how they will affect the overall system.

FMEA makes use of a chart or data sheet such as that shown in Figure 3-2. This chart is not complete but shows enough to illustrate its use. It should be noted that although only two of the failures listed in this example would cause injury to personnel, they are all hazards. Note, too, that failure to secure the workpiece causes an accident of minor severity, and excessive feed causes an accident of minor severity. The probability of the two occurring simultaneously is very small but if they do, the severity becomes major.

In specifying "components," the analyzer must decide what constitutes a unit. An electric motor consists of many parts, but it would seldom be of any value to break it down that way; therefore, the whole motor is considered one component. When a person is part of the system, he or she must be considered a component. Care should be taken in designating the component responsible for a given mode of failure. In this example, it is the operator who failed to secure the workpiece. If the work were held by a clamp and the clamp broke, the component responsible for the failure would be the clamp.

System:___Worker drilling hole with drill press___   Analyzed by:___GAM___   Date:___6-18-80___

Component	Mode of failure	Effect of failure on:			Prob'lity of failure	Severity*			
		Personnel	Hardware	The system		1	2	3	4
Drill	Breaks	None	None	Inoperative	1/1000	x			
Motor	Burns out	None	None	Inoperative	$1/10^6$	x			
Bolt	Breaks	None	None	Inoperative	$1/10^6$	x			
Switch	Sticks "on"	None	Wear	Inoperative	$1/10^6$	x			
Operator	Fails to secure workpiece	None	None	None	1/100	x			
Operator	Uses excessive feed	None	None	Drill burned	1/20		x		
Operator	Fails to secure workpiece & uses excessive feed	Hit by work drill	Broken drill	Inoperative	1/2000				x
Operator	Leaves wrench in chuck & turns switch on	Hit by wrench	None	None	1/500				x

*Classes of severity are: 1. nil   2. minor   3. major   4. catastrophic

Figure 3-2. Example of a Failure Mode and Effect Analysis.

The FMEA does not always include probability or severity, particularly if the purpose of the analysis is only to identify the cause of an accident. If the goal is more than mere identification, however, these data can be very valuable, although frequently difficult to establish. The probability is essentially an expression of how often that particular failure is expected to occur, that is, once in every so many times that the activity is done. If good records of maintenance, first aid cases, and so on, have been kept, quite accurate data can be generated. Otherwise, one must either use probability modeling or just make a guess. Probability modeling is reliable if done properly, but relatively few engineers have the background to do it. A systems analyst can be called in. Even an educated guess has some value.

Severity can be rated on the basis of past experience if that information is available. If not, good judgment should yield acceptable and useful data. Probability and severity rating help identify those things which most need to be corrected. If approvals by management are required for allocation of funds, the existence of quantitative analysis is extremely helpful.

One of the most widely used analytical techniques in system safety is known as fault tree analysis (FTA). Fault tree analysis was developed in the early 1960s at Bell Laboratories as a method to evaluate the safety of an ICMB launch system. Unlike the FMEA, the FTA is what is known as a top-down analysis since it starts with a "top event," which can be taken from the hazard analysis. Once identified, this top event is then analyzed to see what can cause the event to occur. The method used is basically a Boolean logic model that represents the relationships between the various events in a system that can lead to the specified final outcome, which is the top event that has been specified. Each fault tree can have only one top event. The technique is done with the aid of a diagram that starts at the top of the page and branches out toward the bottom, as shown in Figure 3-3. This figure depicts an analysis of one of the failures noted in the FMEA example in Figure 3-2.

In fault tree analysis, a series of events is charted with symbols to indicate the relationships that exist among these events. Figure 3-4 shows the symbols and relationships commonly used. Other symbols have been used for more sophisticated fault tree analyses, but for most purposes this is not necessary.

An "event" is any condition that contributes to the eventual accident, including the accident itself. An event in a fault tree analysis is usually, but not always, an undesirable condition. Whether or not the condition is undesirable often depends on the situation. In the example in Figure 3-3, there are times when it is quite proper and desirable not to have a fixture or not to secure the part to the drill table. It is undesirable when, in that condition, an excessive torque can cause the work to be thrown from the table.

The AND gate is used to connect events when all of the input events must occur simultaneously in order to cause the resulting event. The OR gate is used when any one of the input events alone can cause the resulting event. An INHIBIT gate is used to connect a single input event to a resulting event when the result can occur only if a condition of the input event exists. An INHIBIT gate is not generally used to relate more than one input event to a resulting event, although it is conceivable that such a relationship could arise.

The circle and the diamond both indicate a basic fault or single-point failure. The circle indicates the most fundamental cause, one which cannot be broken down to other input events. The diamond implies that the event may have more basic causes but it would serve no useful purpose to break it down further.

In the example shown in Figure 3-3, only two of the basic faults were considered fundamental. The others could be broken down to find further roots, but were considered to be sufficiently detailed for the purpose of this fault tree. For example, the dull drill could be the result of several different factors, such as use on hard material, too high a speed, or normal wear.

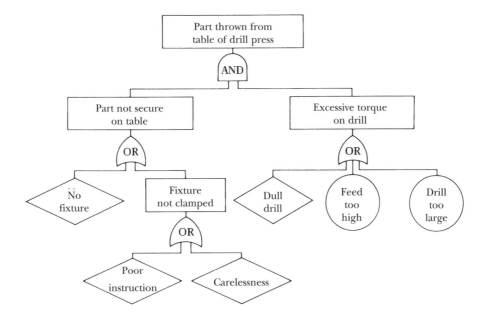

Figure 3-3. Logic diagram for fault tree analysis of accident in drilling operation.

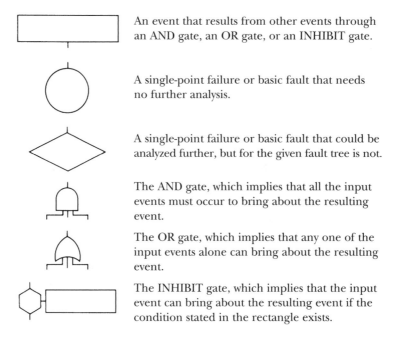

An event that results from other events through an AND gate, an OR gate, or an INHIBIT gate.

A single-point failure or basic fault that needs no further analysis.

A single-point failure or basic fault that could be analyzed further, but for the given fault tree is not.

The AND gate, which implies that all the input events must occur to bring about the resulting event.

The OR gate, which implies that any one of the input events alone can bring about the resulting event.

The INHIBIT gate, which implies that the input event can bring about the resulting event if the condition stated in the rectangle exists.

Figure 3-4. Symbols used to denote the relationships among events in a fault tree.

The fault tree shown here is a relatively simple one. Very complex fault trees can be constructed using many symbols besides the ones shown in Figure 3-4 to describe intricate relationships among a wide variety of events and conditions.

A fault tree may also contain probability and severity evaluations, although the one illustrated does not. Probability can be determined by the same methods used with failure mode and effect analysis. Mathematical modeling techniques such as Boolean algebra or Monte Carlo may be applied. When probability and/or severity evaluations are included in the analysis, these values may be placed within the appropriate symbol, beside it, or on a separate sheet.

Another technique is called management oversight and risk tree, or MORT. MORT is a logic tree that assists in providing the safety engineer with a disciplined method of accident investigation. The biggest difference between MORT and FTA is that MORT utilizes a predesigned tree, intended as a comparison tool, that generally describes all phases of a safety program and is applicable to systems and processes of all kinds. It employs charts similar to the fault tree with a variety of symbols representing lines of responsibility, barriers to unwanted energy, events, priority gates, and assumed risks.

There are three main focal areas in MORT: specific oversights and omissions, assumed risk, and general management weaknesses. The goal is to identify faults and errors and to construct corrective measures. It was developed for use by management as a safety tool but has been recognized as a technique appropriate to a wide range of management problems.

MORT is a collection of many individual techniques, including hazard analysis, accident investigation, and statistical analysis. It is not expected that all of the concepts included in this overall technique will be used for any specific problem. In fact, one of its advantages is that it is versatile and permits the use of separate techniques in any combination that seems appropriate.

The broad goal of MORT is to provide management with a self-analysis from which it can recognize a wide range of errors and see how they are related. This produces a pattern from which management can recognize cause and effect. Once the basic parts of the technique are learned, they can be applied to many aspects of managing an enterprise.

Several comments are in order regarding the use of either a failure mode and effect analysis or a fault tree analysis. The accuracy and value of the finished chart or diagram will be directly proportional to the care taken in making it. If quick identification is all that is needed, a simple analysis will suffice. If the study is to be used to establish acceptable risk, a more thorough analysis is needed. Establishing acceptable risk must involve at least an estimate of probability and severity. Knowledge of just what events are most likely to cause what injuries and what property damage must be obtained before risk can be controlled.

When it comes to providing safeguards to assure that the established risk is met, one can turn to the probability figures generated in the FMEA or FTA. For example, if the acceptable risk for a certain failure is set at one chance in one million, the safeguards built into that system must provide a fail-safe reliability at least as good as one failure per one million operations.

Priorities in Controlling Hazards

When causes of accidents have been determined and their importance has been evaluated, there are several ways to proceed to achieve an established goal. It has been seen that accidents can be caused by people or things. When things—machinery, tools, the workplace environment—are identified as potential causes of accidents, these things or certain characteristics they possess are labeled hazards. People are not referred to as hazards, but perhaps they should be. After all, people design the machinery, the tools, and the conditions in the environment that become hazards, and people misuse all the things that cause the accidents. In one way or another, people are the primary cause of nearly all accidents.

It is frequently valid to ask, "Did the worker cause the machine or tool to malfunction, or did the poor design of the machine or tool cause the worker to malfunction?" Many studies have shown that poorly designed equipment and work environments can cause workers to perform poorly. System safety engineering should have initial input to the system design requirements.

Most of the physical and mental capabilities of people are inherent qualities that can be developed to some extent but not changed very much. We cannot alter the design of people to satisfy a particular job requirement, although we can select those whose abilities and characteristics best suit them to certain tasks. A machine or a tool or an environment, however, can be designed. It can be altered to suit the needs of the task and/or the people involved. In fact, the task itself can be designed or redesigned more easily than people can be altered.

It makes more sense, then, to correct the hazards or, better still, to design things properly in the first place. It is always much faster and less expensive to design and build things properly the first time than it is to redesign and alter things that are already made.

Once hazards have been identified, there are measures that can be taken to mitigate the hazard.

1. Design for minimum hazard (eliminate or reduce).
2. Provide safety devices.
3. Provide warning devices.
4. Provide special procedures.
5. Terminate system.

The priority, or precedence, of these corrective measures is important. Only the first measure can prevent the accident from happening, and only if the hazard is designed out. Providing safety devices lessens the probability of an accident but does not reduce or eliminate the hazard itself. Each step is to be used only if the step above it cannot be accomplished; therefore, step 2 will be utilized only if step 1 cannot be accomplished. Step 5 is always a management decision.

Let us look at an example. The operator of a grinding machine has been given training, is well supervised, and wears safety glasses. In addition, the machine includes guarding adequate to isolate the grinding wheel. What is the hazard? There may be several: the wheel may burst; the workpiece may be thrown from its mounting; certainly there will be particles flying from the wheel and/or the workpiece. Could the hazard be eliminated or reduced? Possibly, if another type of operation could be found that would be as effective as the grinding operation. In fact, even a less-effective method might be acceptable if it eliminated the hazards inherent in grinding. In this example, an acceptable substitute has not been found, and priority 2 has been attempted. Again, isolation cannot be complete or entirely dependable: complete isolation would prohibit the operation. Finally, priorities 3 and 4 have been applied because the first two priorities were found unfeasible.

Another set of terms is sometimes used to distinguish the first two priorities from the last three. Since the first two involve design of the machine, tool, or environment, they are called engineering measures. The last three, involving people, are called management measures. Engineering measures should be attempted first, and then, if they fail to solve the problem satisfactorily, management measures should be undertaken. The fact that employees are using personal protective equipment does not necessarily mean that management has bypassed engineering measures, but it does mean that engineering measures have not solved the problem.

QUESTIONS

1. In 2007 the Company A employed an average of 543 people who worked a total of 1,086,000 hours and experienced the following:

   5 burns (15 workdays, 19 calendar days, lost)
   4 hand and finger cuts (14 workdays, 20 calendar days, lost)
   1 eye injury (12 workdays, 18 calendar days, lost)
   26 recordable cases involving no lost days
   31 first aid cases

   Determine the incidence rate and the alternate incidence rate.

2. In year #2 Company A employed an average of 584 people who worked a total of 1,168,000 hours and experienced the following:

   1 fatality by electrocution
   1 broken arm (16 workdays, 22 calendar days, lost)
   1 finger infection (12 workdays, 16 calendar days, lost)
   2 burns (7 workdays, 12 calendar days, lost)
   28 recordable cases involving no lost days
   32 first aid cases

   Determine the incidence rate and the alternate incidence rate.

3. Referring to problems 1 and 2 above, which year yielded the better performance record for Company A?

4. Reviewing the record of Company A, what would you look for as hazards to be corrected?

5 . What is a *hazard analysis?*

6. True or false, the priority of corrective measures to mitigate hazards from highest (#1) to lowest (#5) is:

   1. Provide warning devices.

   2. Rely on the operator to always do the right thing.

   3. Design out the hazard.

   4. Provide a safety device.

   5. Incorporate special procedures.

7. When constructing a fault tree analysis, what event goes at the top?

8. True or false, a Failure Mode and Effect Analysis is a "bottom-up" analysis.

NOTES

1. Joseph Hrzina. "Single-point Failure Analysis in System Safety Engineering," *Professional Safety* (March, 1980): 20.
2. John M. Watts, Jr. "Probabilistic Fire Models," in *Fire Protection Handbook*, 18th Edition (Quincy, Mass.: National Fire Protection Association, 1997), 11–62 to 11–69.

BIBLIOGRAPHY

Bahr, Nicholas J., 1997. *System Safety Engineering and Risk Assessment: A Practical Approach.* Washington, D.C.: Taylor & Francis.

Brown, David B., 1976. *Systems Analysis & Design for Safety.* Englewood Cliffs, NJ: Prentice-Hall.

Ferry, Ted., 1985. *New Directions in Safety.* Des Plaines, IL: American Society of Safety Engineers.

Grimaldi, John V. and Rollin H. Simonds., 1993. *Safety Management.* 5th ed. Des Plaines, IL: American Society of Safety Engineers.

Hammer, Willie., 1985. *Occupational Safety Management and Engineering.* 3rd ed. New York: Prentice-Hall.

Haight, Joel M., ed., 2008. The Safety Professionals Handbook, Volume II: Technical Applications. Des Plaines, IL: American Society of Safety Engineers.

Hoyos, C. G. and B. M. Zimolong, 1988. *Occupational Safety and Accident Prevention: Behavioral Strategies and Methods.* New York: Elsevier.

Johnson, W. C., 1980. *MORT Safety Assurance Systems.* New York: Marcel Dekker.

Kavianian, Hamid R. and Charles A. Wentz., 1990. *Occupational and Environmental Safety and Management.* New York: Van Nostrand Reinhold.

O'Donnell, Michael P., ed., 1984. *Health Promotion in the Workplace.* New York: Wiley.

Parks, K. S., 1986. *Human Reliability: Analysis, Prediction and Prevention of Human Errors.* New York, Elsevier.

Peterson, Dan., 1989. *Techniques of Safety Management.* 3rd ed. Des Plaines, IL: American Society of Safety Engineers.

Roberts, N. H., W. E. Vesely, D. F. Hansl, and F. F. Goldberg, 1981. *Fault Tree Handbook.* NUREG-0492. Washington, D.C.: Nuclear Regulatory Commission.

Roland, Harold E. and Brian Moriarity., 1990. *System Safety Engineering Management.* 2nd ed. New York: Wiley.

Stephans, Richard A. and Warner W. Talso, eds., 1997. *System Safety Analysis Handbook.* 2nd ed. Springfield, VA: System Safety Society.

4

Work Systems and Ergonomics

Daniel J. Ortiz, MPH, CSP

In Part I of this book, including chapters one, two, and three, the role of management in the prevention of accidents was stressed. In Part II, the human element is to be considered. In Part III, the work environment will be studied, and in Part IV, the equipment used in the workplace will be studied. These four elements make up the work system. No one of these sections will, by itself, prevent accidents. Safety is a team effort, and all four areas must be integrated in order to accomplish the goal of eliminating accidents.

It is interesting to note, however, that people are involved in all of these areas. Management is made up of people. It is for people, primarily, that safety becomes such an important subject. It is people who design and maintain the work environment and the equipment used in this work environment. It is paramount, then, that the behavior of people and the abilities and limitations of people be understood as clearly as possible. Attention will be given to these matters in this chapter. First, we will examine the attributes of people and then consider how people interact with the environment and the equipment they use.

Introduction to Ergonomics

Ergonomics is the multidisciplinary science concerned with optimizing human performance by matching the task to the physical and mental capabilities of the human operator. People may be studied from two broad points of view. In one, the person may be thought of as a machine. In the other, the person may be considered in the context of mental, emotional, or psychological elements—the psychosocial and behavioral aspects of people. In safety engineering, it is important to study the human being in both ways. To conduct an examination from either approach alone is simply an academic exercise, for the mind and body cannot be separated in the living human being.

In many ways, the human body is like a complex machine with three major systems: the skeletal structure of the body resembles the frame and support members of a machine; the muscular system provides and transmits power to the various body members; and the nervous system, including the brain, corresponds to the control system of a machine. In the human body, we find electrical circuits, hydraulic systems, levers, pivots, and complex chemical operations. We have not been able to duplicate all of the intricate machinery of the body, nor do we completely understand how many of these systems function.

As we know, the possible outcome of human system incompatibilities can range from poor product quality and customer satisfaction to injuries and catastrophic system failure. Consequently, where there is a user interface in a process or operation, in order to achieve optimal system performance, it is necessary that human considerations be fully integrated early and throughout the system life cycle. To accommodate the user, it is essential to understand the limits and capabilities associated with human cognition, physical and sensory characteristics that enhance or constrain overall system performance. The safety engineer especially must understand human capabilities and limitations and consider them in all aspects of his or her job.

Information Input

One very important part of human endeavor is the input of data—that is, how the brain receives information and directs various body parts to respond. Except for a few short tasks that are repeated so often that they become habit, every motion requires a signal of some sort and decisions as to what that signal means and what the response should be. A few responses are so well learned that they become automatic.

All other responses result from visual, auditory, tactile, or other form of signal. Our sensory organs must recognize the signal and send the message to the brain. The brain must make a decision based on what it knows about that signal and pass along through the nervous system a message to do something in response. Failure to recognize the signal, for any reason, will result in failure to respond. Similarly, misinterpretation of the signal or lack of knowledge about it will usually result in a wrong response.

The brain can handle only so much information. When signals arrive at the brain too fast or too many at one time, the brain must delay decision. Some must be put on hold until other decisions are made, and some may be lost entirely. This is especially likely to happen if the worker is not well trained to recognize or understand the signal(s). This should tell us that giving too much information will slow the response, while inadequate or vague information will usually result in error. We must design the input data accordingly.

Visual Capabilities

Printed information, scales, and gauges require that the worker not only learn to recognize that information, but also that he or she must be able to see it clearly. Visual acuity is the measure of one's ability to see small details. Acuity is measured as the smallest angle between one side of the character to the other that can be seen, as illustrated in Figure 4-1. Acuity is expressed as the reciprocal of the angle in minutes of arc. If the angle is 2.0 minutes, the visual acuity is ½ or 0.50. Thus, a higher number indicates better visual acuity.

Figure 4-1. Measurement of visual acuity.

The four major factors beyond native capacity that affect visual acuity are luminance (or brightness), luminance contrast (referred to merely as contrast), background reflection, and motion. Other factors that affect the ability to see—but not in terms of visual acuity—include location of the object in the field of vision, adaptation to darkness, color discrimination, changes in focal distance, and changes in direction. To help clarify terms, the term illumination refers to the amount of light that reaches the working surface. Luminance refers to the amount of light given off by some source.

In general, the greater the luminance or intensity of light, the better. However, as intensity increases, reflection from other surfaces or sources also increases. More often, though, too little rather than too much light is the problem. Within a reasonable range of luminance, the contrast between the reflectivity of the detail and that of the area immediately surrounding it is very important. The greater this contrast the easier it is to see the detail.

Background luminance refers to the ability of surfaces outside the immediate work area to reflect light into the worker's eyes. This should be low. Ceilings should be a light color to reflect light downward as much as possible. Walls should also be a light color down to about eye level and a darker color below that level. Floors should be quite dark. The background, including desktop and machine surfaces, should be a moderately dark color and nonreflecting. Green, gray, and buff colors are commonly used to fulfill these requirements.

Relative motion between the worker and the object makes seeing more difficult. The faster the motion, the more difficult seeing becomes. If visual information is important, any such relative motion should be reduced to a minimum. Compensation for motion can sometimes be provided by maximizing luminance and contrast. Relative motion between the viewer and the visual display should be eliminated; where it cannot, the display should be tested under simulated conditions.

The human eye can receive an image from a very large field of vision. But if high visual acuity is required, the field becomes very limited. Acuity diminishes very rapidly as an object moves away from the direct line of sight, as illustrated in Figure 4-2. To be sure a person's visual attention will be captured, objects or signals should be kept within about 30° of the worker's line of sight. Alphanumeric characters and quantitative scales normally require concentration within a few degrees, generally less than 5°.

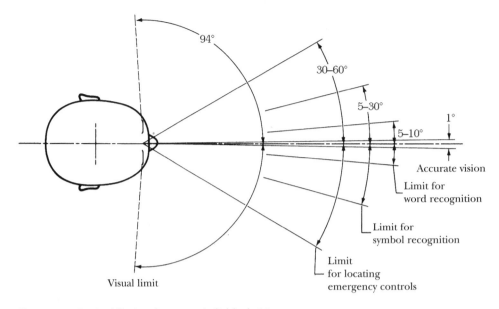

Figure 4-2. Typical limits of a person's field of vision.

Illumination and Color

The important characteristics of lighting for safe and healthful working conditions include the following:

1. Sufficient luminance (brightness)
2. Sufficient contrast
3. Lack of glare
4. Acceptable color.

Some of these characteristics are interrelated. Luminance is the measure of
the amount of light emitted from the source or sources of light.

Illuminance, often called the illumination level, is the measure of the
amount of light that falls on the surface being observed. The amount of
luminance which should be provided depends on each of the other factors
listed above.

$$\text{Illuminance (Lux)} = \frac{\text{candlepower}}{D^2}$$

where candlepower is measured in candellas (cd).

Our eyes are accustomed to an array of frequencies, or color, of light.
If light of some other array of frequencies is reflected toward our eyes, it
produces sensations that we often regard as weird light. Reflected light may
tend to emphasize or diminish certain colors.

Technically speaking, we do not actually "see" an object, but only the
light that is reflected from that object. The qualities of that reflected light are
a result of the frequencies of the light source and of the pigment in the object
absorbing or reflecting those frequencies. For our purpose, we will call this
phenomenon "seeing."

To the engineer, the important issue is how to achieve sufficient illumination, contrast, and color without glare. Ideally, the engineer should be able to do the following four things:

1. Select the best type of light source
2. Select the best luminaries
3. Locate these luminaries in the best place
4. Select the best colors for backgrounds.

The three categories of light sources for artificial lighting are incandescent, fluorescent, and high-intensity discharge. Incandescent lamps are used primarily for localized lighting. The major advantage is the quality of the light—that is, the range of frequencies. The major disadvantage is that, because of the point source of light, it is difficult to control glare.

Fluorescent lamps are available in a variety of color components or frequencies. They provide a broader source of light and are also more efficient in the amount of current used for a given level of illumination. Fluorescent lamps are used almost exclusively by industry and in most other buildings for general lighting. They are even being used in residences. A wide variety of reflectors and diffuser plates makes it possible for fluorescent lamps to provide almost glare-free and shadowless illumination.

High-intensity-discharge lamps include mercury-vapor and sodium lights. Metal halide is another, less commonly used, form. The light from these sources does not provide for color characteristics as well as incandescent or fluorescent lamps, but they are very efficient, have long endurance, and project light that is capable of penetrating long distances. This makes them well-suited for high overhead lamps for general emergency lighting, both inside buildings and outdoors. A major disadvantage, however, is that without an outer cover, these lamps emit very harmful ultraviolet light.

When lighting is designed for industrial buildings, three levels—roof lighting, ceiling or intermediate lighting, and local lighting—are generally considered. Roof-high lighting is lighting at the highest point—that is, roof-support level. Today, this commonly involves mercury-vapor or sodium lamps. In a growing number of industrial plants, the roof support is not exposed. Instead, a false ceiling is hung approximately 3 meters (10 feet) above the floor, leaving a plenum area above the false ceiling, in which case no lighting is usually provided at the roof level.

The ceiling or intermediate level is at the 3-meter level, whether or not there is a false ceiling. Fluorescent lamps are almost always used at this level. Local lighting is that which is provided at the workstation, mounted on the bench or machine within a meter of the point at which the worker is concentrating. Incandescent lamps are usually used on machines because they are smaller in size. Either incandescent or fluorescent lamps are used at a bench or desk.

Cleaning and replacing weak or burned-out lamps should be done on a periodic schedule to assure proper lighting. This is usually the weak point in an otherwise good lighting program. Table 4-1 is a partial listing of the recommended levels of illumination for different tasks, as developed by the Illuminating Engineers Society (IES). These are widely used as guides in industry.

Table 4-1. Recommended Illumination of Selected Areas

	Nominal Illuminance Values	
	in Lux units	in footcandles
Office reception area	30	3
Material handling inside a truck	100	10
Materials handling picking and wrapping	300	30
Simple assembly room	300	30
Rough bench work area	300	30
Simple inspection	300	30
Maintenance	500	50
Difficult inspection	1000	100
Material processing (cutting very fine)	1000	100
Fine bench or machine work	3000	300
Inspection exacting	3000	300

Source: *Illuminating Engineers Society of North America. Lighting Handbook.10th ed. New York: IES, 2011.*

One last step in designing illumination is the background color scheme. Ceilings and upper portions of walls should be a light color but not shiny; flat white, cream, or ivory are commonly used. These colors are good reflectors, so that a large percentage of the light emitted from the lamps is reflected downward. To avoid glare, walls below eye-level height should be a little darker and definitely not shiny. Light green, gray, or buff are the most common colors. Floors should be darker still, but not too dark. Darker shades of green, gray, or brown are commonly used. Within the immediate work area, color schemes can be designed to provide good contrast without glare.

Auditory Capabilities

Hearing is another sensory ability that we use more than we realize. Our hearing ability is unique in many ways. We can discriminate among a variety of sound stimuli and choose what we want to hear to a surprising degree. There are limits to human hearing, of course, and if we wish to convey a message quickly and accurately, we need to know what they are. Normal hearing covers a range of frequencies from about 20 hertz to about 20,000 hertz. Hearing is a fragile sense and is easily damaged. For this reason, noise control is a very important subject in safety engineering and will be discussed in the chapter on noise and noise control. It should be assumed in designing auditory signals that many adult workers have some hearing impairment, particularly in the upper range of frequencies.

If a worker is expected to hear and understand an auditory signal—a bell, a whistle, a siren, or speech—that signal should be easily recognizable among the variety of sounds that make up the background noise. A steady, low-intensity background noise is called *white noise* and is seldom a problem, but louder sounds and intermittent or meaningful sounds distract a person's attention. When one sound tends to obscure another sound, it is called *masking*. Masking may be unintentional and very bothersome when it obscures sounds that need to be heard. But masking may be used intentionally to block out undesirable sounds. All sounds likely to exist must be considered, whether one is designing a signal to be heard or a noise to serve as a mask.

There is often a choice of what type of signal to use to get the attention of the worker and convey a message. Visual signals and auditory signals each have their advantages. The relative characteristics of each are given here.

1. A wider range of codes and variety of information can be expressed with visual signals than with auditory signals, except for speech.
2. Visual signals may be designed so that they can be referred to continuously or repeatedly for a long period of time, whereas auditory signals must be presented in a given sequence and a person cannot refer back to them.
3. Auditory signals are generally better attention-getters and are much superior in most environments for warning and alarm.
4. Auditory signals do not require that the hearer be oriented in any particular direction, as visual signals do, nor are they dependent on lighting conditions.
5. If a great amount of information must be conveyed, an auditory message is less fatiguing than a visual signal.
6. Speech is the most versatile form of signal because it can be changed on the spot as conditions warrant. Speech can also accommodate questions and an exchange of information.

In general, the criteria for the selection of visual versus auditory signals can be summarized as shown below.

Use visual signals if:

1. The message is complex or the worker will need to refer back to it.
2. The message contains spatial relationships.
3. The worker is already burdened with auditory signals.
4. The environment is very noisy.

Use auditory signals if:

1. The message can be made short and simple.
2. The message requires immediate attention.
3. The worker's visual system is already burdened.
4. The environment lacks good illumination.
5. The worker is not likely to be oriented in a given direction.

Use speech if:

1. The message is very complex.
2. The message requires much versatility.
3. The worker is expected to ask questions or offer some feedback to the message.

Another factor to be considered is the confidence a worker places in the message. Because people can refer back to visual messages or stimuli, they often place more trust in them. A written or illustrated message is obviously no more or no less truthful than an auditory or spoken one, yet people frequently accept a picture or printed word more readily and consider it more factual.

Tactile Capabilities

We use our tactile sensory abilities more than we realize. Although it is seldom used to convey messages, the sense of touch does help us to gain information. Nerve endings in the skin, especially at the end of the fingers and at the root of the fingernails, enable us to detect quite small variations in surface height and rather large variations in temperature. From experience we learn to judge smoothness and texture, and, to some extent, we can judge forces by moving the fingers over or against a given surface or edge. A slow back-and-forth motion of the fingers enhances sensitivity to detect smoothness and texture, but fast motion greatly diminishes sensitivity.

The fingers are usually too busy manipulating tools to be used as channels of information. Nevertheless, if the fingers are going to touch tools, controls, and other materials anyway, it can be useful to design those tools and controls to convey simple messages about themselves. The US Air Force has experimented with shape codes to identify the many controls pilots must distinguish and use very quickly. This is a very useful application when visual and auditory channels are already busy with other signals. The tactile sense can and should be used when environmental conditions make visual and auditory signals impractical. The switch used to turn a fan on or off may be made a different shape than the one used to turn the lights on and off. The application of tactile sense is further discussed in Chapter 16: "Tools and Machine Controls."

Human Motor Capabilities

Motor capabilities refer to the movements of body members as well as the force, speed, and accuracy of those movements. These movements are used to guide tools and controls, to exert force on them, and to change the position and location of the body itself—all by the action of muscles. The muscles contract by means of a chemical reaction that is initiated by nerves. Some muscle systems operate involuntarily and without our deliberate control. They include muscles of the heart, the respiratory system, the digestive system, and other internal organs. Some body motions occur without conscious effort and are the result of training. They have become so habitual that we are not aware of directing the action. It is believed that such motions are directed by nerves in the spinal column rather than by the brain. There is certainly much that we do not know about how the nervous system operates. In safety engineering, we need to study these habitual motions, particularly those that may influence behavior in times of emergency. People tend to respond to a signal on impulse or the way they have been accustomed to respond, rather than from a logical consideration of the circumstances. It is extremely important for the safety engineer to recognize situations in which a worker might respond to a signal in a dangerous manner.

For the most part, we need to study those motions that require conscious effort. Practically all human physical work involves hand and wrist movements. The fingers, hand, and wrist constitute an extremely versatile system for doing work, though they obviously have limits of motion, strength, and speed. The hand is normally held in a cupped shape, not in a flat horizontal or vertical position. Any deviation from that cupped position requires effort and time, which, no matter how insignificant, accumulate over many moves. When accuracy and speed of hand and finger motions are required, large deviations from a normal position should be avoided. Handles, knobs, and tools of any kind should be made to fit the hand and fingers that will use them.

In many tasks, the hands and arms repeat a sequence of steps over and over. The object or objects involved, the location, and the purpose of the task may vary from one situation to another, but the types of motions are repeated, as follows:

1. reach for the object,
2. grasp the object,
3. move it to a given location,
4. position the object with respect to another object,
5. use the object in some way,
6. release the object.

The *reach* and *move* steps are similar in that they generally involve arm motion but very little, if any, finger motion. The effort and time for these steps depend on several factors, but the main factors are the nature of the object to be grasped or positioned and the path of motion involved. Two ranges of arm motion have been defined and are referred to as the *normal work area* and the *maximum work area*. These are illustrated in Figure 4-3. The normal work area is that area that can be reached by extending and moving only the lower arm—that is, bending at the elbow with the upper arm essentially stationary. The maximum work area consists of space that can be reached by moving the whole arm without changing the position of the shoulders, back, or legs. The normal work area is obviously more restrictive, but requires less effort and time. In most cases, it is desirable that all work be done within the normal work

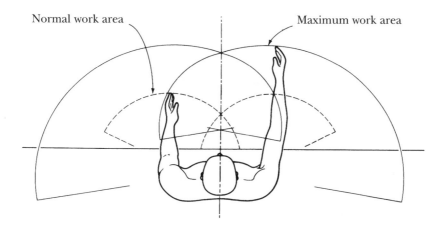

Figure 4-3. Normal and maximum work areas shown on the horizontal plane. The word "area" should be interpreted to include the vertical plane.

area. If that is not feasible, it should at least be kept within the maximum work area. Moving the back is tiring and slow and involves a much greater risk of injury. If the task requires greater strength than can easily be done with the lower arm, then the upper arm should be used.

The *grasp* step involves the fingers and, sometimes, the hands. There are three types of grasps: a pressure grasp, a hook grasp, and a contact grasp. A pressure grasp requires the squeezing of the object with the fingers or the whole hand, thereby exerting a force sufficient to gain control of it. A hook grasp involves using the hand as a hook or platform so that the object can be lifted without squeezing it. The hook grasp is not a common way to grasp an object and usually, if one starts to pick up an object in this way, one quickly closes the fingers around or against the object, and it becomes a pressure grasp. A contact grasp involves touching the object and pushing it along a surface without picking it up. The pressure grasp requires considerably more effort but results in much better control of the object. The contact grasp is very fast and requires less effort but results in little control unless the object is guided in a track of some sort.

In designing work methods, consideration should be given to the nature of the object and what is to be done with it. Although it is well to minimize the effort in grasping, it is often important—and even essential—that the worker have good control of the object. The best way to achieve this may be to leave the object in a fixed position and eliminate the need to grasp it. A fixture or holding device is often used for this purpose.

If an object to be grasped is left from a previous task, it should be left in such a position that it can be grasped most easily. An everyday example is placing a pin in a pincushion (holder) rather than leaving the pin lying on a flat surface, where it is hard to grasp and may be lost. Grasp often requires concentration and may even require visual guidance as well. Because the brain can process only one bit of information at a time, it is impossible to grasp two separate objects simultaneously if either or both grasps require mental concentration. In designing work methods, it is better to avoid simultaneous grasps unless the objects are close together, easily grasped, and intended to be moved together.

The positioning sequence is similar to the grasp in that it usually requires mental concentration or visual guidance. Position includes orienting and aligning one object with respect to another and fitting them together. If this involves tossing a small part into a large box or bin, this step is very simple. If it involves manipulating two close-fitting pieces, then it requires much more concentration and time. As with grasp, it is often difficult or impossible to position two pairs of objects simultaneously. It is equally difficult to grasp one object and position another at the same time. Any activity that requires mental concentration should be done independently of any other activity. To expect a worker to do otherwise is to invite errors and accidents. Table 4-2 is a guide for an engineer to use when designing simultaneous motions of the left and right hands.

Movement of the head and eyes is also required in many tasks. Generally the head is moved for one of two purposes: to help align the eyes in a particular direction or to help achieve body balance. Movement of the eyes should be minimal if it is important that the person see something quickly and accurately. Focusing the eyes becomes very tiring if the change of focal distance is great and/or frequent.

Table 4-2. Relative Ease in Performing Activities Simultaneously with Left and Right Hands

L.H. \ R.H.	Reach	Move	Grasp	Position
Reach	Easy	Generally easy with a little practice	Easy if grasp is easy, but usually difficult	Easy if position is easy, but usually difficult
Move	Generally easy with a little practice	Easy unless destinations are far apart and precise, then difficult	Easy if move and grasp are both easy, but difficult if far apart and/or difficult grasp	Easy if move and position are both easy, but usually difficult
Grasp	Generally easy with a little practice	Easy if move and grasp are both easy, but difficult if far apart and/or difficult grasp	Easy only if parts are together and easy to grasp; usually difficult	Difficult in any case and may be impossible
Position	Easy if position is easy, but usually difficult	Easy if move and position are both easy, but usually difficult	Difficult in any case and may be impossible	Impossible

The feet and legs can sometimes be used to relieve the hands of work, and in some cases can do a better job. The leg muscles are much stronger, but they are slower and not as easily controlled for accurate movement. Since the feet and legs are used to maintain body balance in a standing position, one should avoid trying to operate pedals or to do other work with the feet while standing. In fact, such action can easily throw a person off balance and cause an accident.

The trunk of the body is supported by a very weak bone structure, the spine, which also contains the major nerve system. For these reasons, the back should not be used for lifting or carrying any more than a light load, especially when the spinal column is in a bent or twisted position. The segments of the spine, the vertebrae, are attached to each other by ligaments rather than by bone joints, and the ligaments are quite easily torn. The nerves in the spinal column are partially enclosed by extensions of these vertebrae and are susceptible to damage if the vertebrae are forced out of line. Injury to the vertebrae, damage to the nerves, and excessive stress to the ligaments are among the most common ailments of workers doing physically active work. Much research is being done on low-back pain and injury because they are so common. Figure 4-4 illustrates the construction of the spinal column.

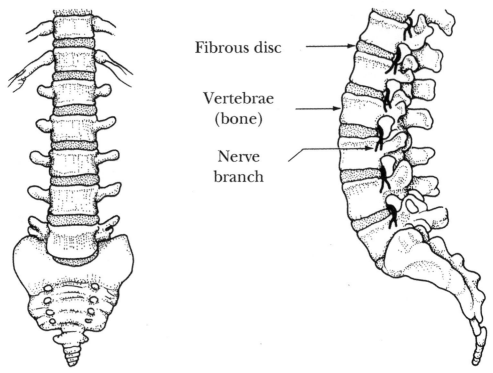

Figure 4-4. Anatomical construction of lumbar (lower-back) region of the spinal column.

Work-Related Musculoskeletal Disorders

The incidence of musculoskeletal disorder (MSD) cases in 2010 increased four percent over 2009. MSDs are the leading cause of disability among people in their working years, costing industry billions of dollars each year. The extent of the problem is revealed in the Bureau of Labor Statistics' analysis of 2010 injury and illness cases involving days away from work (Publication USDL-11-1612).

In 2010, Musculoskeletal Skeletal Disorders accounted for 29 percent of all workplace injuries and illnesses resulting in lost time. The 346,400 MSD cases required a median of 11 days away from work. In 2010, sprain or strain injuries accounted for 40 percent of total injury and illness cases requiring days away from work. Forty-three percent of sprains or strains were the result of overexertion. Eleven percent were the result of falls on the same level. In 36 percent of the sprain and strain cases, the back was injured with a smaller percentage involving the shoulder. Twenty-six percent of the cases involved strains and sprains to the lower extremities, with most of the cases involving the knee. Of all strain and sprain cases, injury to the shoulder and, as an event, repetitive motion required the most days to recover, 21 and 24 days, respectively. Carpal tunnel syndrome cases required a median 27 days to recover, second only to fractures.

Five occupations had MSD case counts greater than 10,000. Of these occupations, nursing aides, orderlies, and attendants had the highest incidence rate of 249 MSD cases per 10,000 full-time workers and also the highest case count. The common factor associated with MSDs for this group is patient or resident handling.

Table 4-3 summarizes the risk factors associated with upper-extremity musculoskeletal disorders and back injuries based on the strength of the evidence

Table 4-3. Evidence for Causal Relationship between Physical Work Factors and Musculoskeletal Disorders (*From NIOSH Publication 97–141*)

Body Part **Risk Factor**	*Strong* *Evidence* **(+++)**	*Evidence* **(++)**	*Insufficient* *Evidence* **(+/0)**
Elbow			
Repetition			X
Force		X	
Posture			X
Combination	X		
Hand Wrist			
Tendinitis			
Repetition		X	
Force		X	
Posture		X	
Combination	X		
Carpal Tunnel Syndrome			
Repetition		X	
Force		X	
Posture			X
Vibration		X	
Combination	X		
Hand-Arm Vibration Syndrome			
Vibration	X		
Back			
Lifting Forceful Motion	X		
Awkward Posture		X	
Heavy Physical Work			X
Whole-Body Vibration	X		
Static Work Posture			X

in the scientific literature. The combination of risk factors, especially force and repetition, has a strong association with hand wrist tendinitis, carpal tunnel syndrome, and elbow tendinitis. A strong causal relationship exists between back injuries and worker exposure to whole-body vibration and heavy, forceful lifting.

Lifting

There is evidence that lifting and forceful movement, awkward postures, heavy physical work, and whole-body vibration are work factors associated with back injuries. Other factors that must be considered when evaluating lifting tasks include the frequency of lifting, the horizontal location of the object away from the body, the vertical location of the object at the start and end of the lift, and the degree of upper-body twisting. Lifting with the back in a hunched position is still common among industrial workers, even though there has been considerable publicity and training on "proper" ways of lifting (see Figure 4-5). It is difficult to illustrate or describe the correct way to lift, because there are so many variable factors. These factors include the size, weight, and shape of the object, where it is located with respect to the body,

Figure 4-5. Flexing of the torso can increase the risk of back injury.

how far it is to be lifted, the surface conditions of the object, and the surface conditions of the floor where the worker is standing.

It is always desirable to keep the object as close to the body as possible (see Figure 4-6). It is usually desirable to keep the spinal column in as nearly a normally erect position as possible and free from twisting. The muscles of the legs should be used to minimize the load on the back muscles. Obviously, the worker should have a firm surface to stand on and a good grip on the object. Lifting should be done with a smooth, continuous motion. Quick, jerky motions should be avoided.

Control strategies should focus on engineering out the exposures as much as possible or practical. This includes converting the manual task to a mechanical task. A lifting device could be provided for lifting any heavy object and when lifting must be done repeatedly for a long period of time. A wide range of lifting devices are available, including cranes, hoists, conveyors, and lift trucks. Improvements can be made to the workstation and layout to prevent torso bending and twisting. If lifting must be done, it should be minimized and kept between knuckle-to-shoulder height as much as possible. A carry

Figure 4-6. Although no lifting technique has been shown to be protective by itself, if lifting must be done from floor level, the load should be handled as close as possible to the spine to minimize the compressive force on the lower back. (From NIOSH Publication No. 2007-131)

should be converted to a push, with the wheels of the mechanical aid as large as possible in order to minimize the force it takes to set it in motion. There is no clinical scientific evidence that back belts (see NIOSH Publications 81-122 and 94-122), lifting technique, and training alone prevent back injuries. Job redesign is the only effective means of protecting workers.

One common type of back injury is the displacement of the fibrous disc between the vertebrae. This happens when the spinal column is put under undue pressure while in a curved position. Commonly, the spinal column is bent forward, and the fibrous disc is forced to the rear, pressing it against a nerve branch and, sometimes, actually tearing the disc.

In 1981, NIOSH published a report titled "Work Practices Guide for Manual Lifting" (NIOSH publication 81-122). The formula that was derived in this application was a back injury risk assessment tool to evaluate manual material handling lifting tasks. In 1991 NIOSH revised the guide (NIOSH publication 94-110) to include additional factors and consider additional research findings. The result was a revised formula for calculating a recommended weight limit (RWL) for manual material handling lifting tasks. According to NIOSH the recommended weight limit is a weight that should cause a compressive force on the discs of the spinal column of no more than 350 kg (770 pounds) and a metabolic rate between 2.2 to 4.7 kilocalories per minute. Working above the biomechanical and physiological limits could increase the risk of injury. Lifting at the RWL would be acceptable to at least 75 percent of the female and 99 percent of male workers. This is important since employees working at jobs that are acceptable to less than 75 percent of the population are three times more likely to sustain a low back injury, according to Stover Snooks and Liberty Mutual.

The RWL is the product of six different multipliers and the load constant. As Table 4-4 shows, coupling (hand-to-object grasp or point of contact) and

Table 4-4. 1991 NIOSH Lifting Multipliers

Factor	Multiplier	
	U.S. Customary	Metric
Load Constant	51lb	23 Kg
Horizontal (HM)	10/H	.25/H
Vertical (VM)	$1 - (.0075 \Omega V - 30\Omega)$	$1 - (.003 \Omega V - 75\Omega)$
Distance (DM)	$.82 + 1.8/D$	$.82 + 4.5/D$
Frequency (FM)	From Table* 4-5	From Table* 4-5
Asymmetry (AM)	$1 - .0032A$	$1 - .0032A$
Coupling (CM)	From Table* 4-6	From Table* 4-6

*__Source:__ *Application Manual for the Revised NIOSH Lifting Equation, NIOSH Publication Number 94-110, 1991.*

H = horizontal distance of the hands in front of the midpoint of the ankles when the lift is started, measured in centimeters or inches,

V = vertical distance from the standing surface to the hands when the lift is started, measured in centimeters or inches,

D = vertical distance from beginning of the lift to the end of the lift, measured in centimeters or inches,

A = Asymmetry in degrees

F = lift frequency (lifts per minute)

Table 4-5. Frequency Multiplier Table (FM)

Frequency Lift/ Min (F)*	Work Duration					
	≤ 1 hour		> 1 but ≤ 2 Hours		> 2 but ≤ 8 hours	
	V < 30*	V ≥ 30	V < 30	V ≥ 30	V < 30	V ≥ 30
≤ 0.2	1.00	1.00	.95	.95	.85	.85
0.5	.97	.97	.92	.92	.81	.81
1	.94	.94	.88	.88	.75	.75
2	.91	.91	.84	.84	.65	.65
3	.88	.88	.79	.79	.55	.55
4	.84	.84	.72	.72	.45	.45
5	.80	.80	.60	.60	.35	.35
6	.75	.75	.50	.50	.27	.27
7	.70	.70	.42	.42	.22	.22
8	.60	.60	.35	.35	.18	.18
9	.52	.52	.30	.30	.00	.15
10	.45	.45	.26	.26	.00	.13
11	.41	.41	.00	.23	.00	.00
12	.37	.37	.00	.21	.00	.00
13	.00	.34	.00	.00	.00	.00
14	.00	.31	.00	.00	.00	.00
15	.00	.28	.00	.00	.00	.00
>15	.00	.00	.00	.00	.00	.00

* Values of V are in inches. For lifting less frequently than once per 5 minutes, set F =.2 lifts/minute

Table 4-6. Coupling Multiplier

Coupling Type	Coupling Multiplier	
	V < 30 inches (75 cm)	V ≥ 30 inches (75 cm)
Good	1.00	1.00
Fair	0.95	1.00
Poor	0.90	0.90

asymmetrical (i.e., twisting) lifting are taken into consideration in addition to the horizontal, vertical, distance, and frequency multipliers. Once calculated, it is compared to the actual load handled. The quotient of the load, divided by the RWL, is the lifting index (LI). Lifting above the RWL, an LI greater than 1, would constitute an increased risk of low back pain and injury that would need to be addressed through some combination of engineering and administrative controls. Under the best conditions, the individual, multipliers will equal one (multipliers can be less than or equal to "1"), so the maximum RWL is 23 kilograms or 51 pounds. This formula is viewed as a more conservative, risk-assessment tool when compared to the 1981 guide. Even so, research is still needed to validate this model to determine whether or not the RWL is protective and a good predictor of back injuries.

$$RWL = 23kg \times HM \times VM \times DM \times AM \times FM \times CM$$

$$LI = Load/RWL$$

Multi-Component Manual Material Handling

There are several limitations to the application of this assessment tool. These are addressed in detail in the NIOSH Publication 94-110, Application Guide. The formula does not apply to one-hand lifts, working in hot environments and lifting for over eight hours. The NIOSH lifting guide provides an analytical procedure to assess multiple lifting tasks providing several examples, including a multi-tiered pallet operation. Even so, manual material handling almost always involves more than one task component, including carrying, pushing and pulling. Lowering arguably exposes workers to the same risk as lifting. The NIOSH guide does not, however, take the other components into consideration. Consequently, it could underestimate the risk to injury for more involved manual material-handling tasks. In the Task Analysis section, we provide a good source for assessing multi-component tasks published by Liberty Mutual, a recognized leader in the field of back injury prevention.

Hand-Intensive Work

As depicted in Figure 4-7, work-related, upper-extremity musculoskeletal disorders (also known as cumulative trauma disorders, repetitive motion disorders, or strains) involve damage to the tendons, tendon sheaths (e.g., synovial and bursal membranes), bones, muscles, and nerves of the upper extremities. Key entry, repetitive use of tools, and repetitive reaching and grasping are activities associated with work-related upper-extremity musculoskeletal disorders (WRUEMDS). They are common in industries where there is hand-intensive work, including meatpacking, poultry processing, automobile assembly, and garment manufacturing. The target regions often affected include the neck, shoulder, elbow, and hand/wrist. Carpal tunnel syndrome, tendinitis, and hand-arm vibration syndrome are three examples of this class of disorder.

Carpal tunnel syndrome is a disorder that results from the compression of the median nerve that provides sensation to the palmar side of the hand. Symptoms include pain, numbness, and tingling of the hands. If left untreated, the muscle that controls the thumb may atrophy. Tendinitis is the result of tendon inflammation, causing dull pain, tenderness, and movement discomfort around the affected area. Vibration syndrome or Raynaud's syndrome is characterized by blanching of the fingers due to constriction of the local blood vessels. This will be discussed further in the section on vibration.

Risk factors include awkward sustained postures, repetition, forceful exertions, and low-frequency vibration. Examples of awkward static postures include a flexed (Figure 4-8) or deviated wrist and shoulder abduction. Frequent reaching to the side, back, and above-shoulder and forearm rotation have also been implicated as risk factors. There is good evidence that worker exposure to a combination of factors is associated with an increased risk of injury or illness.

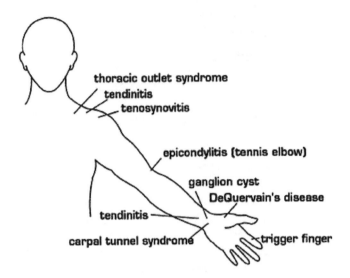

Figure 4-7. Location of common upper-extremity musculoskeletal disorders.

Control strategies should focus on workstation and tool design to reduce or eliminate worker exposure to the risk factors. The redesign effort should allow the individual to assume a posture that is optimal for upper-extremity muscular work (e.g., straight wrist, minimal reaching, and elbows down close to the body). Administrative controls in conjunction with sound engineering controls can help further reduce the exposure. Examples include providing adequate rest periods (for highly repetitive work, 5 minutes every hour in addition to the established break periods) and medical surveillance to detect early symptoms before they become more serious. More research is needed to validate the effect of workplace control measures on reducing the incidence of musculoskeletal disorders. It may take months or perhaps longer following the implementation of a control program before one observes a reduction in the lost-workday incidence rate. It is expected, however, that improvements in work tolerance and worker comfort and performance will be observed much sooner.

The use of fingers and hands and the reach-grasp-move-position sequence were discussed earlier, and it was mentioned that the hand is normally held in a cupped position. It must also be pointed out that grip strength is maximal when the wrist is held in a position in line with the lower arm. This position

Figure 4-8. Flexing of the wrist is a musculoskeletal disorder risk factor.

of the wrist should be maintained as much as possible throughout the work activities. Flexing and laterally bending the wrist reduces hand strength significantly.

Flexing of the wrist is rarely a problem if the flexing is done only occasionally and little muscular exertion is involved. If the effects of pinching and chafing of these nerves are recognized soon enough, they can be relieved by rest or by changing activities. These symptoms are, first, discomfort; then pain; and eventually, disability. Early symptoms include tingling of the thumbs and the next two fingers, usually at night. Frequently, however, the symptoms are often ignored until the condition is serious. In many cases, surgery is required to correct the condition. Carpal tunnel syndrome occurs among office workers as well as factory workers.

Vibration

Vibration that causes audible sound will be discussed, along with noise, in another chapter. Low-frequency vibration can also be disturbing and even physiologically damaging. Although such damage has been recognized for many years, much more research in this area is needed.

Vibration may be transmitted from an external source in such a manner that it is further transmitted through the skeletal structure to all body members. This is called *whole-body vibration*. If the vibration is transmitted to part of the body only, such as a tool's vibrating the hand or arm, it is called *segmental vibration*. Either one can cause damage.

When the whole skeletal structure is vibrating, as when standing or sitting on a surface that is vibrating, the internal organs of the body may not respond to this vibration in the same way as the bone structure. Like any mechanical structure, bodies of different mass can be excited at different natural frequencies. Generally, a greater mass will have a lower natural frequency. When the spinal column and rib cage are excited at some given frequency, the heavier internal organs may tend to remain stationary, thus causing a strain on these systems. In a similar way, we sometimes feel the lower jaw chattering against the upper jaw when the skull is put into vibration.

Differences in frequencies of vibration of adjacent body members are expressed in terms of acceleration ratios. The acceleration ratio of head to shoulder is greatest at frequencies of 30 to 40 cycles per second (hertz). The ratio for lower jaw to skull is greatest at 100 to 300 Hz, and for eyeball to skull it is greatest at 60 to 90 Hz.

Whole-body acceleration ratio peaks occur at 3 to 6 Hz and again at 10 to 14 Hz. Difficult breathing (called "dyspnea") has been reported at 1 to 4 Hz. Chest pain and general discomfort have been reported from 1 to 15 Hz.

Body members can tolerate some vibration without damage. Limits to this tolerance depend on many factors, and this makes it difficult to set limits. Variables include frequency of the vibration, amplitude, and direction of transmission. Tolerance varies considerably from one person to another. The term *tolerance* is also subject to interpretation. Recommended limits are generally aimed at the prevention of physiological damage.

OSHA has not yet adopted standards regarding vibration limits; however, ANSI has published a guide (ANSI 53.18, "Guide for the Evaluation of Human Exposure to Whole-Body Vibration").

Raynaud's syndrome, or "white fingers," is caused by vibration of the hands or fingers. It is aggravated by cold temperatures. Circulation of blood is impaired. This may be a temporary condition but may become incapacitating if allowed to continue for long periods of time. Frequencies ranging from 40 to 125 Hz may bring about this condition.

Among the greatest offenders in whole-body vibration disorders are industrial lift trucks and agricultural machinery. Suspension systems on these machines are not well designed. Work platforms that are attached to vibrating equipment may transmit that vibration. It is the vertical component of vibration that usually causes problems, especially with whole-body vibration.

The first approach to correcting any problem caused by vibration should be an attempt to eliminate the source of that vibration. If that is not possible or feasible, an attempt to interrupt, or damp, the transmission of that vibration should be made. The line of transmission includes any medium through which that vibration may travel from the source to the worker's body, including a cushion or other padding on which the worker is standing or sitting. Such a cushion should be the last resort, to be used only when other attempts have failed to reduce the vibration sufficiently.

Identifying the exact source is essential to any attempt to solve the problem. The source may become quite apparent by listening to or touching various parts of the device. If it is not, it may be necessary to use instruments to measure frequencies. Calculations of the speed of rotation or pulsation of some machine part that matches the measured frequencies will often pinpoint the source. In some cases, proper lubrication of moving machine parts or replacement of worn bearings may reduce or eliminate the vibration and should be tried first. If that fails, a change in speed, if feasible, may eliminate the vibration.

If the source cannot feasibly be changed to reduce the vibration, consider the path of transmission. Common culprits are shafts and panels that have natural frequencies such that they will vibrate at the frequencies being generated by the source. Changing the mountings of these parts to make them either more rigid or less rigid may stop this transmission.

Changes in the design of a seat or standing platform may alter the natural frequencies of those structures so that the offending frequencies will be damped. Vibration-absorbing pads on seats or platforms will sometimes be sufficient. It is important to learn what the offending frequencies are and deal with them, rather than trying to damp other frequencies that are not causing a problem. Damping of higher frequencies that create sound is discussed in the chapter on noise.

Task Analysis Tools

There are many different approaches to task analysis for the recognition and control of risk factors associated with musculoskeletal disorders. They range from computer models and posture tracking systems to simple job observation where the job is broken down to its fundamental components, documenting any observed stressors (job-safety analysis). Video registration and analysis (VIRA) allows a more detailed review of the job, since the task can be observed in slow motion. High-stress postures can be recorded and the amount of time and number of posture changes can be documented and categorized by the reviewer. With VIRA, one gets a more complete picture of the duration of exposure to risk factors.

Manual material handling and upper-limb assessment tools are available on the internet. A sample of these are listed as follows.

Rapid Entire Body Assessment (REBA) and Rapid Upper Limb Assessment (RULA) other ergonomic assessment tools:

- http://ergo.human.cornell.edu/cutools.html
- http://personal.health.usf.edu/tbernard/ergotools/index.html

Liberty Mutual psychophysical tables for assessing lift, lower, carry, push, and pull tasks:

- http://libertymmhtables.libertymutual.com/CM_LMTablesWeb/pdf/ LibertyMutualTables.pdf

Quick checklist assessment and other tools for the state of Washington

- http://www.lni.wa.gov/Safety/Topics/Ergonomics/ ServicesResources/Tools/default.asp

The Occupational Safety and Health Administration (OSHA) and the National Institute for Occupational Safety and Health (NIOSH) Ergonomics web addresses:

- http://www.osha.gov/SLTC/ergonomics/index.html
- http://www.cdc.gov/niosh/topics/ergonomics/

Worker–Machine Systems

Within the whole work system described earlier, we often speak of the worker-machine subsystem. This usually consists of one worker using a tool or a machine to accomplish a given task. It may, however, involve two or more workers or a variety of equipment, but the important thing is the interrelationship between a worker and the equipment being used. The word *machine* is used loosely to mean any machine or tool the worker uses.

A worker-machine system may be very simple, such as a person using a pencil, or it may be very complex, such as a crew operating a ship at sea. Such complex systems may be broken down into simpler units for the purpose of studying that system. Within even the simplest worker-machine system there are two subsystems—the worker and the machine.

In analyzing a worker-machine system, it may be helpful to think of the machine as an extension of the person. People have eyes to see things but sometimes use lenses to help them see small details or distant objects. The telescope and the magnifying glass extend natural human capabilities. Similarly, wrenches and screwdrivers increase our ability to apply torque, motors increase our ability to create and/or control rotation, and welding torches increase our ability to apply heat. The human body is capable of a myriad of different activities; it is the most versatile of all the machines. But it is limited in how much or how well it can do any of those tasks.

A few of the tools that have been devised can be attached to the body and manipulated as if they were a part of it. Most of these are corrective devices, such as contact lenses or even eyeglasses. Most machines are foreign to the body and cannot be adapted to it so easily. When designing a tool or device, we always must consider carefully how well the body will accept it. We should be sure that the skeletal structure can accommodate the device, that the muscular system can manipulate it, and that the nervous system can control it. We should try to ensure that the person can use the machine without undue stress and without damaging the device. If these criteria are not fulfilled, then undesirable results—accidents—are likely to occur. The basic idea in safety engineering is to design machines and tools that are compatible with the capabilities and limitations of the person who will be likely to use them.

One aid in studying this compatibility is referred to as the S-O-R (stimulus-operator-response) model. This is a model of a single activity in which, after some signal is given, the person responds with some bodily activity. There are two interfaces between the person and the machine in this model. In the first, the S-O interface, the person must detect the stimulus and recognize what it means. If the person fails to do either of these correctly, it is very unlikely that a correct response will be made. In the second interface, the O-R interface, the person must respond with some physical activity, such as move a hand with a tool in it or turn a knob. Errors can be made in responding to the stimulus as well as in receiving it. The S-O-R model is usually assumed to represent the correct response for a given stimulus. If an error is made, it is called an *S-O-R error*.

There are three types of errors that can be committed in a worker-machine system. Generally, they are thought of as errors of the worker rather than of the machine, but machines can make errors, too. The three types of errors are:

1. intentional errors,
2. unintentional errors, and
3. errors of omission.

Intentional errors are those that we commit while believing that we are making the correct response, but that are wrong. Unintentional errors are those we commit accidentally. Someone who enters an elevator and wants to go to the fifth floor would deliberately push the button marked "5." If such a person somehow misread the numbers and pushed the "6" button thinking it is a "5," that person would have committed an intentional error. If, however, his or her finger slips onto the "6" button, that would be an unintentional error. An error of omission, of course, is a failure to make any response or to recognize that a response is called for. A person's ability to detect a stimulus, make a decision, and respond depends on many factors. One's mental and physical capabilities certainly pose potential limitations.

When a signal or stimulus is given, people react according to their experience and training. We have done some things so many times that they become routine, and we do not have to think at all about them in order to respond. Other signals require only a little thought in order to respond. Such reactions are referred to as *stereotyped reactions*. Stereotyped reactions can be divided into two categories with respect to the kind of relationship that exists between the stimulus and the response associated with it. One type of reaction involves a well-understood spatial relationship of location and/or motion. The other type involves a conceptual relationship.

The words and sentences in a book are arranged to be read from left to right and from top to bottom. This is the accepted arrangement in our culture and we depend on that relationship when we read. Any other arrangement would be confusing to us. This is an example of a spatial relationship.

The letters of the alphabet are arbitrary shapes that we have come to recognize, and when letters are put together in some sequence, the word or symbol becomes meaningful. The shape of the letter, however, has no significance in itself. We have to learn and memorize the many different shapes. Colors with different meanings are often used for signs or charts and the observer must learn what the colors mean. These are examples of conceptual relationships.

When a person has learned the relationships, he or she can get information very rapidly from a display using those relationships. A young child has to examine each letter in a storybook and then decide what the combination of letters means. An older person who has studied speed reading does not have to look at each letter or even each word; several words are perceived in one glance. The more experience a person gains in using any particular relationship, the easier recognition will be the next time it occurs.

In designing a worker-machine system, one should take advantage, whenever possible, of known stereotyped responses. These enable the worker to make a faster and more accurate response. In choosing displays or designs that will bring about stereotyped reactions, one must consider who will be likely to use that display or device. Not all people have had the same background and training. Many codes and symbols that are widely accepted in one field are meaningless outside that field. Yet, because codes and symbols can convey so much information quickly and accurately, it is often better to use them and train workers to understand them.

Table 4-7 illustrates several common relationships that should be used whenever they are appropriate. A spatial relationship that occurs very frequently in machine and equipment design is the right-hand screw principle. Whenever a rotating motion is translated into linear motion, the relative directions of motion should be the same as if the rotating member were a right-hand screw and the linear-motion member were a nut engaged with the screw. Table 4-8 illustrates several color codes that have been adopted and used quite widely. These may not yet be considered as stereotyped in the minds of people in the general public, but in many industries they are well understood. A designer should try to visualize each step of the worker's operation of the device to ascertain what motions and relationships the worker will expect.

In some cases, people have learned more than one relationship for a given set of circumstances. Consider the arrangement of numerals on a small calculator as shown in Figure 4-9. A person familiar with such a calculator can use it without searching for the keys. Figure 4-10 shows a pushbutton telephone with the same numerals arranged in a different order. Now we have two well-established but contradictory relationships. Similar situations should be avoided, especially if the worker must act quickly and accurately.

How much should we depend on the worker's past experience and training when designing equipment? Does the advantage of codes, symbols, or other special designs outweigh the advantage of stereotyped reactions? Certainly, if a common stereotyped reaction can be used conveniently, it should be, but at times it may be more feasible to achieve this reaction through new training.

A very important factor was mentioned earlier: the brain can process only one thought at a time. The brain develops a hierarchy by which it places

Table 4-7. Common Stereotyped Reactions

Conceptual Relationships

Danger, stop, or hot are implied by the color red.

Safe or go are implied by the color green.

Caution is implied by the color yellow, a yellow or amber light, or alternate yellow and black stripes.

Danger or warning is suggested by a uniformly interrupted sound of a high pitch or by a sound that varies up and down in frequency.

Danger or a failure is expressed by a blinking light.

Spatial Relationships

Words and sentences on a sign are expected to read from left to right; vertical sequence, either up or down, is confusing.

Quantities on a gage or meter, whether expressed numerically or not, are expected to increase from left to right or from bottom to top.

Rotating knobs or scales are expected to increase something (flow rate, temperature, etc.) in a clockwise direction and decrease in a counterclockwise direction.

Handles for controlling the flow of liquids, however, are the reverse of other knobs and increase the flow by turning counterclockwise. Hot and cold water faucets, though, may turn either way.

Any toggle or flip-type switch is turned on by moving it to the right, or by moving it up.

A threaded lead screw moves something away from the knob or handwheel by turning it clockwise and moves it closer by turning it counterclockwise.

A knob to open a door or cover opens it by turning clockwise.

A latch on a door or cover is released by lifting it up.

A swinging door is expected to open in the direction one is travelling if one is in a hurry, no matter which way one is going.

Table 4-8. Standard Color Codes Recommended by the American National Standards Institute[*]

Color	To be used as identification of:
Red	(Danger) Safety containers of flammable liquids, warning signs, (Stop) emergency-stop buttons and bars, (Fire) fire protection equipment and apparatus
Orange	Hazardous parts of moving machinery, opened guards and enclosures
Yellow	Physical hazards which cause tripping, falling, etc.
	Containers and cabinets for flammable materials, corrosive materials
Green	Location of safety equipment such as first-aid kits, stretchers, deluge showers, etc.
Blue	Equipment under repair
Purple	Radiation hazard
Black and White	Traffic and housekeeping information and signs
Black and Yellow	Marking traffic aisleways

[*]NOTE: Recommended in ANSI/NEMA Z535.1-2006 (R2011). Some are required in OSHA, Section 1910.144.

priorities on incoming information. This hierarchy is the product of training and experience, but it may also be subject to the whims of the moment. Regardless of priorities, the brain will accept and process only one bit of information at a time. If two bits of information reinforce each other, they may be processed as one, but otherwise they vie for time in the system.

This characteristic of the worker-machine system is often overlooked. Too frequently a task involves two or more simultaneous activities that both require mental concentration. This situation invariably results in errors, if not accidents. Many tasks in industry have been designed without much understanding of the mental or physical capabilities of the worker. Human factors have been sadly neglected, and our record of accidents and low morale attests to this.

Accidents usually happen when something—whether a tool, a piece of equipment, or a person—fails to do what was intended at a particular time. One common and very useful project a safety engineer can undertake is the study of the reliability of a system such as a worker-machine system. If both the worker and the machine perform their tasks properly, the system will never fail.

Let us consider the ways in which the worker can fail and the ways in which the machine can fail. The S-O-R model provides clues about how the worker can fail, especially if the S-O and the O-R interfaces are studied. The outline below reviews the several areas and means of human failure.

1. Failure to detect signal or stimulus
    a. limitations of sensory capabilities
    b. worker not in proper location to detect stimulus
2. Failure to interpret stimulus properly
    a. lack of education, training, and experience
    b. capability temporarily reduced by stress

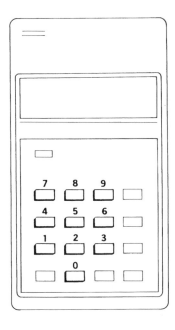

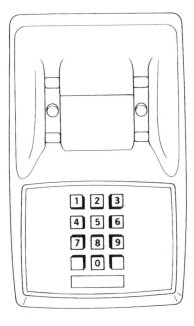

Figure 4-9. Arrangement of keys on a calculator.

Figure 4-10. Arrangement of keys on a pushbutton telephone.

3. Failure to respond properly
   a. lack of training
   b. limitations of physical movement, strength, or speed

Any one of these failures has the potential to cause an accident, and, therefore, anything done to prevent these failures helps to prevent accidents.

The machine can fail as readily as the worker. There are three general categories for machine failure.

1. Inadequate design of machine components,
2. Errors or deficiencies in the manufacture of the machine,
3. Wear and abuse of machine parts.

Using the S-O-R model again, we can find places where the machine can cause a failure of the worker-machine system. Note that a machine failure is not necessarily a broken part. It can be any case in which the machine fails to do what it was intended to do. The following outline shows the possible failures of the machine.

1. Failure to produce an adequate signal to the worker
   a. stimulus is outside the sensory capabilities of the worker
   b. stimulus is masked by interference
   c. stimulus is not given in proper location or direction
2. Failure to respond to a worker's actions
   a. failure of control components
   b. failure in transition of power or motion
3. Failure to do task in acceptable manner
   a. machine not designed to do assigned task
   b. poor quality of components
   c. wear of components.

Worker versus Machine

It should be apparent that people possess some characteristics far superior to any that could be designed into a machine, at least with today's technology. On the other hand, people have some very serious limitations, many of which do not exist or are less severe in machines. Table 4-9 lists some of the major ways in which a person excels over machines, and others in which a machine can be designed to excel over a person.

One rule that should always be followed in assigning tasks to people and to machines is that people should always feel superior to the machine or tool that they use. What makes a person feel superior to the machine may vary depending on the person's background, training, and personal interests.

The author took his class to visit a nearby industrial plant where the students observed what appeared to them to be an extremely boring job. Two men were seated on opposite sides of a circular machine with a narrow conveyor belt traveling near its perimeter. Workpieces were carried on this conveyor through a series of stations at which they were formed, inspected, and then automatically ejected into a bin. Each part traveled 180° so that there were two sets of identical processes. The two workers who picked up the parts and placed them in holding devices on the conveyor could have been replaced by automatic feeding devices. The students asked their guide if the workers objected to this boring job. The guide replied, "Oh, no, they like this job—they feel they've got it made. The job is so easy they don't even have to watch what they are doing. They can sit there all day and talk and see what's going on around them—and get paid for it."

Getting the right person for the job is often as important as designing the job for a person. There are many complex machines that require only an attendant to make sure everything is running properly and to shut them off if something goes wrong. Some people would be completely bored with this job, but others enjoy it, feeling they are the masters of a powerful machine. On the other hand, a worker on an assembly line can feel belittled by and enslaved to a conveyor that moves faster than he or she can work.

Three conditions should exist to make a job satisfying to a worker, namely:

1. the job must utilize the worker's skills,
2. it must be meaningful, and
3. the worker must have real responsibility.

To utilize the worker's skills means that the job must be difficult but within the range of potential competence of the worker, that it must permit a performance that is less than the worker's best, and that there must be feedback to the worker.

Table 4-9. Comparison of Human/Machine Performance for Various Activities

People Excel In:	Machines Excel In:
Detection of low levels of light and sound	Performance of routine, repetitive, or very precise operations
Sensitivity to an extremely wide variety of stimuli	
Perception of patterns and formulation of generalizations about them	Ability to respond very quickly to control signals
Detection of signals in high noise levels	Exertion of great force, smoothly and with precision
Ability to store large amounts of information for long periods of time and to recall relevant facts at appropriate moments	Ability to store and recall large amounts of information in short periods
Ability to exercise judgment where events cannot be completely defined	Performance of complex and rapid computation with high accuracy
Ability to improvise and adopt flexible procedures	Sensitivity to stimuli beyond the range of human sensitivity, such as infrared, radio waves, etc.
Ability to react to unexpected, low-probability events	Ability to do many things at one time
	Insensitivity to extraneous factors
Ability to apply originality in solving problems	Ability to repeat operations rapidly, continuously, and precisely the same way over and over
Ability to profit from experience and alter course of action	Ability to operate in environments hostile to people or beyond human tolerance
Ability to perform fine manipulation, especially where misalignment appears unexpectedly	
Ability to continue to perform even when overloaded	
Ability to reason inductively	

When a machine is already designed and alterations do not seem feasible to adapt it to a given type of employee, it may be reasonable to select an employee to fit the machine. If the machine can be designed for a given job, the needs of the worker should certainly be considered first, even in cases where there is equal capability between worker and machine.

Human Behavior

The reliability of a worker-machine system depends on the reliability of both the worker and the machine. Both are prone to breakdown or failure. Even machines that are mass-produced vary slightly from one another. Yet machines, in general, are far more predictable than people. The physical abilities of people have been studied quite thoroughly and can be predicted fairly well. Physical dimensions and strength have been measured and recorded and are referred to as *anthropometric data,* usually given for a range of sizes as 5th, 50th, and 95th percentile values. Some of this data is presented in the chapter "Tools and Machine Controls" (see Tables 16-1 and 16-2).

Prediction of behavior is not so easy. Certain human characteristics are fairly consistent, but so many variables affect human behavior that prediction is difficult. Psychologists have developed several theories of human behavior, and some of them have been utilized by management. Three of these theories are particularly appropriate to human behavior in work conditions. Douglas McGregor proposed two concepts of worker motivation. His Theory X assumes that workers are not motivated by any satisfaction they might derive from doing a good job, but rather only by pay or other rewards and by the fear of disciplinary action. If management wants the worker to perform in a safe manner, it must provide some tangible reward for doing so and some punishment for failure to do so.

McGregor's Theory Y assumes that the worker is, or can be, motivated by job satisfaction. The task of management is to find ways to make the job satisfying to the worker. Studies and observation indicate that most workers are motivated to some extent by job satisfaction, but that they also need some external incentive to achieve what is expected of them. This raises the question "Has management done a good job in designing the task and working conditions to motivate the worker as much as it can?" This should be continually evaluated.

Another theory, referred to as Theory Z, has been proposed by W. G. Ouchi. This is detailed in his book, *Theory Z: How American Business Can Meet the Japanese Challenge.* Theory Z refers to the relationships among workers and the team effort. Peer pressure, a sense of identification, and camaraderie all have an effect on the performance of each of the workers involved.

Abraham Maslow has presented another theory that is widely accepted. He contends that there is a hierarchy of human needs. Starting at the lowest level, a person concentrates on satisfying those needs, and only when this has been done will he or she move on to the next, higher level. Figure 4-11 shows Maslow's need hierarchy. This diagram might suggest two approaches. First, management might try to place a worker in the type of task best suited to his or her level of need. Second, management might try to provide conditions and opportunities that would encourage the worker to ascend to the next level.

These theories and concepts may not always explain a worker's response to a given situation, but they frequently may help. Theory that is validated over time becomes a piece of the jigsaw puzzle describing human behavior. The more pieces we can fit into the puzzle, the better able we are to design tools, machines, and working conditions for the worker.

Another approach to behavior focuses on the development of attitudes in the mind of the worker. Each of us is the product of all the experiences we have had in our lives. The attitudes developed in childhood remain with us for a long time, probably forever. But attitudes can be changed if a sufficiently strong incentive is provided. Attitudes vary with situations, as well; we do not hold the same attitudes toward all people, all jobs, or all conditions. A worker can have good feelings about his or her work or about other people in spite of antagonistic feelings about the world in general. Those things that affect needs will have the greatest influence on attitude.

In a work environment, the factors that most affect workers' attitude—such as working conditions or the design of tools or equipment—can generally be controlled or influenced by management. The attitude of management toward the worker's welfare can play a major role in shaping workers' attitudes. Management even has some control over the attitude of fellow workers. And, as emphasized in Chapter 1, almost nothing is more important than attitude in determining behavior and safe performance in the workplace.

The Body in Stress

Stress means different things in different contexts, even within one field, such as safety engineering. When we speak of the human body in stress, we are generally referring to the physiological and emotional symptoms that are associated with worry and anxiety over work pressures, family problems, and so on. These symptoms include gastrointestinal disorders, ulcers, cardiovascular

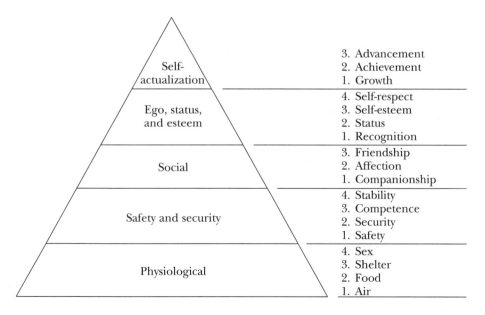

Figure 4-11. Maslow's hierarchy of needs.

disorders, as well as emotional problems of fear, jealousy, and moodiness. There has been a lot of research and publicity on this subject in recent times. Many causes and treatments have been presented and tried with varying success.

Another form of stress occurs when the sensory channels are overloaded with information being transmitted to the brain. The nerve endings in the eyes, ears, and skin have a tremendous capacity for detecting visual, auditory, and tactile signals. It is estimated that up to one billion bits of information per second can be received by these sensory organs. However, only about 0.3 percent of this amount (about three million bits per second) can be transmitted through the nervous system. Of this, only about 16 bits per second flow through the brain as conscious information. An even smaller amount, less than one bit per second, is retained in memory. This breakdown was presented by Dr. K. Steinbuch at the Institute of Radio Engineers' International Congress on Human Factors in Electronics in 1962, not as factual data but as an illustration of the information-input process.

When information comes to the brain faster than the brain can process it, the person is under stress. Only one bit of information can be processed at a time. Other information must be held in some order of priority until it can be processed (time sharing). The longer information waits, the greater are the chances of distortion and error, with concomitant mental fatigue or stress. Applying the principles discussed earlier concerning sensory capabilities will help a great deal in reducing such stress and, thereby, in reducing errors and accidents.

The motor capabilities of the body can be stressed also. Physical fatigue, like mental fatigue, is a complex phenomenon, consisting of three related elements:

1. a feeling of tiredness,
2. physiological changes in the body, and
3. a reduced capacity to do work.

The feeling of tiredness has been found to have little correlation with the capacity to do work. Apparently, it serves as a warning device. Yawning appears to be a mechanism by which the lungs can inhale larger than normal quantities of oxygen. This implies a need of oxygen somewhere in the body, but this is not well understood. The brain, of course, is one organ that requires oxygen, and, in fact, if the brain is deprived of sufficient oxygen, the person quickly dies.

The muscles use oxygen to replenish the chemicals needed for contraction. Blood carries oxygen to the muscles through the circulatory system and carries away the waste products of the chemical reactions, primarily carbon dioxide and water. The bloodstream also transports the nutrients needed for muscular and other bodily activities. The normal basal metabolism is sufficient to carry the oxygen and nutrients needed for light physical activity. When an increase in muscular or nervous activity takes place, the various systems work harder to provide the oxygen and nutrients. Breathing speeds up, the heartbeat quickens, and blood pressure rises. All these changes can be measured with the proper instruments and are sometimes referred to as the physiological cost of doing work.

When oxygen is not supplied as fast as it is needed, the body begins to build up an oxygen debt. The muscles cannot function to their full capacity because lactic acid, formed as part of the chemical reaction in muscular contraction, tends to collect in the muscles and restricts their movement. When the worker stops doing this excessive physical work, the faster breathing, rapid heartbeat, and increased blood pressure continue until the oxygen debt is repaid, and then all systems return to normal.

The position of the arms affects the flow of blood and, therefore, the amount of oxygen carried to the arm muscles. The most efficient flow of blood occurs when the upper arms are nearly parallel with the body and the hands are about at the level with the waist or chest. When the arms are raised above the head, the arm muscles tire very easily. If any of the blood vessels are pinched, blood will not flow as readily and the muscles tire easily. Such pinching may occur when an arm or elbow or finger is bent.

Blood flow is necessary throughout the head, although it is not known exactly how the oxygen is used in the head. Mental activity does cause fatigue, and this type of fatigue seems to be a more difficult condition from which to recover than physical fatigue. Mental fatigue seems to be associated with psychological and emotional stress.

A reduced capacity to do work is another characteristic of fatigue. Workers complain of being tired and seem to lack the capacity to do their jobs. Other factors such as diet, sleep, and worry obviously contribute to the physical condition of a worker, but many cases of loss of the capacity to do work require further explanation. One major factor that has been found to affect a person's apparent capacity to do work is mental attitude. Many researchers and authorities believe that attitude is the greatest single factor. As we have seen in so many other areas, attitude—the product of education and experience, working conditions, and contact with the attitudes of others—is the crucial factor.

Figure 4-12 helps to illustrate how stress causes accidents. A worker's ability to perform a task continually varies due to variations in mental stress, fatigue, and distractions. This is represented by a jagged line varying above and below a "normal" level of activity. The demands of a task also vary from one moment to another. This is also shown by a jagged line varying above and below a "normal" level of demand.

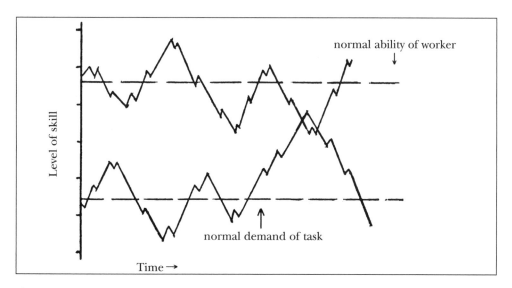

Figure 4-12. Concurrent variations in ability and demand.

Obviously, a worker's ability to perform a task should exceed the demands of the task. Thus, the normal level of ability is shown to be above the normal demand of the task. However, if at some instant a peak demand exceeds the ability of the worker, failure to perform the task properly is likely to happen and be the cause of an accident. In most accidents where this situation occurs, it is most likely to be called *human error*.

To prevent such an occurrence, it is necessary to study carefully the variations of the task and take steps to reduce the variation. It is also important to study the possible variations in the ability of the worker. That may be more difficult to control, but personnel selection, training, and supervision should do much to help.

Safety Training

Training is a fundamental element of any safety and health management system. Safety training is a management responsibility, and the safety engineer may often be called upon both to provide safety instruction and to ensure compliance. In OSHA's General Industry and Construction Industry standards, there are more than 100 paragraphs where worker training is required. The development of training should be based on a needs assessment, identifying goals and objectives, developing learning activities, conducting the training, and evaluating and improving the training program as needed. Safety training begins with the hiring of an employee.

Company policies on safety and health should be made clear at the time of hiring. Observation of safety and health rules and standards should be emphasized as an employee responsibility and a condition of employment. The point is underscored by the fact that OSHA holds the employer responsible for the actions of employees when performance standards, which depend on employees' actions, are enforced. To maintain control over employee actions, the employer must make it known that compliance with these standards is a condition of employment and that violations may result in termination of employment.

A new employee or a person assigned to a new job rightfully expects proper training for the job. In fact, if an employee is injured because of lack of proper training, the employer is held responsible.

One aspect of employee training that is frequently neglected is what to do in abnormal conditions. Accidents usually do not happen when everything is running normally; they happen when something abnormal occurs, and training should include dealing with these out-of-the-ordinary conditions. Fault trees and other techniques of identifying likely sources of failure, along with good record keeping, should help to predict what may go wrong. The employee should be taught to recognize the symptoms that warn of an impending failure, as well as what to do if this occurs.

Very often an unexpected event causes a person to panic and act irrationally. Even if the panic lasts for only a few seconds, the transitional reaction may be enough to turn a minor accident into a catastrophe. The phenomenon of irrational thinking has not been studied enough, but there is no lack of evidence that it exists.

To illustrate this last point, let's consider a common task and, unfortunately, a common accident. A maintenance worker needs to enlarge the diameter of a hole in a piece of sheet metal from 1/2 inch to 5/8 inch. He decides to use a drill and to hold the sheet metal down on the table of the drill press with his other hand. As the drill bit starts to break through the sheet metal, it catches and begins to twist the piece. If the worker were thinking rationally at this point, he would release his grip on the piece and shut off the machine. But, instead, he tries to grasp the piece and gets cut as the sheet metal whirls around. This irrational thinking has been repeated by a great many workers, often with the same results.

Another example illustrates two points. A man driving a car is following another car, a little too closely perhaps, when the car ahead suddenly stops. The driver of the car following anxious to stop quickly, moves his foot to the left, only to realize that his foot is under the brake pedal instead of on top of it. By this time he has hit the car ahead. Quickly he moves his foot to the right and hits the accelerator while a bystander yells, "Hit him again!" Instead of lifting his foot before he moved it to the left, he waited until after he moved it, and then on the next move he pressed down with his foot about the time it was over the accelerator. This sequence of motions sounds funny, but it has happened many times and will, no doubt, happen again.

Does this incident mean that the man was not a good driver? No, this accident has involved experienced drivers. He knew what sequence of motions was required, but when an unexpected event occurred, he panicked for a moment and failed to think rationally.

Another question is raised here, however. Was the designer of the car at fault in designing the brake and accelerator pedals? Since this incident has happened many times and follows a pattern, it should be easy to see that there is a problem. It would be easy to design pedals so that it could not happen again. The more we consider human responses in the design of a worker-machine system, the fewer accidents that will occur.

All too often when a chair, workstation, or tool is introduced to the workplace as an ergonomic intervention, it is not fully utilized or accepted by the intended user. To improve the success of any intervention, the worker should be part of the selection and implementation process. Training is essential to get the full benefit of any ergonomic or safety design change and should occur at the management, supervisory, and worker levels. For the worker it is important to cover the use and maintenance of the equipment and why the change was made. If the equipment is adjustable, then it is important to demonstrate how to make the adjustments and why it is important. Training must be an integral part of a successful hazard recognition and control program and will be accomplished through some combination of class room and on-the-job training.

Horseplay is something that must never be permitted in any work area, especially where there are power tools, switches, and valves. Many accidents have occurred when workers unintentionally, and even intentionally, started equipment while pushing one another around. This is a problem that is often difficult to control. It should be considered a serious offense and cause for reprimand and eventual dismissal from employment. Many companies provide an opportunity for employees to let off steam outdoors or in a game room away from the hazards found in a work area.

Computer Use in the Workplace

According to the Bureau of Labor Statistics, as of October 2003 (their most current publication), more than 77 million workers used a computer as part of their job, representing over half the working population. Computer use in the workplace poses ergonomic challenges associated with prolonged sitting and constrained postures, lighting and vision, and repetitive motion associated with keying and mouse use. Psychosocial and organizational factors may also contribute to the overall computer user stress level. It is easy to see how a poorly designed computer workstation and job can create incompatibilities leading to user performance and health concerns.

Poor lighting conditions and glare are serious and common problems, but ones that can be corrected. Recommended illumination for offices in which paperwork is done is usually about 45 to 50 footcandles (500 Lux). But for areas where video display terminals (VDTs) are used, illumination should be much less—about 25 to 30 footcandles (300 Lux) is the recommended level. Full color-spectrum fluorescent lighting provides better quality illumination than incandescent lighting. One conflict concerning lighting is the requirement for higher illuminance values for viewing hardcopy documents with small text or low contrast. For interactive computer tasks like data entry and report processing, the use of an adjustable task light would be recommended.

Glare, both direct and reflective, is one of the most common problems with VDTs. It is important that all sources of light are controlled so that light does not reflect off the screen into the operator's eyes. The highly reflective screen's surface makes this a difficult problem. Glare filters placed in front of the screen have been tried. They help control glare, but they also reduce the quality of the display. Barriers may be used to prevent ambient light from reflecting off the screen. To control glare, workers may prefer individual workstation lighting. The computer workstation should be arranged so the user is not facing any bright light source, like a lamp or the sun through a window.

Lower-back pain, neck and shoulder pain, and wrist pain are common complaints among computer users. These are usually the result of poor posture and/or the lack of support. These problems are common among other office workers also. Computer operators are often more restricted in their freedom of movement, which adds to the problem.

The arrangement of the computer work area is critical in that there are usually three focal points that need to be coordinated—the keyboard and mouse, the screen, and some source of data or information. Generally, the keyboard is the first to be positioned. The operator's hands and wrist should be three to four inches above his or her elbow. The mouse should be located close to the keyboard to prevent excessive reaching. The screen should be directly—but not too far—above the keyboard. The top of the screen should be about at the same level as the operator's eyes and from 18 to 24 inches in front of the operator's face. A document holder situated close to the monitor can help minimize head rotation and head tilt when going from the screen to hard copy. If the operator must look at the keyboard a lot, it might be better to drop the screen to a level nearer to that of the keyboard to reduce the distance the eyes have to refocus between the two locations.

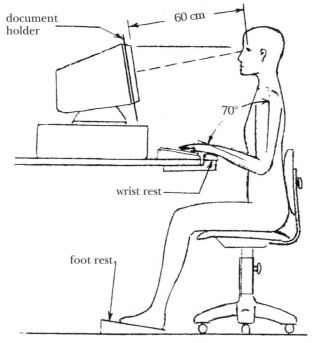

Figure 4-13. Computer workstation.

Figure 4-13 shows a computer workstation for an operator whose task is to type from data. Document lighting should be directed onto the data source but not cause reflected light to shine into the operator's eyes.

Other focal points should be located so that the operator's eyes do not have to make changes in focal distance any more than is necessary. If the operator's wrists must remain in a fairly fixed position for long periods of time, padded rest supports are very helpful. They should be placed under the forearm just behind the wrist.

To accommodate operators of different heights and proportions, the chair should be adjustable. The monitor should be mounted on a tilting and swiveling base. Many of these recommendations apply to general office work stations as well, but they are more critical where video display terminals are used. As in other office jobs, frequent changes in activity are very helpful. Brief rest periods, even while sitting, can relieve tense muscles. Management can sometimes relieve this situation by assigning other tasks periodically that require moving about. For guidance in designing VDTs and computer workstations, consult the American National Standards Institute's standard, HFS100, "Standard on Video Display Terminals."

Providing an adjustable, flexible workstation (computer components and furniture) is one of the most important steps in addressing postural stress associated with day-to-day continuous computer use. Researchers have demonstrated that fully adjustable furniture can result in a reduction of worker discomfort. In one study, users were allowed to make "preferred" adjustments to their workstation. The average computer user adopted a posture resembling one that is used to drive a car, with a slightly reclined back and arms slightly opened or extended (see Figure 4-14.). This "driver's" posture resulted in fewer complaints. Sit/stand workstations have also been shown to improve user

Figure 4-14. Reclined sitting posture.
(From http://www.osha.gov/SLTC/etools/computerworkstations/index.html)

Figure 4-15 Laptop docking station. (From Stanford University EH&S "Ergonomics Guide to Mobile Devises," OHS 12-063- 4/2012)

comfort for some computer tasks. Clearly, a detailed task analysis is required to determine the appropriate interventions. Some questions to answer include: Is the task screen intensive, document intensive or interactive? Is the task mouse and/or keyboard intensive? Are there peak periods of activity?

The use of laptops should comply with the workstation guidelines discussed in this section. This is challenging, especially with use out in the field where any work surface will do. The obvious problem is the monitor and keyboard are attached, making it difficult to place it in a position that accommodates both the visual and postural considerations of the task. When in the office, a docking station could be set up to allow the use of a standard keyboard and monitor. Figure 4-15 shows the use of a special fixture to put the monitor at the right height, using a detached split keyboard instead of the laptop keyboard. More research is needed to determine if laptop use poses the same or greater risk.

In a prospective study looking at the incidence of musculoskeletal symptoms and disorders among computer users, researchers found that more than half of the users developed musculoskeletal symptoms in less than a year at a new job. Among the more than 600 computer users, neck shoulder disorders were more frequent than hand arm disorders. Carpal tunnel syndrome was rare in comparison affecting one out of 100 computer users. Several studies have found statistically significant associations between upper extremity musculoskeletal disorders and certain postures, work activities, and psychosocial factors including those listed in Table 4-10. It is important to note with psychosocial risk factors, and risk factors in general, the results are often inconsistent across studies. "Social support" for example may be a risk factor in one study but not in another.

Table 4-10. MSD Risk Factors Associated with Computer Use

Risk Factor	Symptom/Disorder
Time spent keying	Hand Arm
Non neutral hand/wrist postures	Hand Arm
Head rotation	Neck
Phone use	Neck
Keyboard Height	Neck, Shoulders and Arms
Downward Head Tilt	Neck Shoulders
Head Rotation	Neck Shoulders
Mouse use	Hand Arm
Job Dissatisfaction	Neck and Shoulder
Perceived Time Pressure	Neck Shoulder
Limited job control	Hand Arm
Monotonous work	Neck Shoulder

The workstation guidelines discussed in this section will largely address the posture-related risk factors. Rest breaks to allow muscle and tendons time to recover should be provided hourly for continuous work. This should involve getting away from the seated workstation and doing something else. Alternative style keyboards and devices (e.g., mouse and track ball) can put the hands in a neutral posture. Location of any external devices should be close to the keyboard to limit reaches and users raising their elbow away from the side of the body (See Figure 4-3 "Normal work area"). Head sets to replace the phone and document holders can reduce head rotation. Providing additional resources and adjusting breaks for periods of heavy, intense computer activity could also help to minimize the stress on the upper extremities. In order to be successful, computer operators must be provided formal training on the use and maintenance of any new equipment introduced in the workplace. Training is especially important with alternative keyboards.

There is no solid clinical evidence that any single control intervention will result in the reduction of musculoskeletal disorders associated with computer use. The one consistent finding across studies is the importance of ergonomic training in conjunction with workstation modifications and the use of arm rests. To solve ergonomic challenges in a system where there is a human user interface will require a comprehensive approach that examines carefully the different elements and stressors associated with the job. Human performance can be enhanced by optimizing the relationship between the worker and the environment, technologies, and the organizational structure and design constraints of the task. The design of any system must fully consider the limitations and capabilities of the user.

QUESTIONS

1. What are the three major systems in a machine, and how does the human body compare with such a machine?
2. Distinguish between an intentional error and an unintentional error.
3. Should a designer incorporate into a design signals to which there are recognized stereotyped reactions? Explain why or why not.
4. How does the brain treat signals coming to the brain simultaneously?
5. What is the major difference between McGregor's Theory X and Theory Y?
6. List Maslow's hierarchy of needs in order of priorities.
7. What type of signal would be most effectively employed to warn pedestrians of a moving forklift truck in a machine shop?
8. What are the advantages of an auditory signal over a visual signal?
9. Why is it difficult to do certain types of activities simultaneously with the left hand and the right hand?
10. List the risk factors of carpal tunnel syndrome. Which factor has the strongest evidence of causation?
11. What are the three elements of physical fatigue?

12. CASE STUDY

A material handling task involves lifting a carton weighing 15 kg filled with small parts from floor level to a roller conveyor. The employee begins the task by reading the label on top of the carton then selecting or picking the correct carton for transfer to the conveyor. The carton has a good handhold (i.e., coupling) for gripping. The horizontal (H) location of the carton throughout the lift is 30 cm. The vertical hand location at the start of the lift is 15 cm and at the end, 75 cm. This job is an eight-hour continuous operation requiring two lifts per minute. There is no twisting (i.e., asymmetrical lifting).

Given: The Vertical Multiplier at the start of the lift (VM= 1–.003/15–75/) is **0.82**

a) What is the Horizontal Multiplier (HM =25/H)?
b) What is the Distance Multiplier (DM = .82+4.5/D)?
c) From Table 4-6 what is the Coupling Multiplier (CM)?
d) From Table 4-7 what is the Frequency Multiplier (FM)?
e) What is the asymmetry multiplier (AM = 1–.0032A)?
f) What is RWL for this task?
g) Does this job pose a back-injury risk to the material-handling employees?
h) What would you do to this job to reduce the risk of injury?
i) What is the recommended lighting level for this task?

BIBLIOGRAPHY

Adams, Jack A., 1989. *Human Factors Engineering.* New York: Macmillan.

American National Standards Institute. *Human Factors Engineering of Computer Workstations.* ANSI/HFES 100-2007. Human Factors and Ergonomic Society, Santa Monica, CA.

American National Standards Institute, 2011. *Safety Colors. ANSI/NEMA Z535. 1-2006 (R2011).* National Electrical Manufacturers Association.

Anderson A., Mirka G., Joines S., Kaber D., 2009. *Analysis of Alternative Keyboards Using Learning Curves.* Human Factors 51(1): pp. 35–45, 2009.

Brand J., Office Ergonomics: *A Review of Pertinent Research and Recent Developments. Reviews of Human Factors and Ergonomics,* October 2008, 4(1): pp. 245–282.

Association for Information and Image Management, 1988. *Ergonomics and Safety In the Workplace.* Silver Spring, MD: AIIM.

Booher, Harold R., 2003. *Handbook of Human Systems Integration.* Hoboken, NJ: Wiley-Interscience.

Brammer A. J. and W. Taylor, eds., 1982. *Vibration Effects on the Hand and Arm in Industry.* New York: Wiley.

Eastman Kodak Company, 2004. *Kodak's Ergonomic Design for People at Work.* 2nd ed. Hoboken, NJ: John Wiley & Sons, Inc.

Grandjean, E., ed., 1987. *Ergonomics of Computerized Offices.* New York: Taylor & Francis.

Gerr F., M. Marcus, C. Ensor, D. Kleinbaum, S. Cohen, A. Edwards, E. Gentry, D. Ortiz, and C. Monteilh, 2002. "A Prospective Study of Computer Users: I. Study Design and Incidence of Musculoskeletal Symptoms and Disorders." *American Journal of Industrial Medicine* 41:221–235.

Gerr F., M. Marcus, D. Ortiz, B. White, W. Jones, S. Cohen, E. Gentry, A. Edwards and E. Bauer, 2000. "Computer User Postures and Associations with Workstation Characteristics." *American Industrial Hygiene Association Journal* 61:223–230.

Hagberg M., B. Silverstein, R. Wells, M.J. Smith, H.W. Hendrick, P. Carayon, et. al. *Work Related Musculoskeletal Disorders (WMSDs): a reference book for prevention.* London: Taylor & Francis; 1995.

Hale, A. R. and A. J. Glendon, 1987. *Individual Behavior in the Control of Danger.* New York: Elsevier.

Haskins, Cicilia, ed., 2010. SE Handbook Working Group, *System Engineering Handbook: A Guide for System Life Cycle Processes and Activities.* San Diego: International Council on Systems Engineering.

Houwink A., K. Hengel, et.al. 2009. "Providing Training Enhances the Biomechanical Improvements of an Alternative Computer Mouse Design." *Human Factors* 51(1): pp. 46–55.

Illuminating Engineers Society of North America, ed. David DiLaura, Kevin Houser, Richard Mistrick, Gary Steffy, 2011. *Lighting Handbook.* 10th ed. New York: IES.

Kennedy C., B. Amick, et.al. 2010. "Systematic Review of the Role of Occupational Health and Safety Interventions in the Prevention of Upper Extremity Musculoskeletal Symptoms, Signs, Disorders, Injuries, Claims and Lost Time." *J. Occup. Rehab.* 20(2): 127–162. 2010

Kroemer, K.H.E. and Grandjean, E., 1997. *Fitting the Task to the Human: A Textbook of Occupational Ergonomics.* 5th ed. New York: CRC Press.

Marcus M., F. Gerr, C. Monteilh, D. Ortiz, E. Gentry, S. Cohen, A. Edwards, C. Ensor, and D. Kleinbaum. 2002. "A Prospective Study of Computer Users: II. Postural Risk Factors for Musculoskeletal Symptoms and Disorders." *American Journal of Industrial Medicine* 41:221–235.

Marras,W. S. and W. Karowski, ed., 2006. *Fundamentals and Assessment Tools for Occupational Ergonomics*. 2nd ed. Boca Raton, FL: Taylor and Francis.

National Institute for Occupational Safety and Health, 1994. *Application Manual for the Revised NIOSH Lifting Equation*, NIOSH Publication Number 94-110. Cincinnati: U.S. Department of Health and Human Services.

National Institute for Occupational Safety and Health, 1998. *Assessing Occupational Safety and Health Training: A Literature Review*, NIOSH Publication No. 98-145. Cincinnati: U.S. Department of Health and Human Services.

National Institute for Occupational Safety and Health, 1997. *Elements of Ergonomics Programs*, NIOSH Publication No. 97-117. Cincinnati: U.S. Department of Health and Human Services.

National Institute for Occupational Safety and Health *Musculoskeletal Disorders and Workplace Factors: A Critical Review of Epidemiologic Evidence for Work-Related Musculoskeletal Disorders of the Neck, Upper Extremity, and Low Back*, NIOSH Publication No. 97-141. Cincinnati: National Institute for Occupational Safety and Health, 1997.

National Institute for Occupational Safety and Health *Work Practices Guide for Manual Lifting*. NIOSH Publication No. 81-122, Cincinnati: U.S. Department of Health and Human Services, 1981.

National Institute for Occupational Safety and Health *Workplace Use of Back Belts: Review and Recommendations. NIOSH Publication No. 94-122*, Cincinnati: U.S. Department of Health and Human Services, 1994.

Putz-Anderson, Vern., 1988. *Cumulative Trauma Disorders: A Manual for Musculoskeletal Diseases of the Upper Limbs*. New York: Taylor and Francis.

Rempel D., 2008. "The Split Keyboard: An Ergonomic Success Story." *Human Factors*, 50(3): 385-392.

Salvendy, Gabriel, ed., 2012. *Handbook of Human Factors and Ergonomics*. 4th ed. Hoboken, NJ. John Wiley & Sons.

Saunders, M. S. and E. J. McCormick, 1993. *Human Factors In Engineering and Design*. 7th ed. New York: McGraw-Hill.

Scalet, Elizabeth A., 1987. *VDT Health and Safety—Issues and Solutions*. Lawrence, KS: Ergosyst Associates.

Stanton, N., A. Hedge, (*et al.*), 2005. *Handbook of Human Factors and Ergonomic Methods*. Boca Raton, FL: CRC Press LLC.

Waters, T, V. Putz-Anderson, et. al. "Revised NIOSH Equation for the Design and Evaluation of Manual Lifting Tasks." *Ergonomics*, 1993, Vol.36, No. 7, pp. 749–776.

Woodson, W., B. Tillman, and P. Tillman, 1992. *Human Factors Design Handbook*. 2nd ed. New York: McGraw-Hill, Inc.

5

Personal Protective Equipment

Scott E. Brueck, MS, CIH

A Last Resort

Personal protective equipment (PPE) consists of hearing protection, safety glasses, shoes, gloves, head protection, clothing, and other items that people can wear or carry on their bodies to protect themselves from workplace hazards. Some devices included in this category are not actually worn or carried on the body, but are nevertheless protective equipment. The major criterion of PPE is that it is something that by itself in no way eliminates or reduces a hazard. To some extent it may be thought of as partially isolating the worker from the hazard, but, essentially, PPE merely reduces the effects of exposure to a hazard.

This approach to the control of hazards should never be the primary corrective measure, but rather, as mentioned in Chapter 3, a last resort after all other efforts to reduce or eliminate the hazards have failed. Reasons to resort to PPE last include the following:

1. PPE can be uncomfortable; therefore, many workers are reluctant or refuse to wear it unless there is a strong perception of severe risk.
2. Use of PPE can create a false sense of security and can result in workers exposing themselves more than they otherwise would.
3. PPE must be properly selected, fitted, worn, and maintained to be effective.
4. In most cases, it is difficult to supervise and enforce the proper use of PPE.
5. In some cases, the use of PPE hampers the ability of the worker to do his or her job.

It is important to emphasize that employers can prevent or limit hazardous exposures by first trying to eliminate, isolate, or reduce the hazard. This will not only reduce or eliminate the potential for human injury and illness, but may also help prevent property damage and medical costs associated with injuries and accidents. The safety engineer's primary goal should always be to prevent accidents. Only if it is not possible or feasible to eliminate or isolate the hazard should other measures be considered, and the last of these should be the use of PPE.

When hazards which require the use of PPE are present or likely to be present, OSHA requires employers to conduct a comprehensive assessment of the workplace to identify these potential hazards. Once this assessment has been completed, employers must certify in writing that they have completed the hazard assessment. The written certification must contain the following information:

1. Document title, such as "Certification of PPE Hazard Assessment."
2. Identification of the area(s) or workplace evaluated. For a small company with few employees, this may be as simple as listing the name of the company if the hazard assessment covered the whole facility. For larger employers, it may be preferable to list each department or work area that was assessed as part of the hazard assessment.
3. Date or dates on which the hazard assessment was conducted.
4. Signature of the person or persons who conducted the hazard assessment.

Following the completion of the hazard assessment, employers must provide appropriate PPE to employees. Additionally, OSHA requires that employees be trained about PPE. Training must address the five following areas: when PPE is necessary; what PPE is necessary; how to don, doff, adjust, and wear PPE; limitations of PPE; and proper care, maintenance, useful life, and disposal of PPE.

After the training is completed, employers must document that they have completed the training. The documentation must contain the following information: the name of each employee trained, the date(s) of the training, and the subject of the training. It is important to remember that when new employees are hired, they must be provided with PPE and trained before they actually start working in an area that requires PPE.

When PPE is used to protect employees from a hazard, all persons who may be exposed to the hazard must use the same level of protection. The hazard is the same for a plant manager, engineer, or visitor as it is for the worker. If exposure is unsafe for workers, it is also unsafe for other people having the same exposure. This point is emphasized because, on many occasions, plant management personnel or plant visitors are seen not using PPE in areas where production workers are required to use it. Additionally, it is important to remember that if breaks are taken in the production area, PPE may still be needed.

It is the employer's responsibility to enforce the use of PPE. The employee is also responsible for properly using PPE. Responsibility for observing all safety rules, including PPE, should be accepted by all workers and plant management. Refusal to use appropriate PPE should be grounds for disciplinary action and even dismissal. Many companies accept this in principle but are sometimes reluctant to discipline or dismiss an otherwise good worker. It should be remembered that even the most valuable worker is not very valuable if he or she is seriously injured or killed. Workers who do not wear PPE may also be more unlikely to follow other safety and health requirements, which may endanger themselves or their co-workers.

If PPE is needed for a given job, who provides it? OSHA specifies, in its PPE standard [29 CFR 1910.132], that the employer must provide and pay for PPE. The employer must also pay to replace PPE, unless the employee has intentionally lost or damaged it. However, employers are not required by OSHA to pay for non-specialty, safety-toed protective footwear and non-specialty prescription safety eyewear, as long as the employer allows workers to wear these items outside the workplace. Many progressive, safety-conscious employers realize that these items cost more than the employee would normally pay, and, considering the value of safety to the company, the employer pays either all or part of the cost.

There are many ways to classify PPE for the purposes of discussion. The approach here will be in terms of types of hazards for which personal protective equipment has been designed.

Eye and Face Protection

The National Institute for Occupational Safety and Health (NIOSH) estimates that every day about 2,000 U.S. workers have a work-related eye injury that requires medical treatment. About one-third of the injuries are treated in hospital emergency departments, and more than 100 of these injuries result in one or more days of lost work. The U.S. Bureau of Labor Statistics (BLS) found that almost 70 percent of the eye injuries studied occur from falling or flying objects or sparks striking the eye. Examples include metal slivers, wood chips, dust, and cement chips that are rejected by tools, wind blown, or fall from above a worker. Some of these objects, such as nails, staples, or slivers of wood or metal, penetrate the eye and cause permanent loss of vision. Large objects may also strike the eye or face, or a worker may run into an object, causing trauma to the eye region. Chemical burns to one or both eyes from splashes of industrial chemicals or cleaning products are common. Thermal burns to the eye occur as well. Among welders, their assistants, and nearby workers, UV radiation burns (welder's flash) routinely damage workers' eyes and surrounding tissue.

Safety glasses are the most common and basic form of eye protection against the impact of flying objects, chips, materials discharged from grinding, sawing, cutting, and hammering, molten metal, and potentially injurious light radiation. Today, safety glasses come in a variety of different sizes, shapes, colors, materials, and lens shading. Bifocal safety glasses are also available to workers.

To withstand impact forces and still provide good transparency, three types of materials are used for safety glass lenses: heat-treated glass, chemically treated glass, or high-strength, shatter-resistant plastic. For the latter, polycarbonate is a popular material for safety glasses because it is lighter, thinner, and more impact resistant than glass lenses. Safety glasses made of Trivex®, a urethane-based monomer, are also available. Trivex® is considered to have better optics than polycarbonate. Both polycarbonate and Trivex® lenses provide UV protection. Plastic materials will not break as easily as glass, but they scratch more easily, unless a scratch-resistant coating has been applied, and their optical quality is not as good as glass. Treated glass will break but will not shatter and fly into the eye as untreated glass will.

The American National Standards Institute/International Safety Equipment Association (ANSI/ISEA Z87.1-2010) sets standards for the impact resistance of safety glasses, and any safety glasses provided to employees must be ANSI approved. Ordinary eyeglasses are often worn by workers as safety glasses, but in most cases these glasses do not provide impact protection. However, prescription eyewear can be manufactured to meet impact specifications also. Therefore, when workers need both vision correction and impact protection, make sure that the eyewear is made to meet ANSI impact-protection standards.

In addition to protection from objects flying directly at the eye, workers also need protection from side impact. Therefore, side shields are necessary. Most safety glasses are manufactured with integrated side shields. However, prescription safety glasses, even those meeting ANSI impact requirements, may not have side shields. In those circumstances, it is necessary to purchase separate side shields, which slide into the temple bars of the glasses. Most large PPE suppliers carry side shields for this purpose.

Workers exposed to light radiation from sources such as lasers, welding, or sunlight need eye protection equipped to filter the radiant light energy. Laser safety glasses or goggles must be selected based on the wavelength of the lasers used. Workers conducting welding must wear welding helmets with lenses that have appropriate shading. In its PPE standards, OSHA has published a table of lens-shading requirements based on electrode size, arc current, or plate thickness of material welded (Table 5-1). Workers not actually welding but working in the vicinity of welding operations do not typically need welding helmets, but they should wear appropriately shaded eye protection. Additionally, welding operations should use welding screens or curtains to control the transmittal of light from the welding arc. Workers who spend a considerable amount of time working outdoors in bright sunlight should wear lenses that provide protection from the ultraviolet spectrum of sunlight. If these workers do not need any impact protection, regular sunglasses with maximum UV-A and UV-B protection would be adequate.

In some circumstances, safety glasses will not provide adequate eye protection if workers are at risk from chemical splashes and vapors. Goggles seal to the contour of a person's face and provide eye protection from all directions. Goggles with impact-resistant lenses can additionally be used for exposure to flying particles and are especially useful in extremely dusty environments.

Face shields can help protect to a worker's entire face and neck. Face shields are useful for chemical splash protection and provide additional eye protection when worn over safety glasses or goggles. Face shields also offer some protection from particles but generally do not provide the same level of impact protection as safety glasses. For this reason, face shields should be worn over safety glasses. For exposure to sparks, hot metal particles, and high temperature extremes, face shields made of metal mesh or specially designed plastics can be worn over safety glasses.

Contact lenses are a special subject in a discussion of chemical exposure to the eyes. Previous recommendations stated that employees should not wear contact lenses when working with chemicals that might injure or irritate the eyes. However, lack of sufficient injury data from contact lens use in workplaces with hazardous chemical exposure has prompted many groups, including the American Optometric Association, American College of Occupational and Environmental Medicine, American Academy of Ophthalmology, and the

Table 5-1. Filter Lenses for Protection Against Radiant Energy (OSHA Personal Protective Equipment Standards, rev. [with corrections], July 1, 1994)

Operations	Electrode Size 1/32 in.	Arc Current	Minimum (*) Protective Shade
Shielded metal arc welding	less than 3	less than 60	7
	3–5	60–160	8
	5–8	160–250	10
	more than 8	250–550	11
Gas metal arc welding and flux-cored arc welding		less than 60	7
		60–160	10
		160–250	10
		250–500	10
Gas tungsten arc welding		less than 50	8
		50–150	8
		150–500	10
Air carbon Arc cutting	(light)	less than 500	10
	(heavy)	500–1000	11
Plasma arc welding		less than 20	6
		20–100	8
		100–400	10
		400–800	11
Plasma arc cutting	(light)**	less than 300	8
	(medium)**	300–400	9
	(heavy)**	400–800	10
Torch brazing			3
Torch soldering			2
Carbon arc welding			14

Operations	Plate Thickness-inches	Plate Thickness-mm	Minimum (*) Protective Shade
Gas welding			
Light	under 1/8	under 3.2	4
Medium	1/8 to 1/2	3.2 to 12.7	5
Heavy	over 1/2	over 12.7	6
Oxygen cutting			
Light	under 1	under 25	3
Medium	1 to 6	25 to 150	4
Heavy	over 6	over 150	5

*As a rule of thumb, start with a shade that is too dark to see the weld zone. Then go to a lighter shade which gives sufficient view of the weld zone without going below the minimum. In oxyfuel gas welding or cutting where the torch produces a high yellow light, it is desirable to use a filter lens that absorbs the yellow or sodium line in the visible light of the (spectrum) operation.

**These values apply where the actual arc is clearly seen. Experience has shown that lighter filters may be used when the arc is hidden by the workpiece.

American Chemical Society, to remove previous recommendations against contact lens use. NIOSH also recommends that contact use be allowed, but advises that an eye injury hazard evaluation should be conducted, eye and face protection must still be provided, and workers should be instructed to immediately remove contacts if eye irritation occurs. OSHA specifies that employees should not wear contacts when using acrylonitrile, 1, 2 dibromo-3-chloropropane, ethylene oxide, methylene chloride, and methylene dianiline (NIOSH 2005).

Protective Gloves and Clothing

Workplace hazards that could cause injury to a worker's hands or body include intense cold or heat, burns, hot metals or liquids, sharp or penetrating objects, radiation, infectious materials, and hazardous chemicals. When these hazards are identified in a workplace, employers should attempt to eliminate them through engineering, work practice, or administrative controls. However, if the hazards cannot be successfully eliminated, workers must be protected with gloves or protective clothing.

Many hazardous chemicals can be absorbed into the body through the skin. For some chemicals, particularly organic solvents, skin absorption can be a substantial contributor to total body exposure. Intact skin provides a barrier for absorption of some chemicals; however, cuts, scrapes, and abrasions permit chemicals to enter the body directly. Additionally, workers can develop dermatitis or other skin diseases from dermal exposure to hazardous chemicals.

Many workers and employers are under the assumption that one type of glove or clothing material will protect against all chemicals. However, this is not the case. For example, gloves made of polyvinyl alcohol polymer are protective against exposure to many organic chemicals, but are degraded by water and alcohols. Therefore, different types of materials, including natural rubber or latex, butyl rubber, neoprene, nitrile, polyvinyl chloride, polyvinyl acetate, and polyvinyl alcohol, are available. These materials will vary in their protection against different chemicals. Additionally, some materials have other characteristics that may make them advantageous for specific applications. For example, nitrile tends to be more durable and resistant to abrasions than some of the other glove materials. Some glove or clothing materials are protective against a broad spectrum of chemicals, including Viton®, Teflon®, polyethylene/ethylene vinyl alcohol (PE/EVAL) and polyethylene/polyamide (PE/PA), but may be more expensive, uncomfortable, and may not provide the needed tactile sensitivity for the job. Some workers have allergies to latex materials, so safety and health managers need to look for potential problems related to glove or clothing material if workers are using latex. For each type of glove or protective clothing material, most manufacturers offer different styles and thicknesses from which to choose.

Chemical protective clothing and gloves do not last indefinitely. Once exposed to hazardous chemicals, the protective material begins to degrade, eventually rendering it ineffective. Most chemicals will also permeate through the protective material over time. The time it takes a chemical to permeate completely through chemical protective clothing or gloves is called "breakthrough time." The breakthrough time can vary from minutes to hours, based on the material, chemical, and usage factors. Most manufacturers of gloves and other protective clothing provide breakthrough time and degradation resistance information for their products. Therefore, it is important to review this data so the best protective clothing or glove is chosen for the job and to help ensure that gloves or clothing are not worn longer than the breakthrough time reported by the manufacturer. Other factors to consider in selecting chemical protective gloves or clothing include durability, dexterity, and cost.

To protect against non-chemical hazards, leather or durable cloth material may be appropriate for abrasion resistance. If employees work with knives, glass, sharp metal, or other sharp objects that could cut gloves or clothing, cut-resistant metal mesh Kevlar® can be used. However, these materials may not provide puncture resistance, so employees should still be instructed to handle sharp objects carefully. To protect the forearms, employees can wear gauntlets, which are made from the same material as gloves. Employees working with vibrating tools can wear specialized gloves that are padded in the palms or fingers to reduce vibration at the hands. For poison sumac or other plants that can cause allergic reactions, workers can use barrier creams that contain bentoquatam. Barrier creams work by forming a protective film on the skin that prevents actual contact with the toxic substance. Some barrier creams are also available for exposure to mildly irritating chemicals. Some of these creams form a waterproof barrier, while others are washed away with water but resist oils and other types of chemicals. Caution is urged in using barrier creams for chemical exposure, as these tend to wear off rather quickly, and may not be as effective as necessary for the exposure. Barrier creams should not be used to prevent skin exposures to highly hazardous chemicals.

For protection from extreme heat or cold, gloves or clothing made from aramid fiber offer protection. Other synthetic materials are also available for exposure to heat and cold. Leather can be used for moderate heat and when durability is important. However, for higher temperatures, synthetic materials such as Nomex®, Kynol®, and Preox® could be used. Additionally, gloves and clothing made of aluminized fibers offer insulating and reflective protection against heat.

Regardless of the type of protective clothing or gloves, workers must be properly trained in their use. For example, most chemical protective gloves should be rinsed after use. If chemical residue remains on the gloves, the protective material will continue to degrade and shorten the effective life of the gloves. Workers should also carefully inspect gloves and protective clothing for intact seams and the absence of holes or rips of the protective clothing and gloves should be properly sized to minimize the chance that protective gloves and clothing could be pulled into machinery or equipment.

Head and Foot Protection

Head protection is necessary when there is a potential for injury to the head from falling objects and when work must be done near electrical conductors that could contact the head and cause electrical shock. The most common type of head protector is the hard hat, which consists of a hard outer shell and a harness on the inside to secure the hard hat to the wearer's head and help distribute forces during impact. According to OSHA, hard hats must meet ANSI/ ISEA Z89.1-2009 consensus standards, which have specifications for impact and resistance. Hard hats are grouped into three classes on the basis of electrical insulation resistance. Class G (general) are proof-tested at 2200 volts of electrical charge; class E (electrical) are proof-tested at 20,000 volts of electrical charge; and class C (conductive) provide no electrical charge protection. All three classes provide impact resistance.

Bump caps, as the name implies, are intended to protect workers from bumping their heads on stationary objects, such as overhead pipes or catwalks. Bump caps do not provide impact and penetration protection and should never be used in place of ANSI/ISEA approved class G, E or C hard hats. Often, workers are seen wearing hard hats or bump caps backwards on their heads. If the hard hat is worn backwards, the internal suspension must face forward to provide maximum protection.

Over 100,000 employees experience workplace injuries to their feet every year. However, most of these foot injuries could have been prevented by the use of appropriate foot protection. Proper footwear can protect workers from foot injuries due to falling or rolling objects, punctures, slips, and electrical hazards. Protective footwear provided must meet ASTM F-2412-2005, "Standard Test Methods for Foot Protection," and ASTM F-2413-2005, "Standard Specification for Performance Requirements for Protective Footwear," standards for impact and compression protection. The most common safety shoe has a toe box. The toe box is typically constructed of steel, plastic, or a composite material to protect the toes of the worker. Sometimes, workplace conditions require the use of metatarsal guards. These guards can be contained within a shoe or boot, or can be attached to the outside of the shoe and cover the top of the foot, from the toe to the ankle region. Depending on their work activities, employees may need to wear safety shoes that provide protection from sharp objects, such as nails or screws, that might puncture through the sole of the shoe. If employees have potential contact with electrical sources, nonconducting or insulating shoes must be worn.

Respiratory Protection

OSHA requires the use of respirators when they are necessary to protect the health of workers. An example would be when air contaminant exposures exceed OSHA permissible exposure limits (PELs) or work must be done in a hazardous atmosphere, such as an oxygen-deficient environment. Employers may also choose, and many do, to require the use of respirators even when air contaminant concentrations are less then OSHA PELs. Additionally, there are many chemicals for which no occupational exposure limits have been established. For these situations, good safety practice, manufacturers' recommendations, or research results may indicate that respiratory protection is prudent.

To select appropriate respiratory protective devices, it is important to understand the respiratory system. When we breathe, our diaphragm muscles cause the lungs to expand and create a vacuum, which draws air through the nose or mouth, the trachea and bronchial system, to the lungs. The inside surface of the lungs consists of very thin, single-cell layers of tissue in the form of small pouches called the alveoli. The lungs contain over 300 million alveoli. On the other side of this thin layer of tissue lie thin-walled blood vessels. The blood brings waste products, mostly carbon dioxide, to the alveoli and by osmosis exchanges the carbon dioxide for a new supply of oxygen.

As air moves through the narrow, winding passages of the nose, throat, and bronchial tubes, the larger particles are filtered out by hair, mucous, and impingement against the walls of these passages. Some of these particles are eventually removed from the lungs by coughing. The smallest particles, under 0.5 micron, are usually exhaled and do not stay in the lungs. Consequently, particles ranging in size from 0.005 millimeter and about 30 microns present the greatest problem. Gases are not filtered at all and readily flow into the lungs.

Solid foreign matter can damage the lungs in several ways. Some materials adhere to the surface of the alveoli and prevent transfer of gases. Other materials may irritate the tissue and cause allergic reactions. The most serious problem occurs when air contaminants, such as asbestos, attack the tissue and cause it to harden. This action tends to continue and spread until eventually the lungs are no longer capable of functioning.

Because of the shape of the alveoli, the total surface area of the lungs is enormous and could cover an area the size of a tennis court (100 square yards). A portion of this area can be damaged but if enough alveoli remain, a person would still be able to breathe. Unfortunately, symptoms of lung deterioration do not always become evident until long after damage has been done. Asbestosis or silicosis, for example, are rarely identified until at least five years after exposure, and more often it is 10 to 20 years after exposure. By that time the disease has done serious, permanent damage.

Hazardous particulates inhaled into the lungs can be absorbed or can cause tissue irritation, damage, inflammation, allergic responses, or other respiratory problems. Toxic gases can be absorbed directly into the bloodstream the same way that oxygen is absorbed. The respiratory system has no defense against toxic gases. Some hazardous gases, such as carbon monoxide, bind to the oxygen receptors in the lungs more readily than oxygen and essentially replace oxygen absorption.

Because of hazardous particulates (dust), gases, or vapors in workplace atmospheres, workers may sometimes need respiratory protection To provide this protection, respirators may filter the air (through the use of different filters or absorbent cartridges) or provide breathable air to the worker. Regardless, it is critically important that the respirator fit properly on the wearer's face so that no air contaminants can enter the respirator.

Respirators that filter the air are called air-purifying respirators. There are two basic types of filters that can be used to filter air. Mechanical filters are composed of a fibrous material that removes particles larger than a given size from the air passing through the filter. NIOSH currently classifies particulate filters according to their resistance to degradation from oil mists present in the workplace air. These classifications are "N," not oil resistant (these filters are only for use when no oil mist is present in the air); "R," resistant to oil (should be used for no more than 8 total hours when oil mists are present); "P," oil proof (are not readily degraded by oil mists and can last for a full shift when oil mist is present). For each of these classifications, "N," "R," and "P," there are three levels of filter efficiency that refer to the ability of the filters to remove 0.3 micron-sized particles. The three filter efficiencies are 95 percent, 99 percent, and 99.97 percent. These filter efficiencies are based on the efficiency of the filters for removal of 0.3 micron size particles. When selecting a particulate filter, one needs to choose one based on whether oil mists are present in the workplace and on what level of filter efficiency is desired. It is important to remember that particulate filters will not remove gases or vapors from the airstream.

Chemical cartridges contain an adsorbing medium, such as activated or treated charcoal, that removes gases and vapors as they pass through the adsorbing material. Since one type of cartridge will not protect against every gas or vapor, respirator cartridge manufacturers produce several different types of chemical cartridges. Therefore, when selecting a chemical cartridge for an air-purifying respirator, it is crucial to know what gases or vapors are present in the workplace and then choose a cartridge that will filter that particular chemical. The adsorbing medium in a chemical cartridge can only remove a limited amount of hazardous gas or vapor from the air. The length of time that a chemical cartridge can be used is called its "effective service-life." A chemical cartridge's effective service-life is based primarily on the concentration of air contaminants. However, other factors, such as humidity, temperature, and breathing rate can also affect the cartridge's service-life. The higher the humidity or temperature or the faster a wearer's breathing rate, the shorter the cartridge service-life. A few chemical cartridges have an end-of-service-life indicator (ESLI) on the cartridge to visibly show the respirator wearer when the cartridge is saturated, but most do not. Therefore, OSHA expects companies to have a written change-out schedule for respirator wearers using air-purifying respirators with chemical cartridges. The written change-out schedule is based on the type of air contaminants, expected work effort, cartridge adsorbing media, and other environmental and usage factors. Many respirator manufacturers have online programs to help respirator users determine when the chemical cartridges should be changed. It is important to contact the manufacturer of the respirator chemical cartridges for assistance in determining the effective service life of the cartridges in use.

Four different types of air-purifying respirators are available: mouthpiece, half mask, full face, and powered air purifying (PAPR). The mouthpiece respirator is worn by clenching the mouthpiece between the teeth and holding the nostrils closed with a nose clip. Mouthpiece respirators should only be used to escape from an area with excessive air contaminants. Half-mask respirators cover the mouth and nose of the wearer, and the full-face covers the eyes in addition to the mouth and nose. Both half-mask and full-face are considered negative-pressure air-purifying respirators because, when a person inhales, the air pressure inside the tight-fitting facepiece is less than the air pressure outside the facepiece, which causes the air to flow into the facepiece. Because of this negative pressure inside the facepiece during inhalation, it is important that the respirator fits the wearer's face appropriately. If the respirator does not seal firmly to the face, air can leak into the facepiece. Air leaking into the respirator from around the seal is unfiltered and potentially hazardous. To help ensure that a respirator fits correctly and does not leak, the wearer must undergo a fit test. OSHA requires any person required to wear a tight-fitting respirator to undergo a fit test. The appendices of OSHA's respirator standard (29 CFR 1910.134) contain specific procedures on fit testing.

Powered air-purifying respirators (PAPRs) use a small, battery-powered motor and blower to pull air through the air filters and blow a stream of filtered air over the wearer's face. PAPR respirators are different from negative-pressure air-purifying respirators because under normal conditions of use, the pressure inside the facepiece does not become negative; therefore, any leakage in a tight-fitting PAPR should flow outward. PAPRs can be either tight-fitting full face or loose fitting. Loose-fitting PAPRs have a hood or helmet that fits loosely over the wearer's head but does not seal tightly to the wearer's face.

In situations where the workplace atmosphere is deficient of oxygen, contains a concentration of contaminants at levels that are immediately dangerous to life and health (IDLH), or contains undetermined concentrations of unknown air contaminants, air-purifying respirators will not provide adequate protection. Therefore, respirators that provide breathable air should be worn. These respirators are known as either supplied-air or self-contained breathing apparatus (SCBA). Supplied-air respirators can be either tight fitting or loose-fitting. Supplied-air respirators are attached to a remote source of contaminant-free breathing air which is pumped to the respirators. Air provided to supplied-air respirators must meet grade D air-quality specifications established by the Compressed Gas Association. Steps must also be taken to ensure that carbon monoxide does not enter the supplied-air system and present a health hazard to employees. These protective measures include locating the supplied-air intake in a place where no contaminated air enters the system. Other measures for oil-lubricated compressors include utilizing a carbon monoxide alarm on the downstream side of the compressor or the regular monitoring of the airstream for carbon monoxide.

An SCBA is different from supplied air in that the supply of breathing air is carried by the wearer. A cylinder of compressed breathing air, which can weigh up to 35 pounds, contains a maximum of 60 minutes of grade D air. These respirators are always tight fitting and use a regulator to provide breathing air when needed by the wearer. A system of alarms is used to notify the wearer when the air supply is low. Because of the complicated nature of SCBA respirators, it is extremely important that wearers are well trained on how to use them, the respirators are properly maintained, and repairs are made only by trained technicians.

Selection of respirators must be based on the work activities and the nature of the hazardous air contaminants or atmospheres present in the workplace. It is critically important that the respirator adequately protect the wearer from the hazardous atmosphere to which he or she is exposed. Because different types of respirators provide different levels of protection (for example, a half-mask air purifying respirator isn't as protective as a full facepiece air purifying respirator), the OSHA respirator standard includes a table of assigned protection factors for respirators to help ensure proper selection. When selecting a respirator for use, the assigned protection factor of the respirator should be greater than the measured or expected concentration of air contaminant divided by the occupational exposure limit. The National Institute of Occupational Safety and Health's (NIOSH) has issued the *NIOSH Guide to Industrial Respiratory Protection,* which also provides strategies for proper respirator selection.

Employees who are required to wear respirators in the workplace must undergo a medical evaluation to ensure that they do not have a pulmonary, cardiac, or other medical condition that could be adversely affected by the use of a respirator. Additionally, it is important that respirators are properly maintained, cleaned, and stored to ensure that they remain in proper working condition at all times. The OSHA respiratory protection standard (29 CFR 1910.134) provides detailed guidance for use of respirators. In addition, ANSI has voluntary consensus standards on fit testing, physical qualifications for use, and color coding of canisters, cartridges, and filters.

Protection from Noise

Hearing is arguably the most critical component needed for our communication with other people and our environment, and the ears need to be protected as much as the eyes do. The impact of losing functional hearing on a person's life is difficult to assess, but people with substantial hearing loss tend to feel isolated from their world and often withdraw from interacting with others. Because noise-induced hearing loss (NIHL) usually happens slowly, unlike acute injuries, and it is difficult to imagine what it would be like to be hearing impaired or deaf, people may not value their hearing as much as they should. However, occupational hearing loss is the most common work related injury and hundreds of millions of dollars are spent annually in workers compensation costs related to hearing loss.

The process of hearing begins with sound waves entering the ear and vibrating the eardrum, which is a thin membrane that separates the outer ear from the middle ear. The vibration of the eardrum in turn vibrates a series of small bones in the middle ear and eventually a second membrane that separates the middle ear from the inner ear. This second membrane transmits the vibrations into the fluid that fills a spiral tubelike structure in the inner ear called the cochlea. The cochlea is lined with approximately 20,000 tiny hair cells, which are each attached to nerve endings. Each hair cell is covered with hairlike structures called stereocilia, which are sensitive to a particular frequency like the reeds in a harmonica. By a means not completely understood, the movement of these vibrating stereocilia help convert mechanical stimulation into electrical impulses that send a message via the nervous system and auditory nerve to the brain, where it is interpreted as sound.

Since the inner ear is housed within the bone structure of the skull, sound can also be heard when vibrations are transmitted through our skeletal system and stimulate the inner ear. The transmission of sound through the bone structure of the body, called bone conduction, is nearly as efficient as the route through the outer ear, but very loud sounds can be transmitted this way.

The preferred unit for reporting of noise levels is the decibel, A-weighted (dBA). A-weighting is used because it approximates the way people perceive the loudness of sound. The decibel (dB) unit is dimensionless, and it represents the logarithmic ratio of given sound pressure level to an arbitrary reference sound pressure. The reference level is the minimum sound pressure (equivalent to 0 dB) required to make a sound audible to humans. Decibels are used because of the very large range of sound pressure levels audible to the human ear. Because the dB is logarithmic, an increase of 3 dB is a doubling of the sound energy, an increase of 10 dB is a 10-fold increase, and an increase of 20 dB is a hundred-fold increase in sound energy.

People are capable of hearing sounds within a limited range of frequencies (from 20 to 20,000 hertz), but the human ear does not interpret sound intensity on a linear scale across different frequencies. The threshold of hearing at low frequencies is much higher than it is at midrange frequencies. Our threshold of hearing at a frequency of 1,000 hertz, or cycles per second, is often used as a reference. At frequencies higher or lower than this reference frequency, our hearing threshold is diminished; that is, it takes a greater, or louder, sound intensity to be perceived as an equally loud sound. Loudness is the term used to describe our response or interpretation of sound intensity. Figure 5-1 illustrates the range of frequencies and intensities within which a person with normal hearing ability can detect sounds. Normal hearing ability is defined as that of a young adult with no unusual hearing defects.

Although hearing ability commonly declines with age, exposure to excessive noise can increase the rate of hearing loss. In most cases, NIHL develops slowly from repeated exposure to noise over time, but the progression of hearing loss is typically the greatest during the first several years of noise exposure. NIHL can also result from short duration exposures to high noise levels or even from a single exposure to an impulse noise or a continuous noise, depending on the intensity of the noise and the individual's susceptibility to NIHL (Berger et al. 2003). Noise-exposed workers can develop substantial NIHL before it is clearly recognized. Even mild hearing losses can impair one's ability to understand speech and hear many important sounds. In addition, some people with NIHL also develop tinnitus, which is a condition in which a person perceives hearing sound in one or both ears, but no external sound is present. Persons with tinnitus often describe hearing ringing, hissing, buzzing, whistling, clicking, or chirping like crickets. Tinnitus can be intermittent or continuous, and the perceived volume can range from soft to loud. Currently, no cure for tinnitus exists.

The duration and intensity of noise exposure necessary to cause permanent hearing loss is not known with certainty. NIOSH estimates that

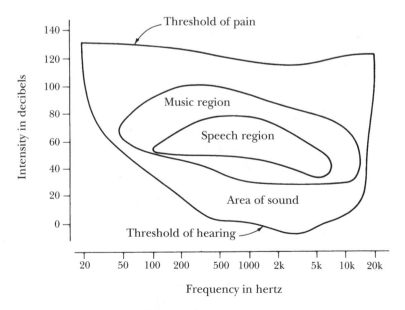

Figure 5-1. Range of audible sounds for human beings.

excess risk of material hearing impairment at age 60 after a 40-year working lifetime with exposure to noise is 8 percent for an average noise exposure of 85 dBA and 25 percent for an average noise exposure of 90 dBA (NIOSH 1998). Therefore, NIOSH has established a recommended exposure limit (REL) for workplace noise of 85 dBA, as an eight-hour, time-weighted average (TWA). NIOSH recommends that hearing protection be worn when TWA exposures exceed the REL. NIOSH also recommends that no one should be exposed to an impulse noise greater than 140 dB, without the use of hearing protection.

OSHA has established a permissible exposure limit (PEL) of 90 dBA and an Action Level (AL) of 85 dBA, both as eight-hour TWAs. When noise exposures exceed the PEL, OSHA requires that employees wear hearing protection, and that an employer implement feasible engineering or administrative controls to reduce noise exposures. The OSHA noise standard (29 CFR 1910.95) also requires implementation of a hearing conservation program when noise exposures exceed the AL. The program must include noise monitoring, employee notification, audiometric testing, providing hearing protectors, training, and record keeping.

Use of hearing protection is the only way to effectively reduce the amount of noise entering the ear canal. There is no satisfactory way to prevent sound from being transmitted to the inner ear via the bone conduction. However, the amount of noise transmitted to the inner ear through bone and tissue is about 40 to 50 dB less than the sound level transmitted through the ear canal (Berger et al. 2003). So noise exposure through this pathway is not significant unless exposures are above 120 dB.

There are two categories of hearing protection: earplugs, which fit into the ear canal to block sound, and earmuffs, which fit around the ears and against the head. Several different types of insert-type earplugs are available:
* Premolded earplugs are made from flexible and soft material, such as silicone or foam, and are usually made in different sizes. Some premolded plugs have flanges to help achieve a better fit. These types of plugs are pushed into the ear canal. Most premolded plugs can be reused, and some can be cleaned with soap and water.
* Foam earplugs are commonly made from closed-cell polyvinyl or urethane foam. Foam plugs must be rolled into a narrow cylinder and then inserted into the ear canal and held in place until the foam expands. These plugs can sometimes be reused if they remain clean, but are generally considered disposable.
* Custom molded earplugs are reusable plugs made by injecting silicone or a similar material with a chemical hardener into the ear canal and allowing it to harden. The result is an earplug that is a replica of the wearer's ear canal. Custom molded earplugs are often considered to be more comfortable than other types of ear plugs, but are much more expensive.

The advantage of so many different types and styles of earplugs is that regardless of an individual's ear canal shape, there should be an earplug that comfortably and effectively fits the wearer. Some other advantages of earplugs include their low cost (except custom molded), portability, and ease of use with other PPE, such as safety glasses and hard hats. However, earplugs can be challenging to fit effectively unless the user is well instructed and motivated. Individuals who need to wear hearing protection should be allowed to try at least three different types to find the hearing protection that is comfortable and provides the best fit.

One of the main advantages of earmuffs is that they are easy to fit. Most earmuffs passively attenuate noise, like earplugs. However, electronic earmuffs with noise-cancelling or active noise reduction technologies are also available. These earmuffs work well at noise cancellation for low frequencies (less than 1,000 hertz), but are not effective for the higher frequencies common in most workplaces.

When using earmuffs, care must be taken to ensure that nothing interferes with the seal of the earmuff to the head. For example, if earmuffs are worn over safety glasses, the temple bar of the safety glasses makes a small gap in the seal of the earmuff, which reduces the amount of noise the earmuff will attenuate.

One of the problems with earplugs and earmuffs is that they reduce desired sounds along with unwanted ones. Most hearing protectors attenuate more noise at the higher frequencies, including the speech frequencies, than at lower frequencies. For this reason some people can have difficulty hearing conversation or other higher-frequency sounds when wearing hearing protection, especially when background noise is relatively low. To address this, some manufacturers make flat frequency response earplugs that attenuate noise relatively equally across all frequencies. To improve communication, earmuffs are also available with built-in communication devices with selectable frequencies like a walkie-talkie. Earplugs and earmuffs reduce noise exposure by attenuating the amount of noise that reaches the ear. The noise attenuation of hearing protectors depends on the type of protectors and how well they fit the wearer. All hearing protector manufacturers publish noise-reduction ratings (NRR) for their devices, which let users know how much noise the protector should attenuate. The NRR values are the result of laboratory testing performed by the manufacturers based on test criteria established by the U.S. Environmental Protection Agency (EPA). The EPA is currently in the process of adopting new methods for determining the noise reduction rating for hearing protection.

Because real-world noise reduction does not always match the results from laboratory testing, NIOSH recommends estimating hearing protector noise attenuation using subject fit data, which is based on ANSI standard S12.6-1997 (ANSI 2008). However, if no subject fit data are available, NIOSH recommends derating hearing protectors by subtracting 25 percent from the manufacturer's labeled NRR for earmuffs and subtracting 50 percent from the manufacturer's labeled NRR for formable earplugs. For dBA noise-exposure levels, an additional 7 dB would need to be subtracted from the derated NRR. An additional 5 to 10 dB of attenuation can be added for use of dual hearing protection (NIOSH 1998). It should be noted that some hearing protection manufacturers now have systems to perform testing of the fit and determination of the actual noise attenuation of hearing protectors on individuals.

An important consideration for selecting hearing protection is to choose hearing protectors that provide enough noise attenuation to reduce noise levels at the ear to a safe level, but that do not excessively attenuate noise. Attenuation of more noise than is necessary, called "overprotection," can lead to problems communicating with co-workers, hearing warning signals, equipment, or other sounds in the nearby work areas.

Personal Protective Equipment for Exposure to Nanoparticles

Engineered nanoparticles are materials that are made with at least one dimension having a size range of 1 to 100 nanometers (nm) (NIOSH 2009). Because of the extremely small size, shape, surface area, electric charge, solubility, chemical properties, and other unique characteristics, they may behave differently than larger particles of the same substance and therefore have additional or greater concerns regarding dermal or inhalation exposure. Some nanoparticles may be able to penetrate the dermal layer of a person's skin under conditions such as flexing of the skin (Ryman-Rasmussen et al. 2006, Rouse et al. 2007) or if the skin has cuts or abrasions. It is important to wear protective gloves and protective clothing when there is a potential for dermal exposure to nanoparticles. Research on appropriate protective clothing and gloves for use against nanoparticles is ongoing. Selecting protective gloves or clothing should take into consideration the nanoparticle properties along with the need for abrasion, cut, or puncture resistance, and the types of liquid chemicals that contain nanoparticles. Nanoparticles may be able to penetrate through some protective clothing materials and through small gaps in seams or closures (Schneider et al. 1999, 2000). Some research suggests that fabrics made from nonwoven materials may be able to protect better than fabrics made from woven materials (Golanski et al. 2009, Golanski et al. 2010). Prior to selecting gloves or protective clothing, it is prudent to check with the manufacturers to determine if PPE that has been tested for use against nanoparticles is available. After removal of protective clothing and gloves, employees should wash their hands to remove nanoparticles that might have gotten onto their skin.

Some research indicates that nanoparticles may be more hazardous when inhaled than larger particles of the similar composition (NIOSH 2009). Engineering controls should be the first method to reduce airborne exposures to nanoparticles. However, if these controls have not been implemented or are in the process of being installed, then respirators may be needed. If concerns about exposure to nanoparticles remain, even after implementation of engineering control measures, using properly fitted and worn respirators can provide additional protection. Currently very few occupational exposure limits (OELs) have been established for nanoparticles. For nanoparticles that do have OELs, a respirator with an appropriate protection factor should be chosen (i.e., a respirator with a protection factor that is greater than the measured airborne exposure divided by the occupational exposure limit). For nanoparticles without OELs, it may be helpful to consider OELs or guidelines for larger particles with similar chemical composition when making respiratory protection decisions (NIOSH 2009). Because the use of engineered nanoparticles is increasing, future research on the effectiveness of protective clothing and respirators should provide additional guidance on choosing the correct PPE and respirators. Consult with OSHA, NIOSH, and product manufacturers for the latest information.

Personal Protective Equipment for Cryogenic Liquids

Cryogenic liquids present unique concerns regarding selection and use of PPE. Because the temperature of these liquids is below −100°F, any skin contact with them or with uninsulated containers or pipes with these liquids can cause instant tissue damage. This is a particular concern for the eyes, which are highly susceptible, and would be damaged very quickly from splashing liquid or a release of vapors or gases. As is the case with hazardous chemicals, contact with cryogenic liquids should be avoided and control measures to eliminate the need for contact should be established. However, if work with cryogenic liquids is necessary, appropriate PPE must be used. PPE includes nonporous clothing, aprons, or coveralls, work boots, insulated gloves, safety glasses or goggles, and face shields. Protective clothing should not have pockets or cuffs in the sleeves or pants, which could catch spilled liquids. Pants should fit over the top of the work shoes. Insulated gloves, designed for protection from splashes of cryogenic liquids, should fit loosely so that the glove can be removed quickly if necessary. The gloves must not be immersed into cryogenic liquids.

Personal Protective Equipment for Ionizing Radiation

The most common forms of ionizing radiation are alpha particles, beta particles, gamma rays, and X-rays. Radiation can affect external tissues, internal tissues, or both, depending on the form, and excessive exposure can cause severe illness or death. Organizations such as the National Council on Radiation Protection and Measurements (NCRP), International Commission on Radiological Protection and Measurements (ICRP), International Atomic Energy Agency (IAEA), and the U.S. Nuclear Regulatory Commission (NRC) have established exposure limits and guidelines or standards for radiation safety. OSHA regulations on radiation safety are published in 29 CFR 1910.1096. The general principle regarding radiation exposure is to keep the exposure dose as low as reasonably achievable (ALARA). Reducing radiation exposure can be achieved by limiting the amount of time spent near radiation sources, increasing the distance from radiation sources, shielding sources of radiation using materials that either block or absorb radiation, and using PPE. It is important to note that PPE can only protect against exposure to alpha or beta particles.

Alpha and beta particles can be inhaled, ingested, or injected and subsequently damage internal body tissues, particularly the lungs, liver, kidneys, and bones. Because inhalation is one of the primary alpha and beta particle exposure pathways, and substantial tissue damage can occur from internal exposure, properly fitting respirators should be used for protection. Minimally, a tight-fitting N95 filtering facepiece respirator can be used for exposure to alpha or beta particles. However, a higher level of protection such as elastomeric half-mask respirators or full facepiece respirators, PAPRs equipped with high-efficiency particulate filters, or SCBA respirators may be necessary. NIOSH also certifies respirators for protection from chemical, biological, radiological, and nuclear (CBRN) agents. As with chemical exposure considerations, the respirator should be selected on the basis of anticipated exposure, work activities, and other respirator usage factors.

Alpha particles are only able to travel a few inches in air and cannot penetrate intact skin. However, cuts or abrasions of the skin should be completely covered for protection against alpha particles. Gloves and protective clothing, which can help prevent contamination of the skin or clothing should be worn for exposure to alpha particles. Several types of gloves and protective clothing materials, such as cotton, disposable synthetic textiles, rubber, or chemically impervious plastics, can be used, and selection should be based on the extent of exposure, type of work activities, and whether conditions will be wet or dry.

Depending on their energy level, beta particles can travel from a few inches to more than 50 feet in the air and are capable of penetrating and damaging the skin and eyes. Unlike alpha particles, which require only a thin layer of protective clothing, exposure to beta particles requires the use of gloves and protective clothing, such as radiation suits, specially designed for protection from radiation energy and the penetrating ability of beta particles. In addition to protective gloves and clothing, safety glasses or goggles, head covering, and shoe coverings may also be needed. For some exposure conditions, it may be necessary to use protective clothing specifically designed for protection from CBRN agents. After removal, any PPE used for radiation protection should be properly disposed of or cleaned.

Gamma rays and X-rays are not particles but are very penetrating waves of energy (photons). Gamma rays originate in the nucleus of an atom, and X-rays originate from the electrons of an atom and can be produced by machines. These types of radiation are external and internal hazards. Protection against exposure can only be feasibly achieved by containing the radiation source with thick layers of lead, concrete, and steel. Protective clothing or respirators do not provide protection against the damaging effects of direct exposure to gamma or X-rays. Leaded aprons or coverings, similar to that used in dental or medical radiology offices, can be used for short-duration exposure low energy (10 keV) scattered X-rays (IEAE 2004).

Personal Protective Equipment for Arc Flash and Electrical Hazards

Electricity is a major workplace hazard and can cause electric shock, electrocution, burns, fires, explosions, and death. According to the U.S. Bureau of Labor Statistics, approximately 4 percent of all workplace fatalities in 2009 were caused by contact with electric current (BLS 2009). The best way to prevent injuries or fatalities from electricity is to completely de-energize and lockout/tagout live electrical equipment prior to working on or near it, refrain from performing work near energized electric lines or circuits, and keep electric-powered equipment and tools properly maintained. Work should not be performed on live or energized circuits unless absolutely required. However, when employees must work on electrical equipment, appropriate safe work practices must be followed, including the use of PPE to protect employees from the risk of electric hazards.

OSHA specifies in its electrical standards that "employees working in areas where there are potential electrical hazards shall be provided with, and shall use, electrical protective equipment that is appropriate for the specific parts of the body to be protected and for the work to be performed" (29 CFR 1910.335(a)(1)(i)). Meeting these requirements requires employers to select and provide the proper PPE, inform employees about what PPE is needed, train employees about using PPE correctly, and ensure that the PPE is properly used. In general, PPE for electrical safety can include insulating gloves and leather gloves; safety glasses or safety goggles; arc-rated face shield or arc flash suit hood; safety shoes or boots; insert-type hearing protection; hard hats; and arc rated flame resistant long-sleeve shirt, pants, coveralls, jacket, parka, or rainwear. Protection can also include the use of insulated tools and insulating blankets or sheeting for live circuits.

The National Fire Protection Association (NFPA) 70E Standard for Electrical Safety in the Workplace has very specific recommendations for PPE that should be used when working around energized electrical conductors or circuit parts (NFPA 2009). Prior to determining what PPE is needed, employers must perform a shock hazard analysis and an arc flash hazard analysis. The shock hazard analysis is used to determine the shock-protection boundaries. These boundaries are based on the voltage of range of electricity to which employees will be exposed, and the boundary distances are specified in Table 130.2(C) of NFPA 70E-2009. Any part of an employee's body that crosses the prohibited approach boundary specified in the NFPA table must be insulated through the use of PPE and insulation of the employee from the energized electrical conductors or circuit parts.

The arc flash hazard analysis is used to determine the flash protection boundary for protection against arc flash and arc blast. Two approaches can be used to help select proper PPE and protective clothing for employees that must perform work within the arc flash boundary of the first approach is to select PPE based on the results of an incident energy analysis. The incident energy analysis allows calculation of the incident energy exposure level based on the employee's distance from the arc source. Using this calculation, PPE and flame-resistant protective clothing with an appropriate arc rating can be selected. The second approach is to select PPE using the hazard/risk category classifications from Table 130.7(C)(9) in NFPA 70E-2009, based on the type of tasks that will be performed on energized electrical equipment. Hazard/risk classifications are numbered from 0 to 4. This table also specifies whether rubber insulating gloves and insulated or insulating hand tools should be used. After the hazard/risk category classification is determined, protective clothing and other protective equipment can be selected from Table 130.7(C)(10) in NFPA 70E-2009 (NFPA 2009). Any employee working within the flash-protection boundary must wear appropriate arc-rated flame-resistant PPE and all parts of the employee's body must be protected.

All PPE used for protection from electricity should meet ANSI or ASTM standards, as noted on NFPA 70E Table 130.7(C)(8). Flame-resistant clothing must be cleaned and maintained according to the manufacturer's instructions and stored to protect it from damage. Inspect all PPE before use to ensure that it is not damaged or contaminated with any combustible material. Additionally, NFPA states that clothing made from acetate, acrylic, nylon, polyester, polyethylene, polypropylene, and spandex should not be used, because they can melt at temperatures below 600° F (NFPA 2009). Because the NFPA 70E standard contains substantial detailed information, it should be referred to for specific guidance on selecting appropriate PPE for work on electrical circuits or conductors.

Personal Fall Arrest Systems

Protection from falling is needed when a person is working on a ladder, scaffold, roof, platform, or other elevated work surface. Protective devices often do not prevent the fall itself but interrupt it so that the person does not get hurt.

A fall-arresting system consists of an anchorage, connectors, a lifeline or dropline, the arrester, deceleration device, and a body harness. The lifeline may be vertical or horizontal depending on the work area involved. It may also include a lanyard between the arrester and the harness, as in Figure 5-2. A body harness distributes forces over several parts of the body. The fall arrester is a mechanism which moves freely up or down on the lifeline as long as there is no load on the arrester. As soon as a load is placed on the arrester, it locks onto the lifeline and prevents any downward motion.

The purpose of a fall-arresting system is to prevent injury to the worker in case he or she slips or loses balance and falls. The system must, therefore, be designed to absorb the energy of the fall without transferring that energy to the worker and causing injury. This requires either that the fall be stopped

Figure 5-2. A fall-arresting system with short lanyard and a harness. Courtesy of Dynamic Scientific Controls © 1994 DSC.

before any appreciable kinetic energy is developed, or that the system absorb that energy. The first alternative requires that the harness worn by the worker must have a very short connection to the arrester to limit downward motion. This severely restricts worker movement and therefore cannot always be used.

The second alternative can be achieved by using a lifeline made of material with a relatively low coefficient of elasticity and a high percentage of elongation. Table 5-2 gives the properties of the two most widely acceptable rope materials for lifelines. Other materials have been, and still are, used but are now considered inferior to these two. These materials include manila rope, polypropylene, and wire rope. Wire rope may be used only when sufficient heat is present to melt plastic wire. Stranded nylon is the most commonly used material. If the length of the lifeline above the arrester is long, however, the total elongation may become so long that the worker will hit something below. In that case, it would be better to have a lifeline with less elongation and insert a shock absorber between the worker and the arrester. This is particularly desirable when the worker needs extra freedom of motion and a lanyard is used between the worker and the arrester. Components of personal fall arrest systems should be inspected before use, for damage, excessive wear, or any deterioration. A fall arrest system that was used in a fall must be discarded and not used again. OSHA has several standards related to fall protection. These should be referred to for additional detailed information (http://www.osha.gov/SLTC/fallprotection/index.html).

Table 5-3 provides a summary of selected OSHA fall protection requirements.

Table 5-2. Typical Properties of Ropes of Various Materials and Sizes

Characteristic	Nylon Rope	Kernamantle Polyester/Polyester
Nominal diameter	16 mm	16 mm
Construction	3-strand twisted	static (solid)
Breaking strength		
dry	49kN	67kN
wet	39kN	66kN
Linear density		
dry	160g/m	183g/m
wet	190g/m	250g/m
Elongation, at 3kN		
dry	8.7%	1.0%
wet	13.8%	1.1%
at 8kN		
dry	20.0%	1.8%
wet	22.7%	1.8%
at 75% of BS	(33kN)	(50kN)
dry	36.6%	8.8%
wet	37.5%	8.8%
Max. Arrest Force*	6kN	18kN

*at mass = 100kg, free fall distance = 1.5m; arresting length of lifeline = 3.5m

Source: Andrew C. Sulowski, P. Eng., Senior Research Engineers, Ontario Hydro.

Table 5-3. Selected OSHA Fall Protection Requirements

	OSHA'S Maximum Allowable Fall Height	
Fall protection is required for:	Construction	General Industry
Sides and edges of floors and platforms	6 feet	4 feet
Openings in walls	6 feet	4 feet
Ramps, runways, and walkways	6 feet	4 feet
Roofs	6 feet	4 feet

Note: Skylights, floor openings, and holes require fall protection at all heights

Personal fall arrest system requirements:

- An anchorage capable of supporting 5,000 lbs. per person or one that has been designed by a qualified person to bear twice the maximum potential force of a falling body
- A lanyard with a locking snaphook
- A safety harness with a maximum fall-arresting force of 1800 lbs.

Training requirements:

- Fall hazards in the work area
- Selection and use of personal fall arrest systems
- Use and operation of any other fall protection systems, such as safety nets, warning lines, or safety monitoring systems

Source: *Tech Guide,* No. 9, published by the Electro-Optics, Environment & Materials Laboratory of the Georgia Tech Research Institute, 1997.

Hazardous Attire and Personal Effects

Clothing, jewelry, and other things workers wear can contribute to or detract from safety as much as the special devices described above. Though the hazards seem obvious, workers often overlook them and come to work wearing unsafe clothing.

Workers should not wear any clothing that could get caught in moving parts on machines or other equipment. Common examples include loose-fitting sweatshirt or jacket, loose shirt sleeves, unravelled threads from a sweater or shirt, and the loose ends of neckties. Hair has also been a cause of many serious accidents. Like clothing, hair can get caught in moving machinery, but, unlike clothing, it cannot be shed quickly or taken off without serious injury. Hair can be a problem in other ways as well. If a worker must use earmuffs, a hard hat, a welding helmet, or anything else that fits over the head, too much hair might prevent a good firm fit and render safety devices inadequate or useless. Therefore, long hair, whether on a man or a woman, must be held close to the head by a cap, hairnet, or some other device. For some jobs, a short hairstyle might be necessary.

Jewelry can also cause problems. Loose or dangling necklaces, earrings, and bracelets, like clothing, can get caught by moving machine parts, or snagged on protruding objects. Even rings have been caught by cutting tools or moving parts on machines. Metal rings and bracelets have also accidentally contacted live electrical parts and carried the current to the wearer. One immediately suspects that those live electrical parts were improperly accessible in the first place, but the jewelry nevertheless added one more hazard.

Finally, improper shoes can be a problem and contribute to the risk of injury. Therefore, open-toed shoes, sandals, and high- and narrow-heeled shoes do not provide the protection or support needed and are inappropriate in a work environment.

Miscellaneous Protective Equipment

Emergency showers and eyewashes are not worn by a worker, but will be discussed here with personal protective equipment. Emergency showers are primarily intended for flushing and washing chemical splashes from the body and clothing. Occasionally they have been used to extinguish a worker's clothing that has caught fire. The ANSI Z358.1-2009 standard for emergency eyewash stations provides specific guidelines on emergency eyewashes and showers.

Several types of emergency eyewashes and showers are available. A fixed, or plumbed, installation is preferable to a portable gravity-fed unit. The handle of an emergency shower must be readily accessible and easily operated. The most common type of handle is a large loop, round or triangular, attached to a pullchain or rod. However, all workers in the areas must be able to reach the handle. If it is low enough for a very short person or someone in a wheelchair, it might get in the way of a tall person. A solution to this problem is a pullcord attached to the wall near the floor, which can be grasped at any height. When activated, an emergency shower should provide at least 15 minutes of continuous water flow at a flow rate of 20 gallons per minute (gpm) and temperature range of 60°F to 100°F.

The tissue in the eyes is readily attacked by corrosive chemicals with permanent damage resulting very quickly. Ten to fifteen seconds of exposure to most corrosive chemicals is considered the maximum amount of time before permanent damage starts to occur, and it is extremely important that flushing of the eyes be done quickly. It is also important to remember that when foreign matter gets into the eyes, the eyelids normally close and tears develop. This built-in defense mechanism is good for small particulate matter, but presents a challenge when one is trying to flush the eyes because it means that the eyes must be held open. The worker's hands should not have to be used to hold open the valve to the eyewash. Usually there is a paddle or flag-type handle that stays open once it is opened. Sometimes a foot treadle is provided instead of, or along with, a hand-operated device. The water in an eyewash should be provided at a temperature of 60°F to 100°F, free of harmful substances, and supplied at an acceptable pressure. Combination eye- and facewash units should provide water at a flow rate of 3 gpm, and eyewash units should provide a flow rate of 0.4 gpm (ANSI 2009). It is also helpful to have some sort of shield to prevent the user from hitting his or her head on the eyewash fountains while trying to wash the eyes.

When there is something in the eyes, the worker is temporarily blinded. Therefore, the location of the eyewash, as well as its handle or treadle, must be well-known and clear of any obstructions.

Proper placement of emergency showers and eyewashes is an important responsibility for a safety engineer. They must be placed within a short distance of any and all points where chemical splashes are likely to occur. Workers should be able to get to an emergency eyewash or shower within 10 seconds of getting splashed by hazardous chemicals. One company that uses many chemicals in its processing operations has set a maximum distance of 7.6 meters (25 feet) from any workstation where liquid chemicals are used to an eyewash and shower unit. The route of travel between work area and emergency eyewash and shower must not involve a door or any obstacle. A 1-by-1-meter square, or a circle 1 meter in diameter, should be marked on the floor immediately in front of the eyewash. Emergency showers and eyewashes are often installed as a combination unit, as shown in Figure 5-3. In many cases, it may be helpful to incorporate an alarm with such a shower or eyewash to alert other people that an emergency exists. This is especially important when hazardous materials are handled in remote areas of the plant. All emergency eyewash and showers should be tested on a weekly basis to ensure proper function. Any problems must be corrected promptly. Weekly testing should also be documented.

Portable and even hand-held showerheads and eyewash units are commonly used in locations where it is not feasible to install permanent units. These portable units usually require the continual use of the hands to support the spray unit, and therefore the eyes cannot be held open. Portable units should not be considered a substitute for permanent or plumbed installations.

Another type of protective device is an alarm. In the past, alarms have been used primarily for fire and evacuation. However, an alarm can also be used to notify people of malfunctions or hazardous situations related to toxic chemicals, lasers, and radioactive materials. Alarms can also be used to alert for violations of security.

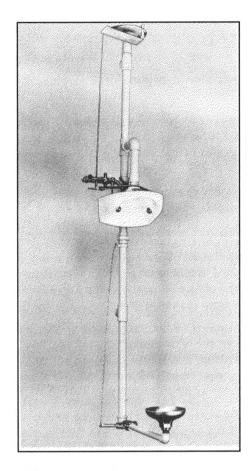

Figure 5-3. Combination emergency shower and eyewash unit. Courtesy Speakman Company.

Another type of warning device available is an alarm to be worn by people in hazardous or isolated locations. If the worker collapses or falls for any reason, the device sounds an alarm audible to the wearer, and then, if the wearer does not respond by shutting it off, it sends a radio signal to a central control. This alerts medical and/or other personnel to the situation of the stricken worker.

PPE is not designed to prevent accidents, but only to reduce the seriousness of bodily harm in case an accident does occur. Misuse of protective clothing, safety devices, and equipment of any kind may actually cause accidents. The one most important thing any worker can carry is a good attitude toward safety.

QUESTIONS AND PROBLEMS

1. What is the basic purpose of all personal protective equipment, and when should personal protective equipment be used?

2. Give several reasons why personal protective equipment, such as earplugs and respirators, should be considered as a last resort.

3. Why might workers need to wear gloves or protective clothing in the workplace?

4. What are some limitations of protective gloves and clothing with regard to chemical exposures, and how can these limitations be addressed?

5. What are three different types of eye protection, and when is eye protection needed?

6. When are workers required to wear respirators?

7. For what purpose is a chemical cartridge or particulate filter respirator used, as distinguished from a supplied-air or SCBA respirator?

8. When is hearing protection required, and what types of hearing protection are available?

9. What is a hearing protector noise-reduction rating (NRR), and using NIOSH recommendations, how much real-world protection would be provided by a pair of earplugs with an NRR of 32 dB in a workplace in which the noise exposure level was measured to be 97 dBA.

10. When is fall protection needed and what does a fall-arrest system consist of?

BIBLIOGRAPHY

American National Standards Institute, 2008. ANSI S12.6–2008, *Methods for Measuring the Real-ear Attenuation of Hearing Protectors.* New York: ANSI.

American National Standards Institute, 2009. ANSI Z358.1–2009, *Standard for Emergency Eyewashes and Shower Equipment.* New York: ANSI.

American National Standards Institute, 2010. ANSI/ISEA Z87.1–2010, *American National Standard for Occupational and Educational Personal Eye and Face Protection Devices.* New York: ANSI.

American Society for Testing and Materials, 2005. ASTM F-2412-2005, *Standard Test Methods for Foot Protection.* West Conshohoken, PA: ASTM.

American Society for Testing and Materials, 2005. ASTM F-2413-2005, *Standard Specification for Performance Requirements for Protective Footwear.* West Conshohoken, PA: ASTM.

Berger E. H., L. H. Royster, J. D. Royster, D. P. Driscoll, and M. Layne, eds., 2003. *The Noise Manual. 5th rev. ed.* Fairfax, VA: American Industrial Hygiene Association.

Ellis, J. Nigel, 2012. *Introduction to Fall Protection.* 4th ed. Des Plaines, IL: American Society of Safety Engineers.

"Fall Protection in the Workplace," *Tech Guide,* No. 9, 1997. Electro-Optics, Environment & Materials Laboratory, Georgia Tech Research Institute.

Forsberg, Krister and S. Z. Mansdorf, 2007. *Quick Selection Guide to Chemical Protective Clothing.* 5th ed. Hoboken, NJ: Wiley.

Golanski, L., A. Guiot, F. Rouillon, J. Pocachard, and F. Tardif, 2009. "Experimental Evaluation of Personal Protection Devices against Graphite Nanoaerosols: Fibrous Filter Media, Masks, Protective Clothing, and Gloves." *Human & Experimental Toxicology* 28:353–359.

Golanski, L., A. Guiot, and F. Tardif, 2010. "Experimental Evaluation of Individual Protection Devices against Different Types of Nanoaerosols: Graphite, TiO2 and Pt." *J Nanopart Res* 12:83–89.

International Atomic Energy Agency, 2004. *Practical Radiation Technical Manual.* Vienna, Austria: IEAE.

National Fire Protection Association, 2009. NFPA 70E, *Standard for Electrical Safety in the Workplace.* Quincy, MA: NFPA.

National Institute for Occupational Safety and Health, 1990. *A Guide for Evaluating the Performance of Chemical Protective Clothing (CPC),* DHHS (NIOSH) Publication Number 90–109. Cincinnati, OH: U.S. Department of Health and Human Services.

National Institute for Occupational Safety and Health, 2009. *Approaches to Safe Nanotechnology: Managing the Health and Safety Concerns with Engineered Nanomaterials,* DHHS (NIOSH) Publication Number 2009–125. Cincinnati, OH: U.S. Department of Health and Human Services.

National Institute for Occupational Safety and Health 2005. *Current Intelligence Bulletin 59—Contact Lens Use in a Chemical Environment,* DHHS (NIOSH) Publication Number 2005–139. Cincinnati, OH: U.S. Department of Health and Human Services.

National Institute for Occupational Safety and Health, 1998. *NIOSH Criteria for a Recommended Standard: Occupational Noise Exposure, Revised Criteria 1998,* DHHS (NIOSH) Publication Number 98–126. Cincinnati, OH: U.S. Department of Health and Human Services.

National Institute for Occupational Safety and Health, 1987. *NIOSH Guide to Industrial Respiratory Protection,* DHHS (NIOSH) Publication Number 87–116. Cincinnati, OH: U.S. Department of Health and Human Services.

National Institute for Occupational Safety and Health. *Workplace Safety and Health Topics: Eye Safety* (Retrieved August 1, 2011). (http://www.cdc.gov/niosh/topics/eye/).

Rouse, J. G., J. Yang, J. P. Ryman-Rasmussen, A. R. Barron, and N. A. Monteiro-Riviere, 2007. "Effects of Mechanical Flexion on the Penetration of Fullerene Amino Acid-derivatized Peptide Nanoparticles through Skin." *Nano Letters* 7(1):155–160.

Ryman-Rasmussen, J. P., J. E. Riviere, and N. A. Monteiro-Riviere, 2006. "Penetration of Intact Skin by Quantum Dots with Diverse Physicochemical Properties." *Toxicological Sciences* 91(1):159–165.

Schneider, T., J. W. Cherrie, R. Vermeulen, and H. Kromhout, 1999. "Conceptual Model for Assessment of Dermal Exposure." *J Occup Environ Med* 56:765–773.

Schneider, T., J. W. Cherrie, R. Vermeulen, and H. Kromhout, 2000. "Dermal Exposure Assessment." *Ann Occup Hyg* 44(7):493–499.

U.S. Bureau of Labor Statistics. *National Census of Fatal Occupational Injuries* (retrieved August 1, 2011). (http://www.bls.gov/iif/oshcfoi1.htm#2009).

U.S. Department of Labor, Occupational Safety and Health Administration, 2000. *Assessing the Need for Personal Protective Equipment: A Guide for Small Business Employers.* OSHA publication number 3151. Washington, D.C.: US Government Printing Office.

U.S. Department of Labor, Occupational Safety and Health Administration. *OSHA Safety and Health Standards for General Industry, 29 CFR 1910.* Washington, D.C.: U.S. Government Printing Office (updated annually).

6

Environmental Controls

Richard T. Beohm, PE, CSP, ARM

Characteristics of the Atmosphere

It would be cost effective and advantageous for companies to integrate safety into their business culture. Safety awareness is very important when working with hazardous chemicals that pose a health, safety, or environmental risk. Chapters 1 and 2 address this importance, its benefits and consequences. Chapter 6 addresses concerns and controls for environmental hazards in the workplace and how to protect workers from the harmful effects of exposure to them. According to the National Safety Council's Environmental Health Center, the chemical risk management problem is as follows:

880,000—The number of hazardous chemicals OSHA estimates are used in the United States.

More than 40 million—The number of employees who are permanently exposed to hazardous chemicals in more than 5 million workplaces.

55,400—The number of illnesses the Bureau of Labor Statistics estimates employees suffered that could be attributed to chemical exposures in 2007.

17,340—The number of estimated chemical-source injuries and illnesses involving days away from work in 2007.

Technological advances in today's world also bring new concerns about possible hazards, such as nanotechnology and biotechnology. There are also concerns with health and viruses and possible epidemics, as well as the hazards of second- and third-hand smoke.

One might well ask how atmospheric conditions relate to safety engineering, but there are several characteristics of atmospheric conditions that affect the safety and well-being of workers. The conditions that will be considered in this chapter include temperature, humidity, and air cleanliness. Illumination and color were discussed in Chapter 4. Noise will be treated as a separate topic.

Temperature and humidity obviously affect a worker's comfort, and comfort, in turn, has some effect on his or her performance. A worker may not be as alert in uncomfortable conditions, although the correlation between the two is not well documented. Extreme conditions certainly cause reduced capabilities.

Air cleanliness refers to two primary qualities of the air we breathe, namely, a sufficient percentage of oxygen and no harmful elements. "Clean" air contains about 21 percent oxygen and no toxic matter. When the oxygen content drops to about 16 percent to 18 percent, a person's thinking becomes unclear and muscular activity slows. The minimum permissible concentration of oxygen is considered to be 16 percent or 18 percent, depending on the authority consulted.

Limits have been established for permissible exposures to toxic materials. These are usually expressed in parts per million (ppm) and are called Threshold Limit Values (TLVs). TLVs will be discussed more thoroughly in Chapter 13 under Hazardous Materials. The effects of toxic materials vary. Most of them do not produce immediate effects but may have severe cumulative effects.

The effect of illumination on a worker's performance has been the subject of studies and experiments for over 50 years and is still unclear. It is known, however, that as luminance drops below some particular level, a person's visual acuity is reduced. The worker is then less likely to see visual signals, warnings, and so on, and this, of course, can lead to an accident.

Color is closely related to illumination and may affect safety in two ways. Color influences a worker's visual acuity, and it may also contribute to a person's feeling of comfort. How much color relates to safety because of its contribution to a person's comfort may be questionable, but its contribution to visual acuity is beyond doubt (see "Illumination and Color" in Chapter 4).

Control of Temperature and Humidity

Comfort zones of temperature and humidity have been defined even though many other factors may influence one's feeling of comfort. The psychometric chart illustrated in Figure 6-1 shows effective temperatures and the comfort zone. The

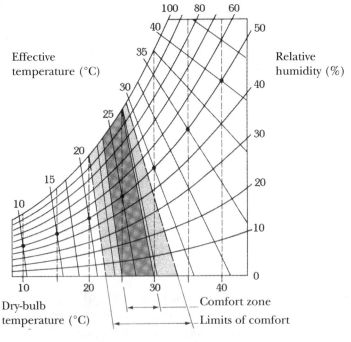

Figure 6-1. Psychometric chart with comfort zone.

comfort zone shifts slightly during the year as people become acclimated to warmer and colder weather; it is also slightly different in northern and southern areas.

Several factors contribute to temperature and humidity in the workplace. It is important to the human body that a fairly uniform body temperature be maintained, and the body has several mechanisms by which to achieve this. These include increasing or decreasing blood flow, excreting sweat to be evaporated on the skin (thus giving off heat), and increasing the effective skin thickness to provide insulation.

Hard physical work increases blood flow and body temperature. If the ambient temperature is not low enough to draw this body heat away, the person perspires to hasten the cooling process. A lower room temperature should be provided, then, for people doing this type of work.

The heat and cold stress equation is:

$$O = M + R + C - E$$

where

O is the oxygen required
M is metabolism
R is radiation
C is convection, and
E is evaporation.

The Heat Stress Index (HSI) is:

$$\frac{E\,(\text{required})}{E_{\text{max}}}$$

where $E(\text{required})$ is $M + R + C$, and E_{max} is the maximum evaporative capacity of the environment.

Circulating cooling fans are useful in hot or humid atmospheres for employee comfort. The total heat load is the sum of the environmental heat and the workload. Adequate ventilation is required by NIOSH standards. The American Society of Heating, Refrigerating, and Air-Conditioning Engineers (ASHRAE) sets the ventilation standards for indoor air quality. The American Conference of Governmental Industrial Hygienists (ACGIH) sets the TLV heat exposure limits to prevent deep body temperatures from exceeding 38°C.

The WBGT Index below is another means of measuring heat stress.

WBGT (Wet-Bulb Globe Temperature) = .7 NWB (Natural Wet Bulb)
+ .2 GT (Globe Temperature) + .1 DB (Dry Bulb)
(outdoors with solar loading)

WBGT = .7 NWB + .3 GT
(indoors or outdoors without solar loading)

There must be a balance of supply and exhaust air to ensure proper functioning of the ventilation system.

Exhaust openings should not be located near windows, roof vents, or air intakes to reduce the chance of reentry into the building. The addition of water vapor to dry air will change the density of air. Ventilation in ducts should be designed to minimize turbulent airflows. Bernoulli's equation applies to the energy within the duct. The upstream run of straight duct should be equal to or greater than 10 duct diameters.

$$TP \text{ (total pressure)} = SP \text{ (static pressure)} + VP \text{ (velocity pressure)}$$

Manufacturers typically rate fans and blowers on the basis of the relationship between airflow and fan static pressure. Ducts are balanced to ASHRAE and ACGIH standards. The brake horsepower (BHP) of a fan motor is calculated below.

$$\text{BHP (of the fan motor)} = \frac{Q \text{ (air flow in cfm)} \times TP \text{ (total pressure)}}{6356 \times \text{Mechanical Efficiency}}$$

Many processes used in industry give off heat, and some give off large quantities of heat. In a few cases the process absorbs heat and thus cools the surrounding area. Many times the heat gain or loss from industrial processes can be used to help maintain the desired ambient temperature, although if the gain or loss is too great, some other means of compensation must be found.

Since it affects the evaporation of perspiration, humidity must obviously be controlled along with temperature. The higher the moisture content of the air, the less additional moisture it can absorb. Air movement hastens the evaporation process, which is why we feel cooler in wind or in front of a fan.

Some work processes add to the problem of humidity. Drying processes, after dipping, spraying, and so on, remove moisture from the work and add it to the air. Many operations involving wood fiber products and natural fiber textiles require the addition of moisture for processing. This is often accomplished by raising the relative humidity of the whole room, thus adding to the discomfort of the workers.

Controlling temperature and humidity is usually not a serious problem for the safety engineer, but it certainly merits some consideration. Air-conditioning is becoming increasingly common in industrial buildings, as it is in other types of buildings. Unless extremes of temperature or humidity are involved, a conventional air-conditioner will control these factors adequately. Commercial units are readily available to provide additional heating, cooling, humidification, or dehumidification. When it is necessary for an engineer to estimate the heat balance for a room, it should be remembered that people contribute to the addition of both temperature and humidity.

Another type of system, a desiccant-based system (EPD), was designed and installed in five of the Georgia Institute of Technology's 1996 Olympic dormitories, as well as several high-rise office buildings in Atlanta, to provide an acceptable Indoor Air Quality (IAQ) complying with ASHRAE 62-89 ventilation guidelines.

The systems provide humidity-controlled, temperature-neutral air using a dual-wheel, total-energy-recovery concept, and they typically produce 2.4 tons of cooling, most of which is latent, for every one ton of cooling input.

A GTRI (Georgia Tech Research Institute) study, headed by Dr. Charlene Bayer, showed that 20 cfm per person was required to maintain total VOCs (Volatile Organic Compounds) below the accepted guideline of 2,500 micrograms/cubic meter and the 1,000 ppm carbon dioxide guideline limit. ASHRAE recommends 15 cfm per person. It also showed that 15 cfm per person with a desired 50 percent relative humidity was well under the mold growth relative humidity level of 70 percent.

The total-energy wheel in the EPD system recovers humidity from the exhaust airstream, thereby maintaining more desirable humidity levels in the space during the heating season. Exceeding 70 percent R.H. for an extended period can lead to odors, allergic reactions, and damage to carpets and wall coverings from mold and mildew. Microbial growth, linked to poor humidity control, can actually generate chemical VOCs and emit them into the indoor environment.

The EPD system allows sensible and latent loads associated with outdoor air to be "decoupled" from the buildings' internal heating-cooling load. Fan coil units process only recirculated room air, and since the outdoor air is pre-conditioned and over-dried by the EPD system, little condensation is experienced at these units.

Methods to improve IAQ and energy efficiency are:

- Build it tight and properly ventilate.
- Maximize the use of air-side economizers.
- Install a total energy recovery system.
- Upgrade air filtration to at least medium efficiency types.
- Improve humidity control.
- Balance the air systems.
- Improve housekeeping in mechanical rooms.
- Seal ductwork properly.

Control of Airborne Contaminants

Technically, air-conditioning includes cleaning the air as well as controlling the temperature and humidity. In practice, this usually doesn't happen, and when it does, it is only a matter of removing dust and other fairly large particulate matter. Clean air means air acceptably free of any substance that may be harmful in any way to human health and comfort.

Toxic airborne contaminants are released by such processes as metallic surface treatment, etching, pickling, acid dipping, metal-cleaning operations, electropolishing, electroplating, and electrolysis plating.

There are thousands of chemicals that could be hazardous to a worker's health. OSHA is now considering revising their 1910.1000 permissible exposure limits.

Definitions

- Permissible exposure limit (PEL): Enforceable regulatory limits from OSHA in the amount or concentration of a substance in the air.
- Recommended exposure limit (REL): Suggested exposure limits set independently by NIOSH.
- Threshold limit values (TLV): Voluntary exposure limits established by the American Conference of Governmental Industrial Hygienists.

OSHA's Air Contaminants Standard, 1910.1000 sets permissible exposure limits (PELs) for over 500 air contaminants and requires in-plant monitoring and analysis. It also sets PELs for a number of dusts and fumes. Air cleaners should be used in the workplace for occupational safety and health as part of the ventilation system. Air filters should be used only for air systems where dust loading is equal or less than 1 grain per 1,000 cubic feet of air. The ACGIH specifies the required capture velocities for the dispersion of various contaminants. The ACGIH ventilation rate is based on:

$$\text{CFM (required)} = \text{lbs. solvent evaporated per min.} \times \frac{387}{MW} \times \frac{10^6}{TLV}$$

Ventilation

The best way to eliminate foreign matter from the air is to avoid generating it in the first place. If that is not feasible, two other approaches may be used without resorting to personal protective equipment. One method involves catching the foreign matter as close as possible to the point of generation and discharging it into a safer place. The other way is to exhaust the air from the entire room and to replace it with clean air. Both methods are ways of diluting the concentration of the foreign matter, since neither will result in capturing 100 percent of it.

These two methods are called ventilation. The former is called local-exhaust ventilation and the latter is called general ventilation. Local-exhaust ventilation is generally much more efficient, provided the area in which the foreign matter is generated is reasonably small. This will involve a much smaller mass of air to be moved and, therefore, smaller ducts, fans, and so on. Furthermore, local-exhaust ventilation will collect the foreign matter, or whatever portion of it that it can, before it is dispersed into the air that people breathe.

General ventilation is employed when the generation of foreign matter is widespread. It is usually more expensive, and it should not be used when the foreign matter is toxic. With general ventilation, a supply of fresh air is needed to replace the contaminated air that is being removed. In local-exhaust ventilation the volume of air removed is generally so small that it can be replaced by the normal circulation of air in the room. The major principles of design are the same for both general and local-exhaust ventilation. Since local exhaust is to be preferred, and since the design of the hood may be more critical in that system, the design procedure that follows will be for local exhaust.

There are four, and sometimes five, components in a local-exhaust ventilation system, namely: a hood, ducts, a fan, a discharge, and sometimes an air cleaner. The hood is the inlet to the system and is one of the more critical parts. The location and shape of the hood determine from what points in the room foreign matter will be removed. The choice of location and shape should be based on the following characteristics:

1. Toxicity of the contaminant
2. Specific gravity of the contaminant
3. Size of the area in which it is generated
4. Air disturbances in the area that will affect the flow of air containing the contaminant

If the contaminant is toxic, movement must never bring it where a worker might breathe it. In fact, this should be a general policy whether the contaminant is toxic or not. The specific gravity indicates whether or not the material can be successfully lifted by a flow of air. If the specific gravity is much greater than 1, and the matter must be lifted to get it into the duct, the air velocity must be high.

If the contaminant is generated in a very small area or at one point, it is fairly easy to collect it before it becomes dispersed in the air. The system will not need as large a hood or as large a volume of airflow. As previously mentioned, it is always desirable to capture the contaminant as close to the point of generation as possible.

When an exhaust system is designed, it is usually assumed that the air around the hood has essentially zero velocity, but this is not always so. Other fans or currents of air from any source may make it difficult to capture and move the contaminated air. If other air movement cannot be eliminated, it may be necessary to extend the hood to form a barrier to it.

Exhaust hoods are used for open-surface tanks, spray booths, welding fumes, foundry dust control, grinding, woodworking, oil mist, melting furnaces, chemistry laboratories, and kitchen cooking equipment.

Laboratory fume hoods should be designed and situated within the lab so that fumes, gases, and particulates are prevented from or minimized from escaping the hood and entering the laboratory space.

Recommended fume hood airflow rates (face velocities) are as follows:

ANSI Z9.5	80–120
ACGIH	60–100
OSHA 29CFR 1910.1450	60 minimum
	150 (carcinogens)
NIOSH	100–150

Care must be taken in labs to prevent cross drafts between hoods. ASHRAE 110 is the standard for testing and management practices for safe laboratory hood systems.

Because of these different factors, a variety of types of hoods is used in industry, as illustrated in Figure 6-2. The simplest and most common is the

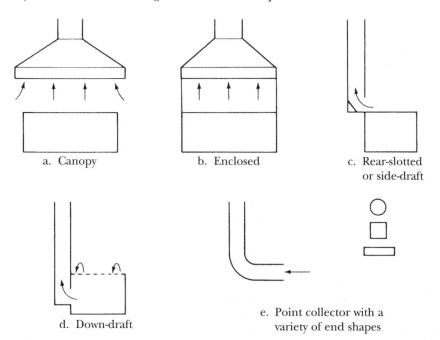

Figure 6-2. Common hood configurations for local-exhaust ventilation.

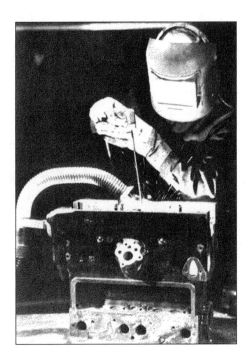

Figure 6-3. Application of a portable exhaust unit with a point collector. Courtesy Eutectic Corporation.

canopy hood. This is not very efficient, however, if the contaminant is heavy or if there are disturbing air currents; nor can it be used for toxic materials if the worker must have his or her head over the bench at any time. The enclosed hood has baffles all around except on the front. The rear-slotted and side-draft hoods have openings at the rear or the sides of the bench and move air horizontally across the bench. The down-draft hood is a more expensive installation because it requires a perforated workbench surface and clear space under the bench, but it is the most effective collection method for heavy contaminants. The point collector is used to capture welding gases and dust or chips from machines. The open-ended duct is brought as close as possible to the source of the generation, as illustrated in Figure 6-3.

The shape of the hood or collector opening determines to a large extent from how far away it will draw contaminated air. The two shapes most commonly used are the plain opening and the flanged opening. The area of their influence and the resulting velocity contours are illustrated in Figure 6-4. Two significant points should be noted from these curves. The first is that a flange at the edge of the opening widens the area of influence of the airflow into the duct. The second is that both of these openings have a very short range of influence. This distance is much shorter than the distance of influence on the discharge end of the duct. Air can be blown much farther by a positive pressure than it can be drawn by a negative pressure by a factor of about 30 to 1.

Capturing contaminants is largely dependent upon the air velocity at some point in front of the hood. Once the contaminant has entered a confined space, a certain velocity is required to maintain an adequate flow; this is called the control velocity. Capture velocity is the same as control velocity, or almost the same, in hoods that enclose the area where the contaminant is generated. Capture velocities and control velocities have been determined

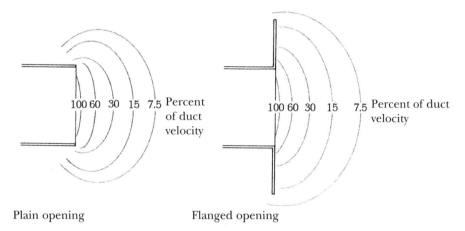

Plain opening   Flanged opening

Figure 6-4. Velocity contours for plain-opening and flanged-opening hoods and ducts.

experimentally and published as minimum or recommended velocities. Much of the data is from *Industrial Ventilation,* a publication of the Committee on Industrial Ventilation of the American Conference of Governmental Industrial Hygienists (ACGIH). The recommended capture velocity for vapors and gases is 50 to 100 feet per minute, whereas for heavy particles it is 500 to 2,000 feet per minute. Recommended velocity in the duct (transport velocity) is 2,000 to 5,000 feet per minute, with an average of 3,500 feet per minute.

In some cases the exhaust rate in cubic feet per minute is given, usually when a certain dilution is needed. Given either the velocity or the exhaust rate, the other can be determined by the relationship

$$Q = CVA$$

where

Q = exhaust rate in cubic feet per minute
C = a constant depending on hood type
V = velocity at a point $\times$ feet in front of the hood in feet per minute
A = cross-sectional area of hood opening

The value of C has been determined experimentally for various hood configurations, as shown in Table 6-1. This constant corrects for variations in turbulence and flow characteristics of the various configurations. It should be noted here that all the data used in ventilation design is in English units, and since it is empirical data it becomes rather meaningless to try to convert it to metric units; therefore, English units will be used.

Several pressure losses occur at various points in a ventilating system. The first one encountered is the hood entry loss. For standard air—that is, air at 70°F and sea-level pressure—this loss in velocity pressure is

$$VP = (V/4005)^2$$

where

VP = velocity pressure in inches of water
V = air velocity in feet per minute.

Table 6-1. Quantity of Air Captured by Various Types of Hoods

Hood Type	Name	Quantity of Air (Q)
	slot (H < 0.2L)	Q = 3.7LVX
	flanged slot	Q = 2.8LVX
	plain opening (H > 0.2L)	Q = V(10X^2 + LH)
	flanged opening	Q = 0.75V(10X^2 + LH)
	booth	Q = VLH
	canopy	Q = 1.4H (2L + 2W)

Slight corrections have to be made if atmospheric temperature and pressure are not standard. The *VP* should be multiplied by the correction factors shown in Table 6-2.

Table 6-2. Corrections for Air Density

Elevation (ft.)	Correction Factors at Temperatures (°F)							
	0	*40*	*70*	*100*	*200*	*300*	*400*	*500*
0 (sea level)	1.15	1.06	1.00	0.95	0.80	0.70	0.62	0.55
1000	1.11	1.02	0.96	0.92	0.77	0.65	0.60	0.53
2000	1.07	0.99	0.93	0.88	0.74	0.65	0.57	0.51
3000	1.03	0.95	0.89	0.86	0.71	0.62	0.55	0.49

Other corrections have to be made to account for friction in the duct, bends in the duct, and changes in cross-sectional area and shape. Obviously the inside of the duct should be as smooth as possible, but some corrosion

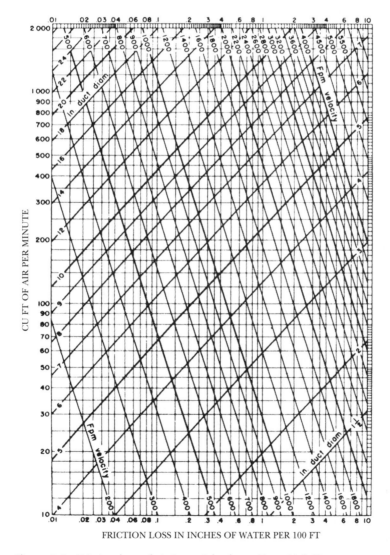

FRICTION LOSS IN INCHES OF WATER PER 100 FT

Figure 6-5. Friction loss of air in straight ducts. From U.S. Department of Health, Education and Welfare, National Institute of Occupational Safety and Health, *The Industrial Environment—Its Evaluation and Control* (Washington, D.C.: Government Printing Office, 1973).

and other defects must be expected. Joints of cut sections are necessary but should be made as smooth as possible. Figure 6-5 is a chart of empirical data relating friction losses to exhaust rate, velocity, and duct diameter. This is based on standard air and straight, clean, round, galvanized-metal ducts with joints spaced at 2.5 feet which is the standard length for duct sections.

A pressure drop also occurs when the flow of air has to change directions or when the duct cross-section changes shape or size. To simplify calculations, such corrections are treated as extra lengths of straight duct that would produce the same loss. Table 6-3 shows the equivalent lengths of straight duct that can be used in place of 90° elbows of varying sizes and bend radii.

Table 6-3. Equivalent Standard Lengths (2.5 feet) of Straight Duct for Friction Losses in 90° Elbows

Centerline Bend Radius	Duct Diameter, D, in inches											
	3	4	5	6	7	8	10	12	14	16	20	24
$1.5 \times D$	5	6	9	12	13	15	20	25	30	36	46	57
$2.0 \times D$	3	4	6	7	9	10	14	17	21	24	32	40
$2.5 \times D$	3	4	5	6	7	8	11	14	17	20	26	32

Multiply by: 0.67 for 60° elbows
0.50 for 45° elbows
0.33 for 30° elbows

When other complications are involved, such as joining two lines of ducts, increasing or decreasing duct size, or using square or rectangular ducts, additional factors must be included in calculations. Data needed for such calculations may be found in the book *Industrial Ventilation,* mentioned earlier, or in the *ASHRAE Guide and Data Book* published by the American Society of Heating, Refrigeration, and Air-Conditioning Engineers (see bibliography).

Another component in a ventilation system is a device to draw the air into and through the ducts and to discharge it. Axial-flow fans and centrifugal fans are the most commonly used items, although compressors and other devices can also be used. There are many types of fans available, each with certain advantages and limitations. Characteristics such as speed, power, efficiency, and physical size must be matched with the requirements of the rest of the system. Fan manufacturers make data about their products readily available and are probably the best source for this information.

In many cases, the air carrying the contaminant cannot be discharged without some treatment. Usable discharge materials such as sawdust may be blown into a bin or a pile outside the building. In other cases, the contaminant must be collected and treated before it can be disposed of. There are several types of devices and methods of treatment to clean air. Among the most common are:

1. Mechanical separators
2. Filtration devices
3. Wet collectors
4. Electrostatic precipitators
5. Gas adsorbers
6. Combustion incinerators

Some of these cannot easily be made a part of a ventilation system, but others can, especially the separators and filtration devices.

There are two general types of separators, impingement separators and centrifugal separators. Impingement separators draw particles by surfaces placed at an angle to the direction of airflow. The particles hit, or impinge upon, these surfaces and fall out, while the cleaned air continues on. In centrifugal separators, a high-speed circular path or vortex is created by forcing the air down through an inverted cone. Centrifugal force throws the heavier particles to the outside, where they slide down the funnel into a bin below, while the cleaned air flows back up through the center into a vertical duct at the top. This is the most common cleaning device used for coarse, heavy, and nontoxic material. Figure 6-6 shows a cyclone separator that operates on the centrifugal principle.

Filters are devices in which a stream of contaminated air is drawn through any one of several types of materials. Solid or semi-liquid particles are caught in the filter, and the cleaned air passes through. The most common type is referred to as a baghouse filter and consists of a series of fabric sacks, much like burlap bags, into which the air is blown. When the bags are so full of particles that they no longer function, they are removed and replaced by new bags. Fine-woven bags are very efficient but clog more quickly than coarser weaves. They are also more expensive to use, not only because the bags themselves are somewhat more expensive, but also because downtime for replacement is greater.

Wet collectors are often referred to as scrubbers. They are essentially like mechanical separators, except a spray or stream of a liquid, usually water, replaces the angled surfaces. The advantages of wet collectors over mechanical separators are that they are continually cleaning themselves, and that they can filter out some gaseous pollutants by absorbing the gas into the liquid spray or stream. The disadvantage is that the discharge of the contaminant is messy to collect and dispose of.

Electrostatic precipitators clean air of contaminants by means of electrodes and a field of charged gas ions. Small solid particles and even mists become charged and attracted to the electrodes. Periodically the frame holding the electrodes is shaken by mechanical cams or levers, releasing the collected

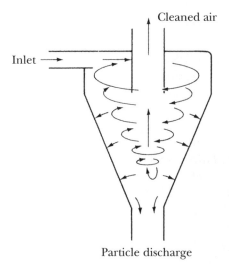

Figure 6-6. Cyclone Separator.

particles. The frame is held in a vertical position and requires fairly low velocities. Therefore, it is more often used in a vertical stack apart from a ventilation system.

Gas adsorbers are not common in most industries. They are expensive and are used for contaminants, such as toxic gases and radioactive materials, that resist other types of cleaners. Gas adsorbers work by means of a chemical reaction in which molecules are attracted to a surface.

Burning under controlled conditions in an incinerator is advantageous for removing pollutants that can be changed from toxic to nontoxic by oxidation. Other uses include removing organic vapors and offending odors. The process must be controlled to effect complete combustion, in which carbon dioxide and water are the principle end products. This method is not easily adapted to a ventilation system and, in fact, may require a bulky installation. Table 6-4 is a summary of these several different methods of removing pollutants from the air, whether as part of the ventilation system or as a separate facility.

If the air-cleaning device is part of the ventilation system, the fan or other air-moving device must be capable of drawing or blowing the air through the cleaning device. No matter how it is done, the final discharge from the ventilation system must be in an acceptable form and place. It would serve no purpose to collect the contaminant and then discharge it back into circulation again.

If the air is not recirculated, then additional air must be brought in to replace it. Local-exhaust systems generally remove such a small quantity of air that a fresh air makeup is achieved through leaks around and through doors, windows, and so on. In general exhaust systems involving the entire volume of air in a room, a fresh air makeup system must be designed to provide oxygen and to avoid a pressure differential.

This suggests another technique often used to deal with contaminated air. To confine a contaminant to a small enclosed area, a pressure differential is created with the higher pressure outside the contaminated area. This assures that if there are leaks, the flow of air will be into, not out of, the contaminated area.

Sick Building Syndrome

Sick Building Syndrome (SBS) is now a major concern for the nearly 70 million employees who work indoors. SBS is defined to be when 20 percent of a building's occupants complain of symptoms such as headaches, dizziness, scratchy throat, dry cough, itchy/watery eyes, nausea, lethargy, fatigue, and the inability to concentrate. OSHA's proposal to regulate indoor air quality (IAQ) is still under review.

Inadequate ventilation is a prime factor. ASHRAE standard 62.1 (2007), Ventilation for Acceptable IAQ Guidelines, requires 20 cfm per person in enclosed spaces to keep VOCs less than 2500 micrograms per meter cubed and 1000 ppm CO_2, but "increasing ventilation up to 52 cfm can prevent the spread of disease among workers in buildings, as well as save the U.S. economy billions of dollars in respiratory illness and SBS costs" (Harlos and West 2010). Chemical contaminants from inside pollutants such as adhesives, carpeting, wood products, upholstery, cleaning agents, tobacco smoke, and office equipment emit VOCs into the air. Formaldehyde (a suspected carcinogen) and carbon monoxide are common VOCs. Pollutants such as car and building exhausts can re-enter the building through poorly placed air intake vents. Buildings over parking garages are especially susceptible to increased carbon monoxide levels.

Another contributing factor to poor IAQ is biological contaminants such as pollen, bacteria, viruses, and mold spores. Water that collects in ceiling tiles, insulation, carpets, and vent ducts acts as an ideal breeding location for those contaminants. Eventually, they spread throughout the building through the ventilation system.

When SBS occurs, the air ventilation system should be checked for possible blockage. All HVAC systems should be designed to meet the building codes. A building inspection should be conducted to find and eliminate the source of the contaminant. Air testing may also be helpful in locating the source.

The building HVAC system should be designed to keep the building in a positive pressure with respect to the outside atmosphere to keep the inside healthy and free of humidity problems. Dew-point temperatures should be assessed for ventilation and moisture control to provide a healthy and productive environment free of black mold, airborne bacteria, dust, and viruses.

Chemical filtration in Air Handling Units (AHUs) removes more air contaminants relative to poor IAQ than carbon-only filters and also reduces allergens in the air. There are also mold-resistant air-duct liners available. Dehumidification systems remove moisture and prevent IAQ problems associated with moisture such as mold, mildew, biological allergens, and odors.

Automatic CO ventilating fans can be used to control CO levels in garages and enclosed buildings. Exhaust fans would activate when the CO level exceeds 100 ppm and cut off when the level drops to 50 ppm.

HEPA filters can be used to control dust and pollen from the outside. Today many public and government buildings have been declared "smoke free," prohibiting smoking.

Contaminated ductwork can be treated by forcing air through a titanium coating and filter, and then exposing it to UV radiation to rid it of VOCs and bacterial contamination.

Nanoparticles

Nanotechnology is an expanding field that involves the manipulation of materials at the nano scale (10 to the -9 meter) that are near the atomic scale in size. It provides new opportunities in a diversity of fields such as medicine, materials science, electronics, and energy. Nanoparticles are defined as those engineered materials that have at least one dimension of 1 to 100 nanometers (nm). The safety and health effect of nanoscale particles could be toxic. They can enter the body through the skin and reach other organs. An effective training program regarding the safe handling of these particles along with adequate industrial hygiene controls is recommended. There is now a NIOSH Nanotechnology Program.

Table 6-4. Comparison of Industrial Air-Cleaning Devices

Device	To Control	Advantages	Disadvantages	Costs	Examples
Mechanical Separators	Medium to large diameter particles	(1) Low initial cost (2) Simple construction (3) Ease of operation (4) Use as precleaners	(1) Low efficiency (2) Erosion of components (3) Cannot remove small particles (4) Large space requirements	Low initial cost	(1) Gravity Chambers (2) Impingement Separators (3) Cyclone Collectors
Filtration Devices	Dusts, fumes	(1) High collection efficiency on small particles (2) Moderate power requirements (3) Dry disposal	(1) High costs (2) Large space requirements (3) Must control moisture and temperature of gas stream	High costs	(1) Fabric Filters (2) Mat Filters (3) Ultrafilters
Wet Collectors	High-temperature, moisture-laden gases	(1) Constant pressure drop (2) Elimination of dust removal problems (3) Compact design	(1) Disposal of waste water may be expensive and troublesome	Moderate	(1) Spray Chambers (2) Cyclone, Orifice, Venturi Scrubbers (3) Mechanical Scrubbers (4) Mechanical-Centrifugal Collectors

Device	Contaminant	Advantages	Limitations	Cost	Types
Electrostatic Precipitators	All sizes of particles—even very small mists which form free-running liquids	(1) High efficiency (2) Dry dust collection (3) Low pressure drop (4) Can collect mists and corrosive acids	(1) Often requires precleaner (2) Large space requirements (3) Cannot collect some high/low resistivity materials (4) High initial cost	High initial costs— low operating costs and low maintenance costs	(1) Single-stage Precipitators (2) Two-stage Precipitators
Gas Adsorbers	Highly odorous, radioactive or toxic gases	(1) Contaminant solvent may be recovered	(1) High equipment and operating costs (2) Corrosion (3) Contamination	High equipment and operating costs	(1) Fixed Bed (2) Regenerative
Combustion Incinerators	Odors, plume opacity, carbon monoxide, organic vapors	(1) Capable of reaching high efficiency operation (2) Catalytic combustion reduces NO_x pollutants	(1) Must burn additional fuel or add catalyst (2) Incomplete combustion can further complicate original problem (3) Catalysts require periodic cleaning and regeneration	Vary widely depending upon application	(1) Direct Flame (2) Catalytic Combustion

Source: U.S. Department of Health, Education and Welfare, NIOSH, *The Industrial Environment: Its Evaluation & Control* (Washington, D.C., Government Printing Office, 1973), p. 645.

QUESTIONS AND PROBLEMS

1. What is the ASHRAE recommended ventilation rate for an enclosed space? What is the advantage to increasing the rate to 52 cfm?
2. Approximately how many hazardous chemicals in the United States does OSHA estimate and how many do they regulate?
3. Name a few health concerns brought about by new technologies.
4. What precautions should a person take when working with nanoparticles?
5. Name several ways to improve IAQ in a building.
6. Name some of the symptoms of a sick building syndrome and what causes them.
7. Name the different organizations that specify fume hood airflow. What does NIOSH recommend?
8. What air velocity does ACGIH recommend to capture vapors and gases? heavy particles?
9. Why are companies striving to obtain LEED certification?
10. What is the velocity in a hood when the face velocity pressure reads .0156 inches of water? (Assume 70 degrees F and sea-level pressure.)
11. Determine the quantity of air (Q) for a canopy hood 3 feet x 6 feet and 4 feet above the work area.
12. Dry bulb temperature should be maintained between what points for an R.H. of 60 percent to control mold growth?
13. What is the pressure loss in inches of water in 10 feet of 8-inch diameter straight duct and two 90 degree elbows with a 2D bend radius and a transport velocity of 500 fpm?

BIBLIOGRAPHY

American Conference of Governmental Industrial Hygienists. *Industrial Ventilation: A Manual of Recommended Practice.* Cincinnati, OH: ACGIH (updated annually).

American National Standards Institute, Inc., 2007. *Fundamentals Governing the Design and Operation of Local Exhaust Systems, ANSI Z-9-2.* New York: ANSI.

American Society of Heating, Refrigeration and Air Conditioning Engineers, 2011. *HVAC Systems and Applications.* Atlanta: ASHRAE.

Burgess, et al., 1989. *Ventilation for Control of the Work Environment.* New York: Wiley.

Burton, D. Jeff, 2011. *IAQ and HVAC Workbook.* 4th ed Bountiful, UT: IVE Books, Inc.

Downing, Chris, 1996. "A Cost Effective Method to Improve IAQ," in *Proceedings of the SESH&E Conference.* Atlanta: Georgia Tech.

_____, 1994. "IAQ and Energy Efficiency Are Compatible," in *Proceedings of the SESH&E Conference.* Atlanta: Georgia Tech.

Guffey, Steven E., 2004. *Industrial Ventilation Design.* New York: Wiley.

Harlow, David and Michael West. 2010. "A Case for Increased Ventilation." *HPAC Engineering*, March 1, 2010.

Hoover, Reynold L. et al., 1995. *Health, Safety and Environmental Control*. New York: Van Nostrand Reinhold.

McDermott, Henry J., 2001. *Handbook of Ventilation for Containment Control*. 3rd ed. Cincinnati, OH: ACGIH.

National Institute of Occupational Safety and Health (NIOSH), "The Industrial Environment: Its Evaluation and Control." Washington, D.C.

Plog, Barbara A., and Patricia Quinlan. 2001. *Fundamentals of Industrial Hygiene*. 5th ed. Chicago: National Safety Council.

7

Walking and Standing Surfaces

Keith Vidal, PE

Kirk E. Mahan, PE, CSP

Injuries from Falls

Why should a safety engineer be concerned about the surfaces on which people stand, walk, and climb? The answer is that more injuries result from falls than from any other single factor except motor vehicle accidents. These falls are caused by tripping, slippery surfaces, unsteady surfaces, and loss of balance for other reasons, and many of them result in death. Figure 7-1 shows some statistics from the 2010 edition of the National Safety Council's

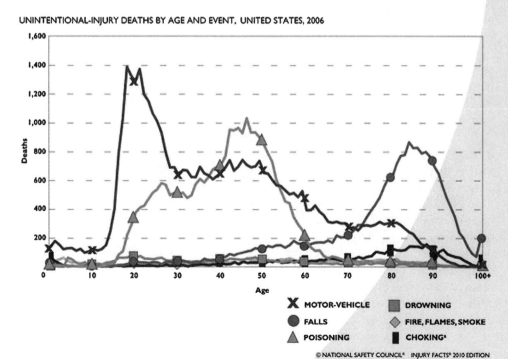

Figure 7-1. U.S. deaths from unintentional injuries, 2006.
Source: National Safety Council, "Accident Facts," Itasca, IL, 2010.

"Accident Facts." The latest data available are for the year 2006. Falls were the second greatest cause of accidental deaths, exceeded only by transportation accidents. This has been the case for many years.

Principles of Safety Engineering

The concept of preventing injures due to known hazards has been addressed in various scientific literature for at least a century, and all of our knowledge regarding avoidance of hazards boils down to a few basic principles, the first of which is to eliminate the hazard through design. The second priority states that protection must be provided when a hazard cannot be eliminated. The third priority involves the use of warnings in instances where a hazard cannot be eliminated through design, or adequate protection cannot be provided. Appropriate signage and/or markings should clearly identify the hazard with appropriately designed icons, markings and/or language that depict the nature of the hazard. Although this is not a substitute for a proper design, it is a recognized means of reducing risk associated with potential hazards. These same principles/concepts should be utilized when considering pedestrian safety on walking and standing surfaces. These concepts are applicable in: the design phase with the choice of appropriate flooring materials; the construction phase; and finally, maintenance, and training of employees.

The Ambulation Process

To better understand how and why people fall, let us review the process by which people walk and maintain balance in an upright position. When we walk, we typically move one leg forward, which shifts our center of gravity forward. Then, in order to prevent falling, we place that forward foot on a firm surface to provide support for the body and move the other leg forward. Several things can happen, however, to prevent this expected sequence from happening. If any obstruction prevents the leg or foot from moving forward, we trip. If either the leading foot or the trailing foot rests on loose or slippery surfaces, it provides insufficient support for the body. If anything causes the legs and feet to move at a faster or slower rate than needed to keep the center of gravity between the points of support, we fall.

These are simple concepts, but they are very basic to an understanding of how tripping, slipping, and losing one's balance can be prevented. Other factors also affect the ability to maintain balance. The center of gravity must always be located between the limits of the points of support. The center of gravity is affected not only by the mass of the arms, the head, and the trunk, but also by any object carried by the body. We learn to compensate for varying locations of the center of gravity, but any misjudgment can lead to a fall.

Human factors also play a significant role in fall incidents. When we are distracted or concentrating on something else, we may not perceive small changes in elevation, contaminants, and other anomalies that initiate falls. The engineer can do little to change our mental habits, but he or she can design the walking surface to minimize the hazards.

One other factor that affects our ability to maintain an upright position is the condition of the labyrinthine organs by which we sense equilibrium, the three semicircular canals in the inner ear. They contain fluid that flows as the head is moved and sensory nerve endings that detect the flow and give us a

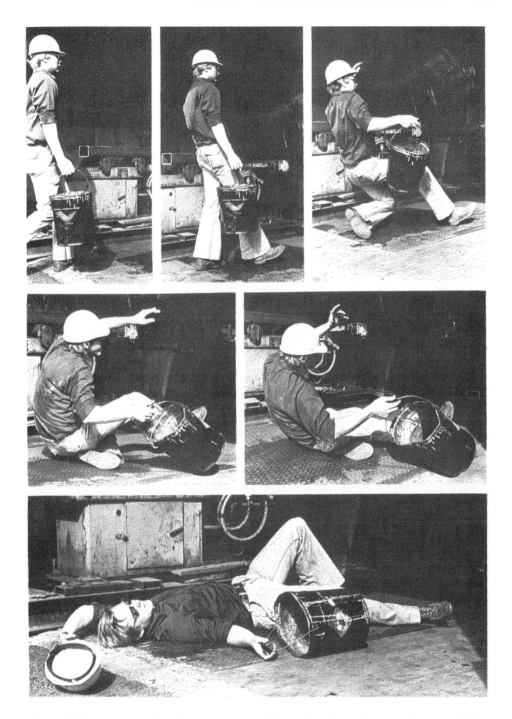

Figure 7-2. Sequence of positions in a typical fall resulting from slipping. Courtesy of Safe Walk, Inc.

sense of whether or not the head is in equilibrium. If this system is damaged or adversely affected by infection or other malady, we feel dizzy and find it difficult to judge how to maintain body balance. Again, the engineer can do nothing to change this situation.

Figure 7-2 shows a sequence of positions in a typical fall resulting from slipping. It is very noticeable that when the leading foot contacted the floor, the force on it resulting from the inertia of the moving body was greater than the resistance offered by the floor. The trailing foot was not strong enough in that position to support the body, so the whole system collapsed.

Human Factors

It is important for facility owners, managers and contractors to implement appropriate safety measures in constructing and maintaining walking surfaces that are expected to be used by persons of all physical abilities and disabilities. Such entities should never needlessly expose persons to dangerous conditions that create slip or trip hazards. If such conditions are created, then those unavoidable and uncorrected dangers must either be guarded, or warnings must be posted.

People generally expect that walking surfaces are uniform and reasonably safe unless something is recognized or perceived as being potentially hazardous. Pedestrian fall incidents commonly occur when the pedestrian encounters an unexpected hazard in the walkway. Unexpected changes in available friction and elevation lead to increased chances of a fall if they are encountered.

It is a well-known fact that people do not look at their feet as they walk, but instead look toward their objective. A person's visual field is constantly changing, and only a small part of the actual visual field is given conscious attention. Objects that fall within the visual scanning field, a cone-shaped area out in front of a person, are given priority. Visual discrimination of objects outside this field (peripheral) is minimal. This leads to the increased likelihood of an accident if a hazard lies outside the scanning visual field, making identification of hazards that much more imperative.

The inability to recognize a hazardous condition can be caused by many factors, including: distractions within the surroundings; widely varying changes in lighting; familiarity of the environment; poor or misleading visual cues or a lack of them; sensory overload; location in the visual field; corrective lenses; and age.

Technology advances have also introduced new sources of distraction: personal music-playing devices; wireless phones; and texting devices are relatively new sources of distraction that put the onus of safe operation on the user. Training of employees as to unsafe and unacceptable practices may be warranted. The best way to keep the general public safe, since their behavior is more difficult, if not impossible, to control is to eliminate hazards altogether.

Design and Maintenance of Level Surfaces

Falls from Trips

One common cause of falls is tripping: one foot is prevented from moving forward by some obstruction. Usually it is the swinging foot that is caught. The obstruction could involve a variety of conditions, as illustrated in Figure 7-3.

Why is it that some people can walk over very rough surfaces and never trip, while others trip over very small obstructions? Some people have poor control over the motion of their legs for physiological reasons. Others have developed poor walking habits and scuff their feet along the floor, a practice that makes them susceptible to tripping. Others don't pay attention to where they are walking and seem quite oblivious to any irregularities in the path.

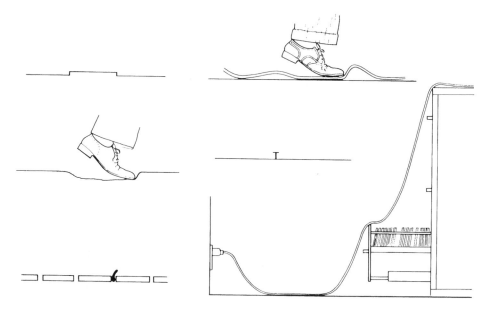

Figure 7-3. A variety of common tripping hazards.

Why is it that people can walk over loose boards, pipes, and all sorts of obstructions at a construction site, yet stumble over an extension cord across the floor in an office? At a construction site we expect to find objects in our path and we are attentive to them. In an office, it is the pedestrian's expectation that the floor is level, smooth, and without obstacles, so we walk accordingly. In this, as in many other situations, our behavior depends more on what we anticipate than on what actually exists.

What constitutes a tripping hazard obviously varies depending on the type of surroundings and the people who will use the area. In many situations, the decision must be left to someone's good judgment, perhaps that of a safety engineer, and should be based on anticipated and forseeable use.

The floor of an office, corridor, stairs, or other clean area is expected to be maintained free of obstructions that could cause tripping. Vertical changes in elevation as small as 0.25 inches (6.4 mm) are acceptable according to ANSI A117.1-1992 "Accessible and Usable Buildings and Facilities" and ASTM F1637-95 "Standard Practice for Safe Walking Surfaces." Changes in elevation greater than 0.25 inches (6.4 mm) are required to be beveled or comply with ramp construction guidelines, depending on the height of the elevational change. If obstructions such as a raised floor board or tile are limited to less than ¼ inch, the probability of a person's tripping would be small enough so that some industrial companies would consider it an acceptable risk.

One of the most common tripping hazards in an office or a laboratory is an extension cord. Presumably an extension cord lying on the floor would be pushed by the shoe and not act as an obstruction. Very often, however, the insulating material has a high coefficient of friction and will not slide if there is any downward force on it. The cord tends to roll but still acts as an obstruction. If the shoe lands on top of the cord, the rolling action makes the surface unstable. There is also a limit as to how far an extension cord will move before it reaches the limit of its free movement. A cord that is attached securely at both ends will move only a given distance and then stop. At this point it becomes an obstruction. This becomes an even greater hazard if the cord hangs just a few centimeters above the floor.

In a manufacturing or maintenance area there are many objects which act very much like the extension cord discussed above. Stub ends of welding rods, screws, balls from ball bearings, and short pieces of round bars are often found on the floor. Trash barrels should always be available in areas where such material is being used or generated.

Other recognized tripping hazards include: mats and runners, walking-surface hardware, stairs (especially short flights of one to three steps), speed bumps, wheel stops, low-lying displays, furniture and fixtures, and gratings. A mat or a runner should be of sufficient density and stiffness to prevent buckling, folding over itself, or rippling from poor storage techniques. Mats and runners should have slip-resistant backing to minimize sliding across a floor they are laid upon.

Another factor in falls, whether a result of a slip or trip, is illumination. Headroom is also a consideration, especially on or around stairs, decks, doorways, and corridors.

Tripping hazards found frequently on construction sites are loose boards, pipes, and debris left in paths of travel. In receiving and shipping areas, it is common to find pieces of strapping and broken boxes and cartons left in pathways. Welding cables and temporary water and compressed gas lines are also frequently found in the way. These practices, among the most common hazards found in industry, should not be permitted and typically will not be found in companies where safety is emphasized. A worker who leaves objects in a path of travel should be reprimanded, and repeated offenses should be considered sufficient reason for dismissal. Managers must also be held accountable for unsafe acts and conditions.

Falls from Slips

Slippery floors are another common cause of falls. Like tripping, the hazard of slipperiness is difficult to assess and measure. Referring again to Figure 7-2, it can be inferred that the heel of the leading foot landed on a surface with insufficient slip resistance. If the stride is not too large and the walker anticipates slipperiness, he may be able to keep his balance by holding his legs rigid in an inverted-V position, although this puts a great deal of stress on the legs and sometimes on the back as well.

Most falls result from an unexpected, localized spot that has lower slip resistance than the surrounding floor. The concept of expectation is important when considering human locomotion; that is, our expectation of consistent floor friction remains constant unless visual or tactile cues tell us that a change in resistance to slipping may be present. It is possible to walk on a slippery surface without falling down if one knows that the surface is slippery. People will adjust their gait and stride for that situation. Prior knowledge of similar surfaces and of conditions on those surfaces dictates our expectations of how that surface will perform under our feet. That knowledge assists us in adjusting to a reasonable walking speed and an appropriate length of stride. Watching people walk on an ice-skating rink illustrates how people will tend to slow down and shorten their strides as they try to walk without falling. In fact, most people try just to shuffle across the ice without actually lifting their feet in order to remain stable.

The study of slip resistance on pedestrian walking surfaces has become a highly debated scientific topic. "Tribometry" is the study of pedestrian slip resistance, and the devices used to measure the coefficient of friction, or slip resistance, are commonly referred to as "tribometers." The traditional definition of the coefficient of friction (COF) relates to the resistance of relative motion between two interacting, interfacing bodies. The static coefficient of friction (SCOF) refers to the resistance to motion between two surfaces where no relative motion (sliding) is involved. The dynamic or kinetic coefficient of friction (DCOF) refers to the resistance to motion once the two surfaces have begun sliding. The traditional equation for COF is:

$$\mu = \frac{F_m}{N} = \tan \alpha$$

where

F_m = tangential or horizontal component of force required to initiate motion (lbf or N),

N = normal or vertical component of force (lbf or N),

α = the slope, or angle, of an inclined surface from the horizontal required to initiate sliding of an object,

μ = coefficient of friction.

Testing under contaminated or lubricated conditions introduces a third interfacing material into the friction model, thereby negating the use of the term SCOF. "Slip resistance" or "slip index," with units of measurement analogous to SCOF, becomes the proper term when such conditions exist. The COF is a unitless value that is generally between zero and 1.0, as is the slip index. A slip index of 0.2 would correspond to a very slippery surface, whereas a slip index of 0.8 would be considered slip resistant. The COF or slip index can be greater than 1.0, but this is usually the result of some kind of interlocking taking place between the surfaces. The slip-index value of 0.5 is generally considered to be the value of slip resistance above which a surface would be considered safe. The ubiquitous value of 0.5 is constantly referred to without consideration as to how it should be measured.

There has been a need to specify a standard foot material for testing because it is necessary to standardize every variable (if possible) except the one we are interested in (i.e., the slip resistance of the surface). The selection of a standard foot material for tests is still a subject of controversy. As of 1999, the generally accepted standard test material was Neolite[1], a material that has been found to have desirable qualities for slip-resistance testing. A prudent safety engineer should refer to the appropriate standard and/or the tribometer's operating manual for recommendations or requirements regarding selection of appropriate foot material for testing.

The term "slip resistance" is preferred, by experts in tribometry, to the term "coefficient of friction" (COF) because, from a technical standpoint, COF can only be measured when there are no contaminants on either of the sur-faces. The surfaces that pedestrians walk upon are rarely, if ever, clean and dry, and the same can be said of footwear. The measurement of the slip resistance of a floor is complicated by several factors, including: interaction between the shoe sole material, as well as the floor material; many of the devices that have been developed, which measure the COF in ways that do not resemble or emulate the dynamics of the human foot in motion; the devices that are commonly used for slip-resistance measurements, which do not correlate with each other.

Many tribometers have been developed over the years that have been the subject of studies and research by academia, bodies that write consensus standards, testing and research facilities, and independent researchers. A nagging problem has been the difficulty in developing acceptable tribometric devices that produce valid, reliable, and reproducible results in a field setting under both wet and dry conditions.

One of the most prevalent issues in the design of tribometric devices is the avoidance of a phenomenon known as "sticktion." Sticktion is associated that devices that are placed on a surface before the application of a sliding force is applied. The interval between the time that the object is placed on the surface and the time that slipping occurs is referred to as "residence time." Sticktion occurs when an adhesion develops between the tribometer's test foot/feet and the surface on which it/they set in the presence of a liquid contaminant (although sticktion can also be evident in dry testing.) Residence times as short as 0.25 seconds can produce disqualifying sticktion. Sticktion leads to erroneously high slip-resistance readings (i.e., a tribometer will seemingly indicate that a smooth, wet surface is more slip resistant than the same surface under dry conditions). The development of new devices that avoid sticktion by applying the components of horizontal and vertical force to the surface simultaneously, thereby avoiding residence time, makes them suitable for testing under wet as well as dry conditions. In the late 1980s and early 1990s several new devices were developed that more closely emulate the dynamics of the human foot and avoid sticktion.

The American Society for Testing and Materials, International, (ASTM) Committee F-13 "Pedestrian/Walkway Safety and Footwear" has responsibility for standards related to the measurement of slip resistance on pedestrian walking surfaces, although there are several other material specific committees that have their own standards for quality control of as-manufactured flooring products/surfaces. It is important to remember that the majority of slips occur on surfaces that are wet or that have some type of surface contaminant, and it is therefore necessary to consider devices that are accepted for use under both wet and dry conditions.

Generally speaking, there are three classes of devices (tribometers) used to measure the slip resistance of walking surfaces: pendulum testers, drag sleds, and inclinable articulated testers.

Pendulum Testers

This type of tester is designed to test dynamic friction only. Several incarnations of this device exist, but they all employ a shoe on the end of a pendulum that swings in an arc of known initial height. The shoe is then allowed to drag across the surface, resulting in a loss of energy and lower swing height. The loss of energy is directly related to the dynamic COF. This device operates quite well under wet or dry conditions, although its detractors claim, among other things, that the slider pad encounters micro skips as it slides across a surface.

Although the test foot on this class of tester strikes the walking surface at an angle, the speed of the test foot dragging across the surface is not comparable to the speed at which the human heel strikes the surface during normal walking. There are no ASTM or ANSI standards that apply to the use of this class of tribometer for testing pedestrian walking surfaces.

Drag Sleds

The classical physics-class experiment using an object of known weight being pulled by a scale is valuable as an academic exercise, but its application in the practice of modern safety engineering is limited to a consideration of dry surfaces only. Drag-sled testers are, as the name implies, weighted objects that are dragged across a surface. This class of device is particularly susceptible to stiction, and, as a result, drag-sled meters are no longer recommended for slip-resistance testing of wet surfaces. Figure 7-4 is a photo of one of many types of drag-sled devices. This particular device is more commonly known as a "horizontal pull slipmeter," or HPS, and its use is standardized in the American Society for Testing and Materials (ASTM) Standard F 609-96 "Standard Test Method for Using a Horizontal Pull Slipmeter (HPS)." It consists of a weighted sled with three replaceable pads (test feet) on its bottom side and a calibrated spring scale mounted on top of the weights. The spring scale is attached by a cord to a pulley that slowly rotates when activated. When sufficient force is applied to the scale, the sled starts slipping on the floor. The force required to overcome this initial friction is a measure of the static coefficient of friction (SCOF). Drag-sled meters do not emulate the

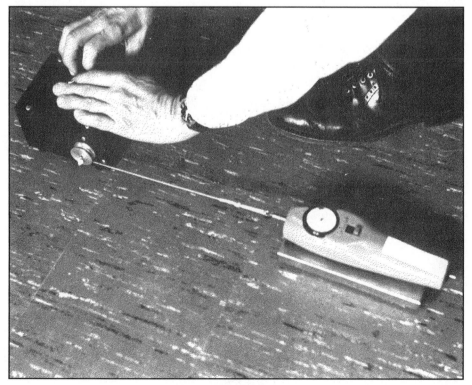

Figure 7-4. A horizontal-pull slipmeter being used to determine the slipperiness of a floor.

dynamics of the human foot and are susceptible to sticktion, not to mention operator-induced variables. As practical tools of safety engineering, their usefulness is limited.

Articulated Inclinable Testers

This class of tribometer uses the trigonometric relationship in the definition of COF to determine the slip resistance on a surface. That relationship is represented as:

$$COF = \tan \alpha.$$

The device often referred to as the standard for measuring static friction is the James Machine. This device measures only static friction and is useful only in a laboratory setting. Another drawback to the James Machine is its susceptibility to sticktion, which makes it unacceptable for use under wet conditions.

In 1974, ASTM issued D2047-74 "Test Method for Static Coefficient of Friction of Polish-Coated Surfaces as Measured by the James Machine." This was the first voluntary consensus standard ever to specify a pass/fail criteria for any floor surface, noting that the standard is specific in its application to polish-coated floor surfaces. The pass/fail criterion of 0.5 in this standard has led to the ubiquitous use of 0.5 as the acceptable value of slip resistance with regard to pedestrian safety.

Another articulated tester that has found its place as the standard test tribometer for evaluating slip-resistant bathing facilities is the Portable Articulated Strut Slip Tester (PAST). This device was originally known as the Brungraber Mark I and its use has been standardized in two ASTM standards: F462-79 (reapproved 1994) "Standard Consumer Safety Specification for Slip-Resistant Bathing Facilities;" and F1678-96 "Standard Test Method for Using a Portable Articulated Strut Slip Tester (PAST)." Although this tester is acceptable for testing floors of bathing facilities, it is not accepted for use under wet conditions on other pedestrian walking surfaces due to its susceptibility to sticktion.

Dr. Robert Brungraber produced a second articulated tester, which he called the Brungraber Mark II, also known as the Portable Inclinable Articulated Strut Slip Tester (PIAST). This device was designed to specifically avoid sticktion by applying the vertical and tangential forces to the test foot simultaneously. The standard for its use is ASTM F1677-96 "Standard Test Method for Using a Portable Inclinable Articulated Strut Slip Tester (PIAST)."

The most recent addition to this class of tribometers is a device commercially known as the English XL, otherwise known in ASTM parlance as a Variable Incidence Tribometer. ASTM Standard F1679-96 "Standard Test Method for Using a Variable Incidence Tribometer (VIT)" covers the operational procedures of this device. This device was also designed to avoid sticktion, and its design comes closer to emulating the dynamics of the human foot by incorporating a test foot that articulates to a flat orientation, much as the human heel does during walking. This device is also acceptable for use under wet/contaminated conditions and has shown exceptional reproducibility and repeatability in multi-user, multi-facility studies.

A Universal Approach to Testing Slip Resistance

The use of several different types of tribometers is problematic in that, when metering a walkway surface, the measurements generated have been found to vary with the tribometer used. This is not unexpected as the measurements are a function of the material being tested, the test foot material and the specific tribometer. In 2011, ASTM International published F-2508 "Standard Practice for Validation and Calibration of Walkway Tribometers Using Reference Surfaces" (ASTM 2011). This standard is accompanied by an adjunct of four tiles or reference surfaces with surface slip resistance properties defined in human ambulation studies. Through the use of the F-2508 standard and the reference tiles, it can be determined whether or not a specific tribometer can produce valid measurements of the slip resistance of a walkway surface. At the same time, a method is provided by which the values of slip resistance generated from tribometric testing of a walkway surface can be related to human ambulation.

The four types of tiles selected as the reference surfaces for testing were polished black granite, porcelain, ceramic, and vinyl composition tile (VCT). In order for the performance of a tribometer to be considered valid, the tribometer must satisfy two criteria: (1) It must be able to statistically differentiate between the surfaces of the four reference tiles; and (2) The tribometer must be able to rank the slip resistance of the tiles in the same order as that generated by the human subject testing. If the tribometer can differentiate between the tiles and rank the surfaces in the correct order, the testing of the reference tiles can provide tribometer-specific slip resistance values for the four different reference surfaces. The values for slip resistance generated during testing of the reference surfaces can be used as benchmarks for the testing of an unknown surface as the validity of the values can be traced back to the human subject testing.

Other Walking-Surface Considerations

A metal plate is often used as a walking surface in certain areas of industrial plants, such as over pits or trenches and around large machines where access is needed below floor level. Overhead balconies, catwalks, raised platforms, and stairs are often made of metal because of strength and rigidity requirements. Flat metal plate is very slippery when wet, especially when oil is spilled on it. Several methods have been used to reduce slipperiness. Raised patterns on the plate help when it is new, but when the raised sections get worn, they can also become very slippery. Some metal flooring has an abrasive material either impregnated into the metal, as illustrated in Figure 7-5, or applied as a coating. The abrasive surface is generally much more dependably resistant to slippage than a raised pattern. Disadvantages to the use of abrasive surfaces are that heavy metal wheels tend to crush and dislodge the abrasive over a long period of time and that they may be more difficult to clean than other types of floors.

Different kinds of floor mats are used to lessen fatigue, to act as sound absorbers, to diminish electric shock, and to reduce slipperiness. Certain types of mats may serve any combination of these functions. Smooth-surfaced mats often become slippery if oil is spilled on them. Mats that are either very thick,

Figure 7-5. Steel plate impregnated with an abrasive material to be used for floor panels. Courtesy of Safe Walk, Inc.

or lightweight and flexible, or not secured to the floor may become tripping hazards.

Slipping and tripping result not only from physical hazards, but also from lack of attention on the part of the walker. It is the latter factor that makes an uneven walking surface acceptable in a sawmill and unacceptable in an office. Slipping and tripping are very common in offices in spite of the comparatively good condition of walking surfaces. The National Safety Council has published a humorous pocket brochure entitled "A Visit to Office Falls," which includes "10 unforgettable events and points of interest along the way." At the end of the brochure a condensed travel itinerary lists the following reminders:

1. Keep file and desk drawers closed
2. Pick up objects on the floor
3. Use aisles—not between-desk shortcuts
4. Wear shoes with moderate heels
5. Carry loads you can see over
6. Wipe up wet spots
7. Walk, don't run, down stairs and use handrails

8. Keep head up and eyes front while walking
9. Use stepladder or step stool for high reaching
10. Avoid tipping chairs

In many industrial plants there are openings in the floor, either for lowering materials to a lower level, for access to utilities or equipment below floor level, or for temporary construction purposes. Since people do tend to be inattentive while walking, someone else must watch out for their safety by placing barriers around such openings. ANSI A1264.1-2007 "Safety Requirements for Workplace Walking/Working Surfaces and Their Access; Workplace Floor, Wall and Roof Openings; Stairs and Guardrail Systems" defines a "floor hole" as an opening measuring less than 12 inches (300 mm) but more than 2 inches (51 mm) in its least dimension in any floor. A "floor opening" is defined as an opening 12 inches (300 mm) or more in its least dimension. 29 CFR Part 1910 Subpart D OSHA (beginning at 1910.21) is another valuable resource regarding requirements on walking surfaces.

At least three types of barriers are commonly used. If the opening is permanent and is located where people might reasonably be expected to walk, a standard railing is required. A standard railing, as defined in the OSHA standards, is made of wood or metal and is capable of withstanding a load of at least 850 newtons (N), or 200 pounds force, applied in any direction at any point on the top rail. It must have a smooth top rail approximately 107 centimeters (42 inches) above the floor and an intermediate rail approximately halfway between the top rail and the floor. This standard railing is required along the side of any raised platform that is 4 feet in height or more, since, in effect, the area around and below such a platform is an opening into which a person could fall.

The design of a fixed railing should take many factors into account, such as height, stiffness, allowable deflection, and the type of loading to which it will be subjected. A railing should not be too rigid, since this places undue stress on its structural members. At the same time, too much deflection is not desirable. The design load of 850 N is an arbitrary but very reasonable load.

If an 850 N load is applied at midpoint on a railing with a span of 2 meters and a maximum deflection of 20 millimeters (0.02 meters) is permissible, assuming a simple beam, the following calculations can be made. The maximum bending moment will be $M = Wl/4 = 425$ N · m. The deflection (d) will be determined by the properties of the rail, assuming a perfectly rigid post, and is expressed as:

$$d = \frac{ML^2}{8EI}$$

where

W = Load
M = bending moment
L = length of span
E = modulus of elasticity of the material
I = moment of inertia of the rail
l = span/2 (midpoint).

From this it is easy to see that the material and cross-sectional shape must be selected to satisfy E and I. That is,

$$EI = ML^2/8d = (425 \times 4)/(8 \times 0.02) = 10{,}625 \text{ N/m}^2.$$

If a pine board is to be used, the value of E is $13.8 \times 10^9 \text{N/m}^2$ (2.0×10^6 psi). Then I must be

$$I = 10{,}625/(13.8 \times 10^9) = 0.770 \times 10^{-6} \text{ m}^4 \text{ or } 77.0 \text{ cm}^4.$$

I is the moment of inertia about the central axis and is expressed as:

$$I = \frac{bh^3}{12}$$

where

> b = width
> h = height

Therefore, $bh^3 = 12I = 12 \times 0.770 \times 10^{-6} = 9.24 \text{ m}^4$.

If a board .04 m (4 cm) thick is used (dimension b), the height h must be:

$$h = \sqrt[3]{12I/0.04} = \sqrt[3]{12 \times 0.77} \times 10^{-6}/0.04$$
$$= 0.0615 \text{ m or } 61.5 \text{ mm or } 6.15 \text{ cm}.$$

A pine two-by-four measures about 4 cm × 9 cm, so such a two-by-four would be quite adequate.

The posts of a railing can be designed in a similar way. The method of fastening the posts to the floor or platform is often the most critical part of the design. Generally, support posts should be installed every 8 feet. One common method involves using a pipe as a post with a pipe flange attached to the floor with screws. A quick look at the location of these screws reveals that when force is applied to the railing, the leverage afforded places tremendous stress on these screws, as well as on the material in the floor holding them, as illustrated in Figure 7-6.

Openings in the floor and edges of platforms must be further guarded if there are people or damageable property below. Additional barriers may be necessary for taller material in storage. A solid or mesh material could be installed up to the height of the midrail or top rail. (See ANSI A1264.1 and 29 CFR 1910 Subpart D.) Toeboards are used to prevent tools and other objects from getting pushed over the edge. Toeboards should be high enough to prevent materials from getting pushed over and close enough to the floor to prevent thin objects from sliding under. A standard toeboard as defined by OSHA is 10 centimeters (4 inches) high and must leave a clearance between toeboard and floor of no more than 6 millimeters (¼ inch).

Toeboards may also be necessary to prevent a person's foot from slipping off the edge of a platform even if it has standard railing. This is not likely to happen if a person is walking or standing reasonably still, but if the work on the platform involves quite a bit of body movement, the person may very easily

Figure 7-6. Moment arms on a pipe railing.

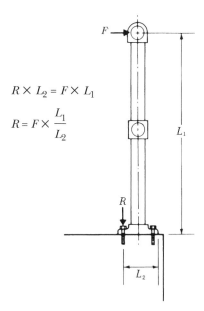

$$R \times L_2 = F \times L_1$$

$$R = F \times \frac{L_1}{L_2}$$

place his or her foot so far out over the edge of the platform that it will slip off the edge. A toeboard will prevent this and should be used on raised platforms where a person is working, especially around and over machinery.

Inclined Surfaces

Although the word "ramp" may not be entirely appropriate for all inclined surfaces, it will be used here as meaning any inclined surface upon which people walk or stand.

The first question concerning the design of a ramp is how steep it should be. Usually when an inclined walking surface is required, horizontal space is also limited, so there is some compulsion to make the ramp as short and, therefore, as steep as possible. Different standards are not in complete agreement about this maximum angle of incline of a ramp. The OSHA standards pertaining to ramps as part of a means of egress permit an incline as steep as one unit of vertical rise for every six units of horizontal distance, a ratio of 1:6. The ANSI A117.1-1992 "Accessible and Usable Building Facilities" calls for a maximum incline of 1:12, only half as steep as that allowed in the OSHA standards.

Military standard MIL-STD-1472D recommends a maximum and a preferred range of angles of incline for ramps, stairs, and ladders, as shown in Figure 7-7. Even the 7° angle of incline shown here exceeds that recommended by ANSI for use by disabled persons.

The ratio 1:12 is commonly being used in industry today because such ramps are often required by the Americans with Disabilities Act (ADA). Industry has found, too, that the ratio 1:12 is much more feasible for the use of hand and powered trucks.

The surface of a ramp must obviously be as free of tripping and slipping hazards as can feasibly be attained. Usually a rubber or abrasive material is used on permanent ramps. Railings are recommended at least on one side and preferably on both sides of the ramp. However, such railings, or handrails as

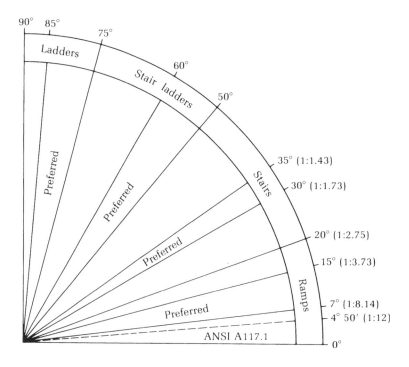

Figure 7-7. Military Standards (MIL-STD-1472D) for angle of incline of ramps, stairs, and ladders, with ANSI A117.1 Standard superimposed.

they are called in the ANSI standard, should be 80 to 82 centimeters (32 inches) high rather than the 107-centimeter height prescribed for standard railings. The surface of the ramp should be wide enough to accommodate whatever traffic may travel on it. The minimum width for use by a person in a wheelchair should be 82 centimeters, and preferably 90 centimeters, although that is not specified in the ANSI standard. When a ramp is used as a path of egress, it must also satisfy the standards that apply to that use, which will be discussed in Chapter 8.

Negotiating a long ramp can be very tiring for a person in a wheelchair or on crutches, or for someone pulling or pushing a hand truck. For that reason it is recommended that a continuous ramp be no longer than about 9 meters (30 feet). If this is not feasible, a level landing at least long enough to allow the vehicle to rest should be provided every 9 meters.

Landings or level flooring must be provided at both top and bottom ends of the ramp and at any door or other obstacle. If a door swings out toward a ramp at either the top or bottom end, a level place is needed long enough to back a vehicle away as the door swings out. The ANSI standard recommends a landing 1.8 meters (6 feet) long by 1.5 meters (5 feet) wide for this purpose.

Design of Stairs

A continuous series of four or more steps constitute stairs as the term is generally used. There are several standards pertaining to the design of stairs, found in section 1910.24 of the OSHA standards. The preferred angle of incline for

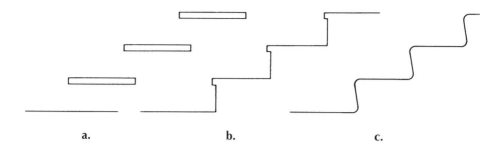

Figure 7-8. Three designs of stair treads and risers.

stairs is 30° to 50°, according to OSHA, and, at least as importantly, the angle of incline must be uniform throughout the length of the staircase. Stairs that are either too slight in incline or too steep are difficult to traverse without continual concentration.

There are some advantages and some disadvantages to open risers on stairs. An open riser permits a larger tread area because one tread overlaps the next, as shown in Figure 7-8a. This design is usually less expensive and permits easier cleaning. But since the person climbing the stairs can see through the opening, it may create a distraction. Too large an overlapping may also invite tripping. Figure 7-8b illustrates the most common type of riser and tread. One limitation of this design is the tripping hazard, although the probability of tripping is low. The design shown in Figure 7-8c is recommended in the ANSI standard because it allows a relatively smooth transition from one step to the next, something especially needed by people with disabilities.

The design of the tread edge, or nose, is very important to the overall safety of people using the stairs. One of the most common causes of falls on stairs is the inability of the person to see the edge of the tread. This problem is more serious when the person is descending, but it can also occur when he or she is ascending. The edge should be accentuated, either by proper illumination or by color on the edge itself. One thing to be avoided is having the floor and tread covering match so well that it is difficult to locate the edges of the steps, or even to perceive that there are edges at all, as illustrated in Figure 7-9. Diagonal patterns on the wall and the presence of handrails help draw attention to stairs.

Walking up or down stairs requires continual concentration, and anything that distracts a person's attention, such as an abrupt change in scenery or environment at the top of the stairs or part way down, is likely to cause an accident.

The surface on the treads of steps should have a high coefficient of friction at all times. Any work process that might involve spilled liquids, or solid objects for that matter, should be located away from steps to reduce the probability that the step's surface will become slippery or unstable.

Stairs of four or more risers must have a handrail, but short flights of one to three risers should also, as they provide a visual cue to the change in elevation as well as the obvious function that they serve. A handrail on both sides is preferred, but if only one is provided, it should be on the right side descending. This is because a person is more likely to be hurt in a fall when

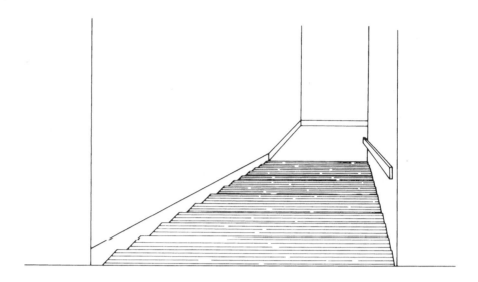

Figure 7-9. Stairs with poorly defined edges of treads.

headed down the stairs than when headed up. The term "handrail" is used rather than "railing" because it serves a different purpose, and the height is different as well. A railing has to prevent a person from falling over it, so it should be at least as high as the person's center of gravity. A handrail, on the other hand, is used to steady a person's balance while he or she is walking along an uneven surface and should be at a height to be grasped conveniently with the hand. This obviously varies, but a height of 81 centimeters (32 inches) is considered a good median.

A very important characteristic of a handrail is the cross-sectional shape. Ideally it should be nearly round and about 5 centimeters (2 inches) in diameter. This permits the maximum contact area between the hand and the handrail. Too many handrails in public stairways are made to be aesthetically pleasing but are very inefficient functionally.

The military standards refer to another form of stairs called stair ladders, which are constructed like stairs but are so steep that the use of the hands is required to ascend or descend. They have flat treads and open risers. Although other standards do not cover them, stair ladders are frequently used in storage areas, in basements, and to provide access to overhead platforms and walkways. They should not be used where there is frequent travel. OSHA standards refer to inclined fixed ladders as being "for special hazard only." Stair ladders have handrails on both sides and should have an angle of incline not less than 50° or more than 60°.

Spiral stairways present some unique problems and should be avoided if possible. A spiral or winding stairway must have an inside radius of at least 1.5 meters (5 feet) in order to assure that the angle of incline will never exceed the acceptable angle of 50°. Spiral stairways must have a handrail offset to prevent walking on those portions of the stairway where the tread width is less than 6 inches. Even then, any spiral will produce a nonuniform stair tread width, which requires even greater concentration than usual.

People walking up or down stairs usually look downward or straight ahead. They seldom notice overhead obstructions. Therefore, an overhead clearance measured vertically from the edge of each tread must be at least 2.1 meters (7 feet). If stairs or other walkways are used infrequently or by only a few people, lower clearances are acceptable if sufficient warning is given, such as strips of fabric hanging a few meters ahead of an overhead obstruction.

Fixed and Portable Ladders

Ladders are restricted to angles of incline of between 60° and 90°, with angles between 75° and 90° preferred. A person ascending or descending any ladder is expected to use his or her hands to help maintain balance, not only because the angle is so steep, but also because it is not feasible to have treads broad enough for the feet alone to maintain balance. This is basic to the design considerations for ladders of any type.

People should never carry objects in their hands while on ladders. Both hands should be used on the ladder, and necessary materials should be raised or lowered by some mechanical means, such as a rope. A major cause of falls from ladders is misuse of the ladder, as distinguished from poor design. When a person moves from one step to the next, he or she should be especially careful to keep the center of gravity as well centered as possible between the hand(s) and the foot that is in contact with the ladder at any instant. The sides and rungs of a ladder should not be slippery, but even when they are in good condition, they do not provide much resistance to the body's moving sideways. Even when an individual-rung ladder is designed with rungs cast in a concrete structure, the ends of the steps or rungs should have a barrier to prevent the foot from sliding off the end.

OSHA standards require that a fixed ladder be strong enough to withstand a load of at least 200 pounds (90 kg) concentrated at any point on the ladder. Other sources suggest this as a base figure and use a safety factor of 5. Table 7-1 lists some of the dimensional requirements for fixed ladders as described in 29 CFR part 1910.27. Other requirements are also described in that section. Similar requirements for ladders, as well as for stairs, are described in military standards, MIL-STD-1472D.

Certain factors associated with the use of a ladder are as important as the ladder itself. Any ladder, fixed or portable, should extend above the landing or platform surface it serves. This extending portion of the ladder provides greater stability when a person is getting on or off the top of the ladder. Fixed ladders must extend 42 inches above the upper landing. A grab bar, firmly secured to the structure, may be helpful. Portable ladders should be set up on firm, level sufaces and secured in place. When long portable ladders are used, they should be secured at the top and bottom. This reduces the possibility that the ladder will tip or slip.

While the recommended angle of incline for a ladder is between 75° and 90°, a portable ladder would obviously be very unstable at or close to 90°. An angle smaller than 70° to 75° not only makes the ladder awkward to climb but also increases the chance that the bottom of the ladder will slip. As the

Table 7-1. Dimensional Requirements for Fixed Ladders from 29 CFR 1910.27

Fixed Ladders Dimensions in inches		
Ladder Feature	Minimum	Maximum
Rung diameter, wood	1 1/8	
metal	0.75	
Rung length	16	
Distance between rungs		12
Clearance on climbing side of ladder,		
for pitch of 76 degrees	36	
for pitch of 90 degrees	30	
Clearance behind ladder	7	
Clear width either side of center	15	
A cage or well is required around ladders over 20 feet		
Cage extension above ladder	42	
Height of opening at base of cage	7 feet	8 feet
Width of cage	27	

angle increases, the center of gravity of the ladder and the person climbing it moves farther out from the vertical, thus increasing the danger that the system will fall backwards. A 75° angle is ideal, or a ratio of about 4 units of height to 1 unit of horizontal distance.

Fixed ladders are usually installed in a vertical position. If a person loses control on a fixed ladder, there is greater likelihood that he or she will fall *off* the ladder rather than on it. For this reason, a cage or a ladder safety device is required on vertical ladders more than 20 feet in height. A cage must permit easy movement on the ladder but restrict how far away from the ladder the body can move.

A ladder safety device is often referred to as a fall arrester. Fall arresters, whether for ladders or platforms, require a harness and are covered in Chapter 5 under personal protective equipment.

Specifications for the construction and use of ladders are described in sections 1910.25, .26, and .27 of the OSHA Standard.

Temporary and Movable Elevated Platforms

The raised platforms discussed earlier in this chapter were considered to be reasonably permanent and were therefore treated the same as floors. There are also several common types of movable surfaces on which people stand to perform work, including the following:

1. Scaffolding supported from the ground and commonly used in construction, and maintenance work. Frequently constructed with sections of metal pipe or tubing, these platforms can be easily assembled, disassembled, and moved.

2. Hanging platforms or suspended scaffolds supported from a roof or wall. They consist of wood or metal planks with a railing and are raised and lowered by pulleys.
3. Platforms attached to a hydraulic lift or scissors mechanism. They are mounted on a powered truck or portable platform.

These platforms are illustrated in Figures 7-10, 7-11, and 7-12.

There are many other types of temporary platforms are used in the workplace. All platforms should be assessed for hazards prior to use. Platforms should be in stable condition and secured in place to prevent unintentional movement or overturning. They must also have a railing and/or fall-arrester device. The supporting structure—the ground, wall, or roof—must be stable and strong enough to support the load.

29 CFR 1910.28 and 1926.451 provide additional requirements for scaffolding. 1926.451 covers the design and use of all types of platforms. Material and personnel hoists and elevators are covered in 29 CFR 1926.552.

Figure 7-10. Supported Scaffolding

Figure 7-11. Suspended Scaffolding

Maintenance of Walking and Standing Surfaces

The design, construction, and installation of walking and standing surfaces are certainly important, but improper maintenance often creates unexpected hazards. Preventative maintenance is critical for keeping surfaces in a safe condition; however, these issues must also be addressed during the design phase. Design has a great influence on subsequent use. If facilities are designed for the convenience and benefit of the worker, then the worker will be more likely to use them as intended. If they are not convenient, they may be used in an unsafe manner.

There are several factors that should be controlled either by design or by management, or both. Here is a list of things to be considered.

1. Keep surfaces free of oil and other liquids.
2. Keep surfaces free of loose objects, especially small cylindrical objects.
3. Keep surfaces free of larg obstructions, which may become tripping hazards.
4. Keep surfaces properly illuminated.
5. Keep handrails clean, free of slivers and sharp edges, and always readily accessible.

Figure 7-12. Scissor Lift

Despite the best efforts made to keep oil and other liquids off walking and standing surfaces, frequently some is tracked or spilled. As soon as liquids begin accumulating on a surface, it should be removed. There are a number of absorbant materials commercially available.

Small cylindrical objects, such as screws, bolts, welding rods, or pieces of wire, many create unstable walking or standing surfaces. They are commonly found in industrial plants, especially in assembly and maintenance areas. Ideally, these things should never get on the floor. If the items are scrap pieces, they should be put into barrels or bins conveniently located for that purpose. Other items that are still usable occasionally get spilled. Good training helps reduce spillage, and good supervision ensures that spilled parts are collected promptly.

Large objects that fall on or are placed on a walking or standing surface can create tripping hazards. These include such things as pieces of lumber, boxes, pallets, or tools. Sometimes these items fall or get pushed on walking or standing surfaces, but frequently they are left there inadvertently. There are two design features that can help reduce this hazard. First, good storage and handling facilities can be designed and provided to reduce the possibility of objects falling or getting pushed onto the surface. Second, borders of the area designated for walking or standing should be clearly marked by brightly colored tape or paint. This type of marking helps remind workers that the area is to be kept free of objects.

Proper illumination is another important, but often overlooked aspect of walking and working surface safety. Recommended levels of illumination for a wide range of conditions have been developed by the Illuminating Engineers Society (IES).

If a handrail is provided to aid in maintaining balance and stability, it must have a surface that can be grasped securely. It must not be slippery or in any other way prevent a secure grasp. This involves not only the shape, as discussed earlier, but a clean, nonslip surface.

The width of aisles, corridors, and stairs used as part of the means of egress is specified in detail in the Life Safety Code and discussed in Chapter 8, "Egress and Life Safety."

The space needed for standing and moving about at a given work station depends on the activities involved at that work station. Refer to Chapter 4, "Work Systems and Ergonomics," for additional design information.

QUESTIONS AND PROBLEMS

1. What is the difference between a slip/fall versus a trip/fall versus a stumble?

2. Describe the differences one might expect of the dynamics of a person that trips versus a person that slips?

3. Explain why we can walk over stones, tree roots, and other obstacles on a mountain trail and then trip on a loose board on the steps at home.

4. What minimum change in elevation is recognized in the standards as being capable of causing a person to trip? What standards recognize this threshold?

5. Calculate the COF or slip index of a surface where a 20 pound object begins to slide with the application of 13 pounds of force. What would be the corresponding angle of an inclined surface using the answer obtained from the first portion of this question?

6. What types of tribometers are unsuitable for measuring slip resistance on wet walking surfaces? Why are they unsuitable?

7. What are two methods to safeguard a climber on a fixed ladder that is more than 20 feet in length?

8. What is the recommended angle of incline range for a ladder? Why should a portable ladder not be used at an angle of incline at or near 90 degrees?

9. List several types of temporary or movable platforms.

10. List several factors to consider when maintaining walking and working surfaces.

NOTES

1. Neolite is a registered trademark of the Goodyear Tire and Rubber Co., Shoe Products Division, Windsor, VT 05089. Neolite can be obtained from Smithers Scientific Services, Inc., 424 W. Market St., Akron, OH 44303. Specify "Neolite (Break-in Compound)," RMA Spec. HS-3; Size 36 by 44 inches; 6 irons; Color, Natural 11; Specific Gravity 1.27 + or − 0.02 m; Hardness Shore A 93-96.

BIBLIOGRAPHY

Americans with Disabilities Act of 1990 (ADA)

American National Standards Institute. ANSI A117.1-09 – *Specifications for Making Buildings and Facilities Accessible to, and Usable by the Physically Handicapped,* NewYork:American National Standards Institute.

_____. ANSI / ASSE TR-A1264.3-2007 *Technical Report: Using Variable Angle Tribometers (VAT) for Measurement of the Slip Resistance of Walkway Surfaces,* Des Plaines, IL: American Society of Safety Engineers; New York: American National Standards Institute.

American Society of Safety Engineers and American National Standards Institute. ASSE/ANSI A1264.1-2007 – *Safety Requirements for Workplace Walking/Working.*

Surfaces and Their Access; Workplace, Floor, Wall and Roof Openings; Stairs and Guardrails Systems, Des Plaines, IL: American Society of Safety Engineers; New York: American National Standards Institute.

_____. ASSE / ANSI A1264.2-2006 – *Provision of Slip Resistance on Walking/Working Surfaces,* Des Plaines, IL: American Society of Safety Engineers; New York: American National Standards Institute.

American Society for Testing and Materials Intl. ASTM F1637-10 *Standard Practice for Safe Walking Surfaces,* West Conshohocken, PA: American Society for Testing and Materials, Intl.

_____. ASTM F1694-09 *Standard Guide for Composing Walkway Surface Investigation, Evaluation and Incident Report Forms for Slips, Stumbles, Trips, and Falls.*West Conshohocken, PA: American Society for Testing and Materials, Intl.

_____. ASTM F802-83(2003) *Standard Guide for Selection of Certain Walkway Surfaces When Considering Footwear Traction.* West Conshohocken, PA: American Society for Testing and Materials, Intl. This standard has been withdrawn, but is still available through ASTM, Intl.

_____. ASTM F2508 - 11 Standard Practice for Validation and Calibration of Walkway Tribometers Using Reference Surfaces. West Conshohoken, PA: American Society for Testing and Materials, Intl.

Bakken, G., H. Cohen, J. Abele, A. Hyde, C. LaRue, 2007. *Slips, Trips, Missteps and Their Consequences (2nd ed)*. Tucson, AZ: Lawyers & Judges Publishing Company, Inc.

Chang, Wen-Ruey, Theodore K. Courtney, Raoul Grongvist, Mark Redfern, eds., 2003. *Measuring Slipperiness: Human Locomotion and Surface Factors*. Boca Raton, FL: CRC Press.

DiPilla, S., 2010. *Slip, Trip, and Fall Prevention: A Practical Handbook* (2nd ed). Boca Raton, FL: CRC Press.

English, W., 1989. *Slips, Trips, and Falls: Safety Engineering Guidelines for the Prevention of Slip, Trip, and Fall Occurrences*. Oakton, VA: Hanrow Press.

English, W., 1996. *Pedestrian Slip Resistance: How to Measure It and How to Improve It*. Alva, FL: William English, Inc.

English, W., 2003. *Pedestrian Slip Resistance: How to Measure It and How to Improve It* (2nd ed). Alva, FL: William English, Inc.

International Code Council and American National Standards Institute. A117.1-09, *Accessible and Usable Buildings and Facilities*. Washington, D.C.: Intl. Code Council, New York: American National Standards Institute

National Safety Council. 2010. *Injury Facts*. Itasca, IL: NSC.

Public Law 101-336, 7/26/90. Federal Register Vol. 56, No. 144, Chapter A4.5.1, Friday, July 26, 1991 Rules & Regulations. (http://www.ada.gov).

Templer, J., 1992. *The Staircase: Studies of Hazards, Falls, and Safer Design*. Cambridge, MA: MIT Press

RECOMMENDED READING

Government References

ADA *Standards for Accessible Design*. http://www.ada.gov/stdspdf.htm.

Air Force A-A-59124 Canc Notice 1 Deck Covering, Lightweight, Nonslip

Air Force A-A-59166 Coating Compound, Nonslip (For Walkways)

Americans with Disabilities Act of 1990 (ADA)

MIL-D-17951D, Military Specification: Deck Covering, Lightweight, Nonslip, Abrasive Particle Coated Fabric, Film, or Composite and Sealing Compound (5 Jun 1975) [S/S by Mil-Prf-17951E]

MIL-D-18873B Deck Covering Magnesia Aggregate Mixture

MIL-D-3134 1, MIL-D-3134J (Navy) (Amendment 1), Military Specification, Deck Covering Materials (12 Sep 1989)

MIL-PRF-24667C, Performance Specification, Coating System, Non-Skid, for Roll, Spray, or Self-Adhering Applications (22 May 2008) [Superseding DOD-C-24667 and MIL-D-23003A and MIL-D-24483A]

MIL-W-5044C, Military Specification, Walkway Compound, Nonslip and Walkway Matting, Nonslip (25 Aug 1970) [S/S by A-A-59166 and A-A-59124]

NAVY MIL-D-23003A CANC Notice 2 Deck Covering Compound, Nonslip, Rollable

Public Law 101-336, 7/26/90. Federal Register Vol. 56, No. 144, Chapter A4.5.1, Friday, July 26, 1991 Rules & Regulations (http://www.ada.gov)

RR-G-1620D Fed. Specification—Grating, Metal, Other Than Bar Type (Floor, Except for Naval Vessels)

United States Access Board Technical Bulletin: Ground and Floor Surfaces http://www.access-board.gov/adaag/about/bulletins/surfaces.htm

Walking-Working Surfaces and Personal Protective Equipment (Fall Protection Systems); Proposed Rule. 05/24/2010 - 75:28861-29153, http://www.osha.gov/pls/oshaweb/owadisp.show_document?p_table=FEDERAL_REGISTER&p_id=21518

ASTM, Intl. Standards Committee

F-13, Pedestrian/Walkway Safety and Footwear

F1646-12 – Standard Terminology Relating to Safety and Traction for Footwear

F802-83(2003) – Standard Guide for Selection of Certain Walkway Surfaces When Considering Footwear Traction. This standard has been withdrawn, but is still available through ASTM, Intl.

F695-01(2009) – Standard Practice for Ranking of Test Data Obtained for Measurement of Slip Resistance of Footwear: Sole, Heel, and Related Materials

F1240-01(2009) – Standard Guide for Ranking Footwear: Bottom Materials on Contaminated Walkway Surfaces According to Slip-Resistance Test Results

F2048-00(2009) – Standard Practice for Reporting Slip-Resistance Test Results

F1694-09 – Standard Guide for Composing Walkway Surface Investigation, Evaluation and Incident Report Forms for Slips, Stumbles, Trips, and Falls

F2508-11 – Standard Practice for Validation and Calibration of Walkway Tribometers Using Reference Surfaces

Proposed new standards (2012) under the jurisdiction of ASTM Committee F13 – Pedestrian/Walkway Safety and Footwear

WK33603 New Guide for Walkway Auditor Qualifications (Technical) Ballot F130112 Item 017.

WK33621 New Guide for Selection of Walking Surface Treatments When Considering Aggressive Contaminant Conditions Ballot F13 (12-02) Item 051.

WK34033 New Practice for for Snow and Ice Control for Walkway Surfaces Ballot F13 (12-02) Item 052.

ASTM Committee D21 on Polishes

D2047-04 - Standard Test Method for Static Coefficient of Friction of Polish-Coated Flooring Surfaces as Measured by the James Machine

ASTM Committee C21 on Ceramic Whitewares and Related Products

C1028-07e1 - Standard Test Method for Determining the Static Coefficient of Friction of Ceramic Tile and Other Like Surfaces by the Horizontal Dynamometer Pull-Meter Method

ASTM Committee F15 on Consumer Products

F462-79(2007) – Standard Consumer Safety Specification for Slip-Resistant Bathing Facilities

National Fire Protection Association Codes and Standards

NFPA 101: Life Safety Code
NFPA 1901: Standard for Automotive Fire Apparatus

American National Standards Institute (ANSI) Standards

ASSE/ANSI A1264.1-2007 – Safety Requirements for Workplace Walking/
Working. Surfaces and Their Access; Workplace, Floor, Wall and Roof
Openings; Stairs and Guardrails Systems
ASSE/ANSI A1264.2-2006 – Provision of Slip Resistance on Walking/Working
Surfaces
ANSI/ASSE TR-A1264.3-2007 – Technical Report: Using Variable Angle
Tribometers (VAT) for Measurement of the Slip Resistance of Walkway
Surfaces
ANSI A117.1-80 – Specifications for Making Buildings and Facilities Accessible
to and Usable by the Physically Handicapped
ANSI A137.1 – 2008 – American National Standards Specifications for Ceramic Tile
ICC/ANSI A117.1-03 – Accessible and Usable Buildings and Facilities

Underwriters Laboratories (UL)

UL 410 - Slip Resistance of Floor Surface Materials

Model Building Codes - Stairs and Means of Egress

International Code Council (ICC) – International Building Code – Sections on
Means of Egress

Stairs and Egress

Pauls, J. and various others, http://web.me.com/bldguse/Site/Stairways.html,
as well as http://web.mac.com/bldguse/Site/Downloads.html, for many
excellent papers/presentations on stair safety and safe egress as well as many
excellent references.

Papers and Publications

Visit www.sciencedirect.com and search "slip fall" and the like, for peer reviewed
publications on slips, trips, and falls.

8

Egress and Life Safety

Kevin L. Biando, PE, MSFPE

Richard T. Beohm, PE, CSP, ARM

Definitions

The word *egress* has greater significance in the field of safety engineering than is implied by a dictionary definition. In this field, egress implies leaving a building or structure under conditions of emergency or duress. It is not the same as the exodus at the end of the workday, although some of the same elements are present. People are leaving for reasons of safety, not just for pleasure. Human behavior is different, and this behavior is a major factor in the design of the means of egress.

The term *life safety* as used in connection with egress refers to the movement of people from a hazard within a building in a manner that does not result in any physical harm to any of those occupants. Such a hazard is most often fire, but it could be any kind of hazard that would require evacuation of the building.

The term *means of egress* also takes on particular significance because of the conditions of departure. The means of egress is whatever route people take to get from wherever they are inside a building to a point outside or away from the building. For convenience, the means of egress is broken down into three separate and distinct parts, referred to as the *exit access*, the *exit*, and the *exit discharge*.

The exit access is that portion of the means of egress that leads to an exit. Normally, this part of the route is entirely within the building, but if a person had to travel through a courtyard into another building in order to get completely outside, the courtyard would be part of the exit access. Some codes may give the impression that only the more common routes of travel—the aisles and corridors—need be considered as the exit access. This is not the intention. More people travel through a given aisle or corridor, but a person usually has to get to the aisle or corridor from a point within a work area. These passageways are also part of the exit access.

The exit itself is the portion of the structure that separates the inside of a building from the outside. This may sound like a complicated way of saying that an exit is a door, but often it is not as simple as that. In some structures, there may not be a door. Many buildings today, especially multistory buildings, have enclosed stairs of a fire-rated construction and fire-rated doors. The whole stairwell, including walls, doors, and stairs, becomes the exit. Since fire is usually the hazard, the walls and doors must be fire rated to withstand heat, smoke, and toxic gases for a given period of time, such as 1 or 2 hours.

A person is not completely isolated from the hazard inside the building until he or she is out of the possible reach of that hazard. The third portion of the means of egress, the exit discharge, is a path or open area leading to a place considered entirely safe from the building and its hazardous contents, or into an area of safe refuge, as in the case of some high-rise buildings. (The safe refuge usually involves a "defend-in-place" philosophy as opposed to "exiting the building.") The exit discharge may be a sidewalk, an open lawn, or any area free of obstruction through which a person can go to get completely away from the building. The term *public way* is often used to describe this haven from danger. It should be remembered that an open space between two buildings is not a public way if a person has to go back into one of the buildings to reach a place of complete safety.

The ability to egress from a building also depends on the type of occupancy—that is, the type of activity that goes on in the building and the type of people who occupy it. The Life Safety Code®[1] identifies the following classifications of occupancy:

1. Assembly
2. Educational
3. Health Care
4. Detention and Correctional
5. Hotels and Dormitories
6. Apartment Buildings
7. Lodging or Rooming Houses
8. One- and Two-Family Dwellings
9. Residential Board and Care
10. Mercantile
11. Business
12. Industrial
13. Storage
14. Day Care
15. Special Structures and High-Rise Buildings
16. Mixed Occupancies.

In each classification, the type of activities and their inherent hazards, the grouping of people, and/or the age and capabilities of the occupants combine to create unique situations as far as evacuating the building is concerned. Some specifications included in the codes are applicable to all classifications, and others apply only to a particular classification.

The Life Safety Code sets requirements for the design, construction, operation, and maintenance of buildings and structures to provide for safety to life from fire. It also serves as an aid for other similar emergencies.

The protection of occupants is achieved by the combination of prevention, protection, egress, and other safety features. Chapters 1–10 to all structures and include "Means of Egress," "Fire Protection Features and Buildings Service,"

[1]*Note:* The Life Safety Code® and 101® are registered trademarks of the National Fire Protection Association, Quincy, MA.

and "Fire Protection Equipment." Chapters 11–42 involve the requirements and provisions of individual occupancies.

The design of a building usually requires a Life Safety Design Analysis which addresses such features as building-code requirements, exits, exit-egress paths showing travel distances, common paths of travel, dead ends, fire extinguishers, exit capacity, and the occupant load at each exit. This is usually a written narrative which accompanies a Life Safety Plan or Drawing and should be completed for all new and existing buildings.

In addition to the Life Safety Code, several other codes, generally called building codes, have been developed and are in common use. In addition to fire safety, some of them also cover electrical safety, plumbing specifications, mechanical requirements, structural integrity, and so on. OSHA has adopted certain portions of the Life Safety Code (LSC). The majority of states and municipalities have adopted the Life Safety Code, as well as other NFPA codes, and a particular building code. There are some differences among these codes, but they are similar in most respects. All of these codes are revised approximately every three years, when the current codes appear to be insufficient or when new information makes current codes obsolete or inadequate. There are now about 300 NFPA codes.

There are some 50 million Americans with disabilities. This number is expected to grow as more Americans reach 65 and experience some kind of disability. The federal government and individual states have enacted accessibility laws such as the Americans with Disabilities Act of 1990, the Architectural Barriers Act of 1968, and the Uniform Federal Accessibility Standards to guarantee individuals with disabilities the right to independence, freedom of choice, and the opportunity to participate fully and equally in American life.

ANSI A117.1 and ADAAG (ADA Accessibility Guidelines) provide the technical design requirements for buildings that exceed the Life Safety Code in various aspects of egress and exits. Accessibility requirements should be addressed in the Life Safety Design Analysis along with life-safety/fire-protection and building-code requirements.

The Life Safety Code is primarily the code used to specify design criteria for safe egress and exiting from buildings in a hazardous condition, usually fire. Building codes are mainly concerned with containing a fire and preventing its spread to adjoining buildings and structures.

The Life Safety Code and other NFPA-adopted codes are usually enforced by the State Fire Marshal's Office. The building code is usually enforced by local building officials. The Fire Prevention Code is enforced by local fire marshals and/or fire departments. NFPA codes are consensus codes, as they are developed by private groups and voted upon by the members of such groups. NFPA codes are also ANSI standards. The major building and life-safety codes currently available include the following:

Life Safety Code®, NFPA 101®, The National Electric Code, NFPA 70
    National Fire Protection Association
    One Batterymarch Park, Quincy, MA 02169

Also available as ANSI A9.1 (LSC) and ANSI C.1 (NEC) from the
    American National Standards Institute
    1430 Broadway
    New York, NY 10018

International Building Code (IBC)
International Code Council
500 New Jersey Avenue, NW
6th Floor
Washington, DC 20001

Differences among the building codes have led to a recent attempt to bring uniformity to these codes. Representatives from BOCA, SBC, and the UBC met and developed what is referred to as the International Building Code.

NFPA codes are called *prescriptive* codes in that they are followed step by step in the design process to achieve the desired fire-safety objective of the code.

Performance-based design, which uses computer modeling, is becoming increasingly popular and accepted. It provides an equivalent level of safety in reducing loss of life, injury, and property (direct and indirect) due to fire, explosion, and related perils by using alternative design methods as variances to adopted prescriptive codes and regulations. Fire-protection engineers often use such codes as NFPA 101, 101A (Alternative Approaches to Life Safety), and 92B (Smoke Management Systems), as well as computer programs, for their performance-based design.

Some of these computer tools, such as ASET and ASET-A and B by the National Bureau of Standards, are used to calculate safe egress time. Some other programs used to calculate fire behavior and egress are FPETOOL (NIST), FAST, CFAST, LAVENT, WPI/FIRE, EVACNET 4, EESCAPE, and many others. These programs use algorithms based on fire-dynamics formulas. Two programs using agent-based models are SIMULEX and EXODUS.

The Fire Dynamic Simulator (FDS) is a model by the National Institute of Standards and Technology (NIST) of fire-driven fluid flow. FDS and EVAC were developed at the VTT Technical Research Centre in Finland, and is an evacuation model that simulates human egress.

Human Behavior in Emergencies

The question is often asked, "Why is it necessary to label exits when it is obvious that the door is an exit?" One employer was heard to say, "If I had an employee who didn't know enough to know that's an exit, I'd fire him!" Unfortunately, things that may seem very obvious under normal conditions are not so obvious under conditions of emergency. Furthermore, what may appear to be an adequate exit may not be. Fire doors, which will be discussed later, are usually identified as such only by an identification plate mounted on the door.

There are several reasons for this lack of clarity, the most common one being the irrational thinking of people in a state of emergency. We become accustomed to a normal set of conditions, but do not train ourselves to respond to other conditions. If we are given enough time to react to a change, we can usually arrive at a logical way to respond. When we do not have enough time, or believe we do not, we take shortcuts in our thinking and often arrive at an irrational response.

Another reason we may not recognize something obvious in an emergency is that we may be suffering from the effects of smoke and hot gases.

Smoke lowers visibility, and heat limits where a person may go. A more serious restriction is created by the gases a fire produces and by the lack of oxygen. When smoke is inhaled, it reduces the ability to breathe by coating the surface of the lungs with carbon particles and by reducing the oxygen content of the inhaled air. If the air being breathed is too hot, it will damage the lining of the lungs.

As previously mentioned, the air normally contains about 21 percent oxygen. But fire uses oxygen also, and as the oxygen is depleted, harm to people increases. When the oxygen level is reduced to 16 percent to 18 percent, mental activity slows down, and at 14 percent, the brain is severely impaired, thinking is very irrational, and the person feels extremely tired. At 8 percent to 10 percent, all bodily functions become impaired, and death occurs after only a few minutes of exposure to this atmosphere.

Lack of oxygen is usually not as serious a problem, however, as the inhalation of toxic gases, such as sulfur dioxide, hydrogen sulfide, hydrogen cyanide gas, hydrogen chloride, and, especially, carbon monoxide (CO). CO is extremely toxic and is more readily absorbed into the blood than oxygen, thus very quickly replacing oxygen throughout the body. Concentrations of as little as 0.2 percent to 0.4 percent can cause death after prolonged exposure.

Burns can be serious, too, causing pain, disfigurement, and even death. But the number of deaths and injuries caused by burns is relatively small, probably less than 10 percent, compared with those caused by toxic gases and lack of oxygen.

It becomes clear that in planning a path of egress, one must realize that people using that path in times of emergency may need help in finding and following it. A person in panic often thinks any door is an exit. Doors opening into a closet or a boiler room may appear to provide a way out of the building, especially when they are located at the end of a corridor or at the intersection of corridors.

Several studies have been made of people in emergencies. Sometimes people pay no attention to others and run to what they believe is the way out, quite oblivious of anyone else. In other cases, they become part of the crowd. Much has been written about mob psychology, and this form of reasoning, or lack of reasoning, applies to a crowd trying to egress from a building in an emergency.

It is very difficult to know how people will react in an emergency, but several things can be done to reduce irrational behavior. Well-designed and appropriately placed signs help direct traffic. The following signs, which may seem unnecessary under normal conditions, can be very useful in emergencies:

1. "Exit" signs with legible arrows at:
   a. all intersections of aisles
   b. corners of aisles
   c. any other place where the way is not obvious
2. "Not an Exit" sign on all doors that could be mistaken for an exit door
3. "No Exit" sign at entrance of dead-end aisles.

The National Fire Protection Association's NFPA 170, "Standard for Fire Safety Symbols," provides symbols for fire prevention and other emergency situations. Included are symbols for an emergency exit, what is not an exit, a warning to use stairs in case of fire, and others.

The ISO 7010 Safety Sign and Symbol meaning "exit" is green/yellow showing a person exiting through a door. Egress is represented by the word

NFPA 170 symbol for emergency exit.

NFPA 170 symbols for emergency exit route.

Figure 8-1. ISO 7010 Safety Sign and Symbol meaning "exit" top, and "egress" pictorial bottom. Reprinted with permission from NFPA 101®-2009 *Life Safety Code®*, Copyright © 2008, National Fire Protection Association, Quincy, MA. This reprinted material is not the complete and official position of the NFPA on the referenced subject, which is represented only by the standard in its entirety.

"exit," a pictorial of a person running, and an arrow. NFPA 170 signs are shown in Figure 8-1.

Whenever any sign is used, it must be clearly visible and legible. There should be no point along any path of egress at which an exit, or a sign indicating the direction to an exit, is not visible and legible. These signs should be illuminated at all times with normal or emergency power and conform to the Life Safety Code.

Features of Building Design

The design of all the building features that affect the egress of people from a building should aid that egress and never impede it. Several such building features serve other purposes as well and in some cases call for design characteristics that would hinder egress. Usually designs can accommodate all uses. The Life Safety Code and the various building codes have been developed to guide such design. The codes establish specifications which, for the most part, are minimums or maximums. An engineer or architect should not design just to meet the codes, but should analyze the problem at hand and design to provide adequate protection or performance for that situation.

Practically all paths of egress involve one or more doors, which are often the most critical part of the system. Such doors should meet the expectations of a person who is in a hurry to get out and perhaps even in a state of panic. Some of the characteristics of any door in a path of egress include the following:

1. Latching mechanism
2. Door type and direction of swing
3. Locks
4. Door width

Fire barriers, smoke doors, and exit doors should be self-closing/-latching doors that can be opened using a single motion to open the door. The Accessibility Code A117.1 section 4.13 and ADAAG specify door width, thresholds, clear width, latches, and side-of-door spaces. It requires that the force necessary to open the door be no more than 5 pounds.

Panic hardware involves a self-closing device with no latch or with a latch operated by pushing a bar or plate. It does not require the use of fingers or even a hand. It opens when any part of the body or any object pushes against this bar or plate.

Doors in the means of egress generally should be of the side-hinged or pivoted-swinging type. There are some conflicting principles regarding the direction in which certain doors should open. With some exceptions, any door serving as an exit for 50 or more occupants should swing outward—that is, in the direction of egress. Doors opening onto an exit access require special discussion. Figure 8-2 illustrates several possibilities. If people are moving quickly through an aisle and a door opens out into the aisle, a serious accident could occur, particularly if the door opens like door No. 1. Door No. 2 is not quite as dangerous, because people walking along the aisle would walk into the face of the door at an angle rather than the edge of the door. Door No. 3 is more desirable. Doors No. 1 and 2 are permitted if the reduced width of the aisle is still within acceptable limits. Door No. 3 is more desirable if there are fewer than 50 occupants in that room.

Consideration must also be given to the people inside the rooms. If there are only a few people and no serious fire or explosion hazard in the room,

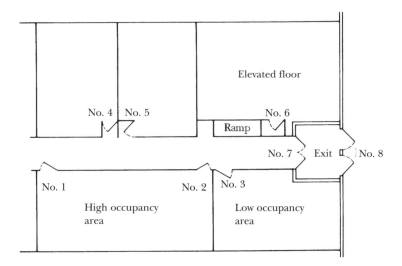

Figure 8-2. Several door-opening configurations.

the configuration of door No. 3 is acceptable. If there are many people or a serious hazard, the door must swing in the direction of exiting. A recessed door, as illustrated by doors Nos. 4, 5, and 6, is the best type to use. It is also a good idea to avoid emptying two rooms through doors located directly opposite each other on an aisle.

The codes also prohibit the use of any kind of lock on a door in a path of egress that prevents or hinders a person from opening it from the inside. Delayed-egress locks and access-controlled egress doors are allowed in some occupancy types. A lock that is incorporated into the latch and permits opening by the latch is permissible. This can be unsatisfactory as far as security is concerned, but when people's lives are at stake, safety must prevail over security.

The door opening should always be wide enough to accommodate the number of people exiting through it. Essentially it should be as wide as the corridor leading up to it. The width of doors and corridors will be discussed in more detail under the heading Egress Planning. It will just be noted at this point that no matter how well doors are designed and installed, they still hinder the free flow of people. As long as rooms with walls are needed, doors are needed, but movement of people would be much faster if there were no doors.

Walls are another important structural feature in designing for life safety. Walls hinder movement of people, but they also hinder movement of smoke, gases, and the heat of a fire. Thus, if designed properly, they can be a help as much as a hindrance. Almost any kind of wall or barrier will help to prevent the spread of smoke and gases. When a fire threatens the safety of people in a building, it is the smoke and gases, not the heat, that constitute the major hazard. Heat and flame injure or kill very few people in a fire. Over 90 percent of all fire deaths and serious injuries are caused by suffocation. For this reason, it is more important to protect people from the smoke and gases of combustion. Property, on the other hand, is destroyed or damaged more often by heat and flames.

Barriers are needed to prevent or hinder the movement of smoke and gases to other parts of the building. Accordingly, smoke doors and barriers are used to divide a large building into smaller smoke areas. They often include wired glass in non-hazardous locations to allow an accessible view to the other side.

Fire walls and fire doors are constructed of metal or metal-covered solid core or other materials that will provide a fire-rated protection. Construction of fire doors is governed by NFPA 80, "Standard for Fire Doors and Fire Windows." The design of any fire door or fire wall is tested to assure that it meets certain specified criteria. They must retain structural integrity and not allow passage of flame or toxic gases for a stated period of time. Standard hourly ratings for walls and partitions are ½, ¾, 1, 1½, 2, 3, and 4 hours. Standard fire-door ratings include ⅓, ½, ¾, 1, 1½, and 3 hours. Any opening through the door or wall for pipes or wires must be sealed or "fire-stopped." A fire door must have some sort of closing mechanism that will assure that the door is closed when a given ambient temperature is reached. An identification plate or label is attached to each fire door to help identify its fire-resistance rating.

Stairwells and other areas considered to be part of the exit must be enclosed by fire-resistive construction (i.e., fire walls and doors) with a rating of one or two hours, depending on the number of floors they serve. When a person enters the stairwell, he is safely separated from the rest of the building for a certain period of time. Many stairwells in high-rise buildings are pressurized so that smoke and toxic gases that might otherwise flow into the stairwell will be forced out, providing a safe atmosphere for egress.

Ramps may be used as a part of the path of egress as long as they meet the criteria discussed in Chapter 7 and in the Life Safety Code. Stairs can be used to a limited extent, but since they obviously impede the flow of people, they should be avoided as much as possible. When stairs are used in this way, they should be wide enough to accommodate the traffic, structurally sound, and made of a fire-resistant material. Open risers should generally be avoided.

The use of fire escapes as a means of egress is not recommended and is not permitted in new construction (with a few exceptions). Permanent fixed ladders may be used to provide access to unoccupied roof spaces or to provide escape from platforms, boiler rooms, towers, and other spaces. Elevators should not be used for emergency exiting. They are usually recalled to the street floor for fire department use.

Fire escapes have saved occupants of burning buildings and have aided fire fighters in getting to vantage points to fight fires. However, many fire escapes are of inferior construction, and many people are reluctant to use them. Other means of egress, such as enclosed or "smoke-proof" stairs, are far superior and are required in all new construction.

Egress Planning

In planning facilities for getting people out of the building quickly in a fire or other emergency, the following factors must be considered:

1. What are the capabilities of the people individually?
2. What are the capabilities of the people moving as a group?
3. Where are the people normally located with respect to potential hazards?
4. What is the longest distance anyone will have to travel to get to an exit?
5. What restrictions or potential hazards exist along each path of egress?
6. Does each person have an alternate route of egress?

The capabilities of individuals vary widely. Even in industry, there are likely to be some people in wheelchairs, on crutches, or restricted by some physical limitations. These restrictions are often not visible, such as heart disorders, arthritis, impaired hearing, and so on. Some people with restricted movement may need assistance in getting out of the building in emergencies. Some people depend on elevators to move from one floor to another, but elevators are not to be used during a fire. A *mobile occupant* is defined as one who can move by his or her own effort at a rate of 30 meters (100 feet) per minute.

All of this points to one conclusion: people must be evacuated from a burning building as quickly as possible. How much time can be allowed? This varies, obviously, depending on many factors, but guidelines have been set. An evacuation time of 3 minutes from a residence is considered to be the maximum acceptable time. For most industrial and office buildings, the maximum acceptable time is 10 minutes. In high-rise buildings, evacuation takes much longer. For this reason, high-rise buildings should have *areas of refuge,* entire floors or rooms within the building that are well protected from fire and easy to get to.

Many new and renovated facilities, including the US Department of Defense (DOD), are now requiring a Fire Emergency Voice/Alarm Communication System (ECS), or Integrated Fire Alarm Mass Notification System (FA/MNS), to alert and guide building occupants to safety should there be a terrorist or on-campus shooting incident or a warning of a natural disaster such as a tornado, hurricane, or flood. NFPA 72 2010 and the DOD UFC 4-021001 (2008) address specifications for Fire Alarm/Mass Notification Systems.

The *SFPE Handbook of Fire Protection Engineering* contains several chapters dedicated to egress of occupants and the effect of fire and smoke on human behavior. Significant research has been conducted on crowd movement. This is highlighted in the *SFPE Handbook* along with a time-based egress analysis with basic assumptions and calculations. Some of the related topics are discussed below.

The flow rate ($ft^3/ft^2 \times min$) of the occupants, whether down the corridor, through a doorway, or on the stair, depends upon the number of people and the speed of the occupants. Density (number of occupants per sq. ft.) is indirectly proportional to the speed of the occupants. As the number of people per square foot increases, the speed decreases. As an example, the speed for occupants in the corridor decreases from 250 ft./min. for densities less than 0.05 persons/sq. ft. to less than 60 ft./min. for density of 0.30 persons/sq. ft. Additional crowd movement parameters can be found in the *SFPE Handbook.*

The *SFPE Handbook* provides hand calculations for a timed egress analysis. Important factors include width (i.e., corridor, aisle, door, and stairs), density, and speed. At densities less than 0.05 persons/sq. ft., occupants move at speeds independent of others. Occupants are at a standstill when densities reach 0.35 persons/sq. ft. For densities in between these extremes, speed (S), can be calculated with the following equation:

$$S = k(1 - 2.86D)$$

where k is 275 for corridors, aisles, ramps, and doorways, and is variable depending on the riser height and tread depth for stairs.

Specific flow is speed times density ($S \times D$), with maximums given for each type of exit route (corridor, aisle, door, and stair). Calculated flow is specific flow times effective width. Effective width is the actual width available for the exiting occupants (i.e., for corridors it is equal to the corridor width less obstructions, less boundary layer width—given as 8 inches for corridors).

Transition points are any points where the speed of the occupants may decrease, such as where the exit route becomes smaller or wider, where corridors enter the stairs, or where two corridors merge. The complete set of equations along with an explanation and a sample problem can be found in the *SFPE Handbook.*

Several egress computer models have been developed in recent years that are used by many fire-protection engineers. Such models include ASET, EESCAPE and Rescue model, and EVACNET 4. Each model has its assumptions and limitations, and it is important to know and understand these when using them.

A General Services Administration (GSA) study measured the forward movement rates and resulting discharge rates for different concentrations of people on stairs, as shown in Table 8-1. The discharge rate is per unit width, which in this case was taken as being 0.56 meter (22 inches). It is seen that a space of 0.28 square meter (3.0 square feet) per person yields the highest discharge rate, and this is commonly used as a basis for planning evacuation via stairs.

The Life Safety Code sets maximum distances any person is required to travel from the normal work position to an exit. Travel distance depends on occupancy type and, in most occupancies, whether the building is provided with a fire sprinkler system. As an example, travel distance in existing health-care occupancies may be increased from 150 feet to 200 feet if the facility is sprinklered. Similarly, travel distances in new or existing business occupancies may be increased from 200 to 300 feet for fully sprinklered buildings.

The limiting factor in most facilities is the *common path of travel*, which is defined by the Life Safety Code as "that portion of exit access that is traversed before two separate and distinct paths of travel to two exits are available. Paths that merge are common paths of travel." The Life Safety Code provides requirements for common paths of travel based on occupancy type and sprinkler protection.

The following example pertains to a two-story, 100×200-square-foot office building. The occupant load of each floor is $100 \times 200/100$, or 200 persons. There will be two remote, enclosed stairways at each end of the building, so that 100 persons will egress to one and 100 will egress to the other. If the corridor width is 60 inches clear width, then $60/.2$ gives a 300-person capacity, which is more than enough for the given building. If the stairwell doors are 36-inch-wide doors, then the capacity through each door would be 160 persons (32 inch [clear width]$/.2 = 160$). And, if the stairs are 44 inches wide, then the exit-stair capacity would be 146 persons $(44/.3)$.

Intersections of aisles create problems for traffic, especially if one or more of the corners is occupied by a partition, cabinet, or other large object that obscures visibility around the corner. Moving vehicles create a major problem at intersections during normal conditions. When there is an emergency, vehicles should be moved off the aisles and stopped. But blind corners still hamper people trying to move quickly.

Several solutions have been tried, with varying degrees of success. The best answer, if feasible, is to remove the obstructions. Widening the aisles or cutting back the corners where it is feasible to do so also helps. The use of flat or convex mirrors mounted on each corner of the intersection provides better visibility but creates reverse images; that is, one must look to the mirror on the left to see what is coming from the right, as illustrated in Figure 8-3. Furthermore, the mirrors are easily hit by trucks, and maintenance costs are high. Figure 8-4 shows the use of windows at blind corners, a helpful but limited solution. Another option is the use of hemispherical mirrors mounted overhead at the middle of the intersection, as shown in Figure 8-5. The image is always in right orientation and is complete. However, people have to be trained to look up as they approach intersections.

Dead-end aisles also cause problems. They tend to create the illusion that the aisle continues around the corner. Furthermore, if a fire or other hazard exists at the entrance to the dead-end aisle, there is no alternate route of escape.

Table 8-1. Relationship Between Concentration of People on Stairs and Rates of Movement and Discharge

Concentration of People on Stairs (sq. ft. per person)	Forward Rate of Movement (ft. per min.)	Discharge Rate (persons per min. per exit unit width)
2.0	0	0
2.5	53	39
3.0	75	45
3.5	82	43
4.0	94	43
4.5	106	43
5.0	117	43
5.5	129	43
6.0	139	43
6.5	143	40
7.0	147	39
7.5	150	37
8.0	152	35
8.5	154	33
9.0	156	31
9.5	157	30
10.0	158	29
11.0	158	26
12.0	158	24

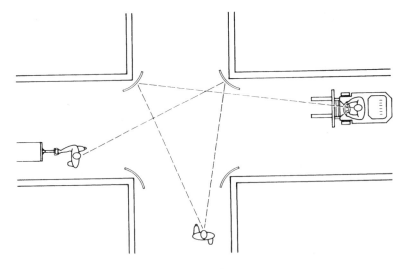

Figure 8-3. Use of mirrors at intersections.

Figure 8-4. Windows at blind corners offer limited but helpful visibility.

The Life Safety Code specifies maximum lengths for dead-end conditions. Dead-end requirements vary from 20 feet to 50 feet, depending on occupancy type and sprinkler protection. Additionally, some occupancies may not have dead-end requirements, while others (e.g., high-hazard occupancies) do not permit any dead-end condition.

The Life Safety Code provides criteria for determining the actual required egress capacity for stairways and level components (i.e., doors, corridors, etc.). With the exception of board-and-care facilities, nonsprinklered health-care occupancies, and high-hazard areas, 0.3 inches per person for stairs and 0.2 inches per person for level components and ramps must be provided. For example, consider 300 occupants attempting to exit a facility. At least 60 inches of door or corridor width (300 persons × 0.2 inches per person) must be provided if the occupants will all exit via a corridor or the front entrance of a building. If these same occupants need to traverse a stairway, 90 inches of stair width is required (300 persons × 0.3 inches per person).

The Life Safety Code specifies minimum widths for all egress components. With several exceptions, the minimum egress width of any means-of-egress component must be at least 36 inches. Existing buildings, aisles in assembly occupancies, and industrial-equipment areas are some of the exceptions where the minimum egress width may be reduced.

To determine the number of people involved, one must calculate the *occupant load.* This is not necessarily the number who will use the path of egress; rather, it is based on a density figure specified for each class of occupancy. The occupant load for industrial occupancy is calculated on the basis of 1 person per 30.5 square meters (100 square feet) of gross floor area, although several exemptions and waivers exist.

In planning paths of egress, one should analyze each floor separately; then, starting with the top floor, analyze the cumulative effects as people are added to the flow of traffic at each floor level of the stairs. It should be assumed that no one will use an elevator during a fire. The time may come when all elevators

Figure 8-5. A 360° hemispherical mirror at an aisle intersection. Courtesy Global
Marketing Corp.

have an emergency power supply, but that is not the case today. Many elevators
now in use would very quickly become inoperative in a fire, thus trapping the
occupants inside.

Every person should have a way to reach an exit that is free of obstructions. In fact, there should be at least two routes to the exits that are remote
from each other, particularly in hazardous occupancies.

People on below-ground floors must travel up stairs to get out of the
building. Generally there are many fewer people in these locations. There is
no particular problem for people on below-ground floors if they can enter a
protected stairwell at that same level. If a person must traverse unprotected
stairs, he or she may be exposed to more gases, smoke, and heat at ceiling level
than at floor level. Carbon monoxide has a specific gravity of 0.967 and will
rise in air. Some other gases are heavier than air and will settle to the floor, or
below (if there are openings in the floor).

In planning exits, it is very helpful to organize the needed data in some
form in which it can easily be assembled and reviewed. Rolf Jensen and Associates,
Inc., have designed such a form, which is shown in Figure 8-6. Instructions for
using this form follow.

Step. 1. Identify each area in the building. (Column 1)

Step. 2. Identify the occupancy classification according to the code being used.
(Column 2)

Step. 3. Determine the area in square feet for each location. (Column 3)

Step. 4. Determine occupancy load factor as given in the code—that is, the persons per square foot used for exit calculations for a given classification. (Column 4)

Step. 5. Determine the population, for calculation purposes, by dividing the area (Column 3) by the occupancy load factor (Column 4). (Column 5)

Step. 6. Whenever people must move from an upper floor or a lower floor to the floor being studied, or when people must move to this area from another part of the building on the same floor to get out of the building, the number of people so involved must be added to the calculated population of this floor. The number added is shown in Column 6, and the total population thus calculated for each area is shown in Column 7.

Step. 7. The various codes give an exit capacity required by that code for determining the total exit width needed. As previously mentioned, this is expressed as so many inches per person. In many cases, a different number will be specified for horizontal corridors and doors and for stairs. Take the measured minimum clear width of the corridor in inches, including corridor doors, and divide by .2 to determine the number of persons it can safely handle. The stairs' capacity is determined by dividing the minimum clear width of the stairs/stair door by .3. Enter this in Column 8.

Step. 8. Determine, then, the total width in inches or centimeters for corridors, doors, stairs, and horizontal areas by multiplying the total population (Column 7) for that area by the appropriate code requirement number (Column 8). Enter these in Columns 9, 10, or 11.

Step. 9. From the building plan, or the exiting building, determine the actual width in inches or centimeters, and enter in Columns 12 through 14.

Step. 10. The codes also specify a maximum allowable travel distance along the path actually traveled from any work location to an exit. This distance, along with the actual distance, should be entered in Columns 15 and 16.

Another form, shown in Figure 8-7, is useful in studying exit routes from a small building with many rooms, as shown in Figure 8-8. Each room is identified by a number or letter code, which is entered horizontally and vertically on the form. Then it is assumed that a person is in the first room indicated in the vertical column and that a fire occurs in one of the rooms indicated in the horizontal row. How many escape routes are available to that person without going through the room where the fire is located? This number is entered in the space below the room where the fire occurred, and the procedure is repeated for every room in the building. When the form is completed, "0s" and "1s" indicate rooms that need to be modified to provide one or more escape routes; rooms with "0s" require special attention. In the example, two sets of figures are given because the use of windows as exits is questionable. The first figure includes windows as exits; the second does not.

Rolf Jensen and Associates, Inc.

Exit Calculations

Building _____   Page _____ of _____ pages

Code Reference _____   Date _____

1	2	3	4	5	6	7	8	Exit width (inches)						Max. travel dist. (ft.)	
								Required			Provided				
								9	10	11	12	13	14	15	16
Loc. in bldg.	Occup. class.	Area (sq. ft.)	Occ'y load factor (Persons per sq. ft.)	Pop'n Prov'd	Pop'n added from above, below	Total pop'n	Factor	Horiz.	Doors	Stairs	Horiz.	Doors	Stairs	Req'd	

Figure 8-6. Form for calculating exit requirements. By permission of Rolf Jensen & Associates, Inc., Deerfield, IL.

Room occupied	Escape routes with fire in:									
	L&D	B₁	B₂	B₃	Bt	K	H	W	Sh	G
L&D		3/2	3/2	3/2	3/2	2/1	3/2	3/2	3/2	3/2
B₁	3/1		3/1	3/1	3/1	3/0	3/0	3/1	3/1	3/1
B₂	3/1	3/1		3/1	3/1	3/0	3/0	3/1	3/1	3/1
B₃	2/1	2/1	2/1		2/1	2/0	2/0	2/1	2/1	2/1
Bt	2/1	2/1	2/1	2/1		2/0	2/0	2/1	2/1	2/1
K	3/1	3/2	3/2	3/2	3/2		2/2	3/2	3/2	3/2
H	5/1	4/1	4/1	4/1	4/1	4/?		5/1	5/1	5/1
W	1/1	1/1	1/1	1/1	1/1	?	1/1		1/1	1/1
Sh	2/2	2/2	2/2	2/2	2/2	2/2	2/2	2/2		1/1
G	3/2	3/3	3/3	3/3	3/3	2/2	3/3	3/3	2/2	

Figure 8-7. Form for analyzing escape routes from a building.

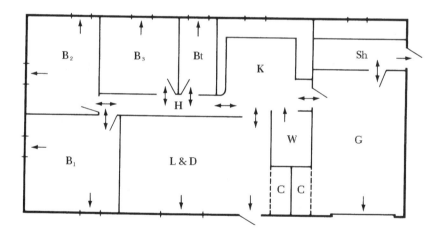

Figure 8-8. Plan of small residential building.

Confined and Hazardous Work Areas

Several industrial jobs require that the worker be in an area that is small, has a restricted means of exit, and, in many cases, has a limited supply of fresh air. Aside from the hazards of the task being performed, this work area itself involves several hazards. The supply of oxygen may have to be supplemented either via a hose from an outside tank or by a self-contained breathing apparatus carried on the worker's back. Since both require the worker to wear a mask of some sort, they are considered personal protective equipment and were covered in Chapter 5.

These confined spaces include storage tanks and bins, tanks on vehicles, and frames and structures that need work done on the inside. Most provide only one way out. In vertical tanks, the worker often has to be lowered and raised by a hoist of some sort. This must be considered part of the worker's means of egress.

The type of work done in such confined spaces frequently involves vapors of chemicals that have been stored there, vapors of chemicals used to clean the tank, or welding equipment. These jobs are generally hazardous because of the toxic or flammable vapors, heat, smoke, or sparks.

It is strongly recommended that anyone working in a tank or other confined space be accompanied by another worker who stays just outside as a monitor. The monitor should maintain continuous visual and/or voice contact with the worker inside, and must be able to help get the worker out of the confined space in case of emergency.

The worker on the inside must have proper personal protective equipment, often including breathing apparatus with the supply of air or oxygen outside. Metering devices that indicate how the worker is breathing will greatly aid the monitor if the worker cannot easily be observed by the monitor. A self-contained breathing apparatus carried on the worker's back provides better mobility but it does not provide any means of monitoring the worker. The OSHA Standard for Confined Space is 29 CFR 1910.146.

Mines constitute a special type of confined space, and this book will not attempt to cover the safety needs of miners. A great deal of work has been done in the study of mining safety, and several sets of standards and many legislative acts exist to protect miners. In spite of all the work and laws, mining remains a hazardous occupation. Toxic gases, loss of oxygen supply, and the ever present risk of collapse of any means of egress make mine safety a complex issue.

Many tasks in the construction of buildings, especially multistory buildings, place workers in positions from which it would be difficult to egress in case of an emergency. Work progress should be planned so that at no time will a worker be put into a position from which he or she cannot reach a safe area quickly and easily. We want to avoid the situation of the painter who paints himself into a corner from which there is no exit, or of Dagwood on the roof when Blondie loans the ladder to a neighbor.

Case Studies

A great many deaths have occurred when people were unable to escape from a burning building. One of the most tragic cases was the loss of 119 people, dead from the Winecoff Hotel fire on December 7, 1946, in Atlanta, Georgia. Another was the loss of 164 people, mostly children, when a circus tent caught fire and collapsed on the crowd inside in Hartford, Connecticut, in 1944. A look at the history of tragic fires shows that many of them took place in hotels or nightclubs. Very few lives have been lost in fires in industrial buildings. It is interesting to note that, in many cases, especially the nightclub fires, a practical design had been replaced by a more aesthetically appealing design. The decorative furnishings were the major sources of fuel and caused the fire to spread rapidly.

One fire that deserves particular attention occurred in Sao Paulo, Brazil, on February 24, 1972, in the 31-story office building known as the Andraus Building. Although it was considered a modern structure, there was only one stairway serving the entire building, and it was not enclosed in fire-resistive construction. Elevators normally serving the occupants of the building were not to be used during a fire. Doors into the stairwell were not fire-rated doors. There was no emergency lighting and no fire alarm.

The fire started on the fourth floor and quickly spread to the fifth, sixth, and seventh floors, aided by open windows, strong winds, and combustible ceiling tile. The fire did not travel above the seventh floor and the wind kept the air suitable for breathing above that level. People above the seventh floor could not have used either the elevators or the stairs, but were safe where they were. However, people on the burning floors had no means of escape except the windows; 16 people died in that fire. A record number of 450 people were rescued. Immediately after the Andraus fire, an attempt was made to enforce more stringent controls for high-rise buildings.

Another tragic fire occurred on May 28, 1977 in Southgate, Kentucky, in a nightclub known as the Beverly Hills Supper Club. One hundred and sixty-four persons died and about 70 others were injured because they could not get out of the building fast enough. Briefly, the major contributing factors to the tragedy were the following:

1. Lack of sufficient exits for the population
2. Lack of sufficient egress corridors
3. Lack of direction for egress
4. Lack of fire walls, fire doors, and smoke doors
5. Lack of an alarm system
6. Excessive crowding of people in available space
7. Lack of sprinkler or other extinguishing systems
8. Excessive use of combustible materials
9. Violations of safe electrical-wiring practices

Some other interesting facts about this building and the development of the fire came to light after the event. Several long-standing violations of building codes were known to exist by a number of people in positions of authority and responsibility, but nothing had been done to correct them. No instructions or plans for evacuating the building existed. The building had been changed and enlarged many times, and the result was a maze of oddly shaped rooms, narrow corridors, and few visible means of egress. In reviewing the incident, it seems remarkable that only 164—only 5 percent—of the more than 3,000 people inside the building were killed. This disastrous fire was the worst multiple-death fire in the United States since the Coconut Grove Night Club fire in 1942 in Boston, Massachusetts, claimed 492 lives.

On September 3, 1991, a fire in a poultry-processing plant in Hamlet, North Carolina, took the lives of 25 people. Another 54 were injured. About 90 people were in the plant at the time. The building was made of noncombustible material, but the interior had been modified several times without appropriate changes in the means of egress. The fire started in an unsprinklered part of the building when heat from a gas-fired cooker ignited hydraulic oil from a ruptured line and spread rapidly with very dense black smoke and hazardous products of combustion. A regulator that should have stopped the flow of natural gas failed because of the intense heat, thus adding fuel to the fire.

Workers had not been trained in emergency evacuation, and directions to exits had not been posted. Many of the workers had to depend on their sense of touch to find their way to a door. There were several outside exits, but many of the workers had to pass through one or two rooms to reach them. A few workers found their way to one exit that they found was bolted. Another exit was blocked by a truck. The door beside the loading dock was found to be inoperable. Most of those who died were trapped in areas with no means of egress and out of reach of rescuers.

A fixed fire-suppression system in the area where the fire started would have, at least, contained the fire and possibly extinguished it. Adequate corridors and exits were needed. Proper training of workers would have prevented some loss of life, also. Each of the doors normally used by personnel was found to be deficient in some way and could not pass requirements for use as a means of egress.

A tragic fire at the Station Nightclub in West Warwick, Rhode Island, on February 20, 2003, resulted in 100 deaths and over 200 people injured. Most of these were young people. The following factors contributed to the deadly disaster:

1. Overcrowding
2. Spray-on foam insulation causing extremely rapid fire spread and heavy smoke
3. No automatic sprinklers
4. Delayed evacuation and panic
5. Inadequate exits and egress paths
6. Pyrotechnics that ignited the combustible interior

The Rhode Island Station nightclub fire resulted in stricter code requirements and code enforcement by state and local officials for assembly occupancies. The fire also resulted in a number of civil lawsuits and criminal arrests, due to blatant lack of concern and enforcement of fire safety requirements.

QUESTIONS AND PROBLEMS

1. How is an exit distinguished from an exit discharge?
2. What is the purpose of occupancy classifications?
3. What is the purpose of building codes, and how do they differ from the Life Safety Code?
4. What is the Common Path of Travel and why is it important to keep it to a minimum?
5. Explain where and how exit signs are to be placed.
6. How should an engineer or architect design a building to be fire-safe?
7. List the characteristics of doors that you would expect to encounter as you leave a building in an emergency.
8. Calculate the number of exits and the required minimum exit capacity for a floor with 300 persons.

9. How fast is an adult in reasonably good physical condition expected to move when walking toward an exit in an emergency?

10. With the use of the Life Safety Code, determine the required width of an aisle in a large factory that is likely to be used by 160 persons.

11. How wide must stairs be in an industrial building to accommodate 160 persons?

12. Using the Life Safety Code, determine how wide corridors, doors, and stairs should be in your classroom or office building.

13. List several ways a panic situation can be avoided during an evacuation from a nightclub.

14. What could have been done to the building structure of the Andraus Building to prevent the loss of lives from fire?

15. List several Life Safety design features for a new nightclub that would prevent a disaster such as the Rhode Island Nightclub fire.

16. List several things you can do to ensure your workplace has a safe means of egress and exits from the building you work in.

17. Name several advantages of having automatic sprinkler protection.

BIBLIOGRAPHY

Hale, A. R. and A. J. Glendon, 1987. *Individual Behavior in the Control of Danger.* New York: Elsevier.

National Fire Protection Association, 2010. *NFPA 80, Fire Doors and Windows, and Smoke-Control Door Assemblies.* Quincy, MA: NFPA.

_____. 2010. NFPA 72, *National Fire Alarm and Signaling Code.*

_____, 2012. *NFPA 101, Life Safety Code.*

_____, 2009. *NFPA 170, Fire Symbols.*

_____, 2012. *NFPA 252, Fire Tests of Door Assemblies.*

_____, 2012. *The Life Safety Code Handbook,* Quincy, MA: NFPA.

Society of Fire Protection Engineers. *The SFPE Handbook of Fire Protection Engineering,* 4th Ed., 2008, Bethesda, MD: SFPE.

The Americans with Disabilities Act Accessibility Guidelines (ADAAG) www.access-board.gov/adaaglhtml/adaag.htm

Uniform Federal Accessibility Standards (UFAS) www.access-board.gov/ufas/ufas-htm/ufas.htm

9

Fire Prevention and Suppression

Kevin L. Biando, PE MSFPE

Richard T. Beohm, PE CSP, ARM

Setting Goals

Safety engineers have two areas of concern in regard to fire: the prevention of unwanted fires and the suppression or extinguishment of unwanted fires and continued operations. For the sake of simplicity, the term *fire* will be used in this book to mean unwanted fires. In many situations, fires are beneficial and there is no reason to prevent or suppress them. But when fires are causing, or threatening to cause, personal injury or damages to property or the environment, they are considered unwanted.

Building design can lessen the potential for unwanted fires. We have the means to design, construct, and use any type of building and its contents in such a way that a fire cannot occur. Fires in less confined areas may be less easily controlled by design, but vehicles and even outdoor storage facilities can be designed to reduce the probability of fire. Such design and utilization may require sacrifices in convenience and cost. As in other areas, we must weigh the benefits against the costs and set the limits of acceptable risk.

In any design, we should establish goals as to acceptable risk and then design to meet them. With regard to fire, we might establish goals such as the following:

1. We will not tolerate loss of life or hospitalization of any occupant of the building due to fire.
2. We will tolerate no more than $75,000 damage to property as a result of fire.
3. We will not tolerate a fire's spreading beyond the room of origin.

Goals may vary from one situation to another, but these are the kinds of goals that should be established. The building and its contents can then be designed to ensure that the goals will be met. Note that meeting such goals requires a consideration of the use of the building and its contents as well as its design and construction.

Chemistry of Fire

Fire is a chemical reaction in which a material, called a fuel, is oxidized rapidly, producing heat. Fire cannot occur unless certain conditions of the fuel and the oxidizing agent exist. The fuel must be in the form of a gas or vapor in order to mix properly with the oxidizer, which must also be a gas or vapor. The temperature of the mixture must be sufficiently high to cause the chemical reaction to take place.

Fire has been schematized both as a triangle and as a pyramid. The fire triangle consists of the three components—fuel, oxidizer, and thermal energy—needed to start the chemical reaction. Sometimes the chemical reaction itself is included as one of the components, producing a four-sided body in which all four components contact each other—that is, a tetrahedron. The fire pyramid becomes useful when fire suppression is discussed, since a fire can be extinguished by removing any one of the three sides.

There are several characteristics of fuels that are important in safety engineering. One is the ability of a given fuel to mix with an oxidizing agent. In only a few isolated cases is the oxidizer anything other than oxygen from the air. Oxygen is sometimes made available by the decomposition of water or another chemical containing oxygen. Fluorine and chlorine will react with some types of fuels to start a fire, but in this discussion it will be assumed that air is the medium by which an oxidizer is made available to the fuel.

The fuel must mix with the air within a certain range of proportions in order to burn. The limits of this range are called the lower flammability limit and the upper flammability limit (LFL and UFL). They are expressed as the percentage of fuel vapor, by volume, to air. Unfortunately, some literature has referred to these as explosive limits (LEL and UEL). Explosive limits may be the same as the flammability limits under some conditions of temperature and pressure, but quite different under other conditions. These limits are well demonstrated in the engine of an automobile. If the fuel/air mixture is too lean, it will not burn, and if there is too much gasoline, the engine becomes flooded. For a further illustration of this, as well as of the fact that fuels must be in the form of a gas or vapor, a spark device can be inserted into a container of gasoline. When a spark is discharged, nothing happens to the gasoline, but if the spark is discharged on the surface of the gasoline, the gasoline vapor immediately ignites.

Several temperature characteristics of fuels are important, especially with liquid fuels but to some extent with solid fuels as well. The ignition temperature is the lowest temperature at which a fuel, either solid or liquid, will ignite without an external source of ignition. Solid fuels will ignite from an outside source of ignition after exposure to heat for varying lengths of time, but the ignition source must provide a temperature no less than the ignition temperature at the surface of the fuel.

Three other temperatures are associated with liquid fuels. The flash point is the lowest temperature at which a liquid fuel will form an ignitable vapor. The fire point is the lowest temperature at which vapor will be formed rapidly enough to sustain a fire. The boiling point is the lowest temperature at some given pressure at which a liquid begins to evaporate or boil. The flash point of a liquid fuel is usually a more critical temperature than the fire point. An external source of heat sufficient to ignite a flammable mixture will usually be available to sustain that ignition until it reaches the fire point, at which the fire generates enough heat to be self-sustaining. The fire point is usually a few degrees above the flash point.

Liquid fuels are classified according to flash points and boiling points. Table 9-1 gives the several classes of combustible and flammable liquid fuels. Note that the term *flammable* implies a more hazardous condition than the term *combustible*. Any material which is capable of burning is called combustible or flammable. A material which will not burn under ordinary conditions is noncombustible. The term *inflammable* should never be used because of the implications of the prefix "in-." In past years, several accidents have occurred because people thought the label "inflammable" meant that the material would not burn.

It is often important to know the vapor density and the specific gravity of a fuel. Combustible vapors with a vapor density greater than 1.0 will tend to sink in the air and collect in low places such as basements, pits, and shafts. Liquids with a specific gravity less than 1.0 will float on water if they are insoluble in it. Properties of several common combustible materials are given in Table 9-2. It should be noted that the materials wood, aluminum, magnesium, potassium, and titanium ignite very easily and burn very rapidly when in dust form. More complete tables can be found in several handbooks and references listed in the bibliography.

Under certain conditions, a fuel will ignite without an apparent outside source of heat; this is called *spontaneous ignition.* It occurs when the fuel is loose enough to contain air within it, and slow decomposition of the material raises the temperature to the ignition point. Spontaneous ignition happens most frequently when a pile of small pieces of solid material settles, and the increased pressure and temperature bring about ignition. Large piles of coal, heaps of moist hay stored in a barn, oily rags, and accumulated piles of grain dust or wood dust are places where it is likely that spontaneous ignition will occur.

Ignition of a fire can be prevented by not allowing the three essentials of the fire triangle to exist simultaneously. The absence of any one makes fire impossible. The fuel can be any material that produces heat when it reacts rapidly with an oxidizer. The oxidizer most frequently is the oxygen in the air, but it can be any other material acting to reduce the fuel in an oxidization-reduction reaction. Some materials contain oxygen as an element of the compound and may release that oxygen when subjected to heat. When the oxygen content of the air is reduced to about 16 percent, it will not support

Table 9-1. Classes of Flammable and Combustible Liquids

Flammable Liquids	Flash Point	Boiling Point
Class IA	<22.8°C (73°F)	<37.8° (100°F)
IB	<22.8°C (73°F)	≥37.8° (100°F)
IC	<37.8°C (100°F)	
Combustible Liquids	**Flash Point**	
Class II	≥37.8°C (100°F) <60.0°C (140°F)	
IIIA	≥60.0°C (140°F) <93.3°C (200°F)	
IIIB	≥93.3°C (200°F)	

Table 9-2 Properties of Some Combustible Materials

Material	Sp. Gr.	Flash Pt (degrees F)	B.P. (degrees F)	Flammability Limits (% vol.)
				UFL LFL
Acetone	.79	0	133	12.8–2.5
Acetylene		gas	sublimes	100–2.5
Benzene	.88	12	176	7.8–1.2
Ethlene glycol	1.11	232	388	15.3–3.2
Fonnaldehyde		NA	–6	73–7.0
Gasoline	.72–.76	–45	102	7.6–1.4
Isopropyl alcohol	.79	53	181	12.7–2.0
Methane	.87	40	232	7.1–1.1
Nicotine	1.01	203	482	4.0–0.7
Propane			–44	9.5–2.1
Turpentine	.86	95	309–338	?–0.8
Vinyl acetate	.93	18	162	13.4–2.6
Aluminum	2.70	NA	4221	NA
Zirconium	6.51	NA	6471	NA

Source: NIOSH Pocket Guide to Chemical Hazards, 2011.

combustion of most materials. Furthermore, a fire will not be ignited without sufficient heat to initiate the chemical reaction.

By the same reasoning, a fire once started can be extinguished by removing any one of the three items of the fire triangle or by repeatedly interrupting the chemical reaction itself. Later in this chapter, the several most common methods of extinguishing, or suppressing, a fire will be discussed.

Flame Growth

All fires start at a point of ignition and, if satisfactory conditions exist, expand in some predictable manner. If a fire starts in a building, and assuming that these essentials are continuously available, the fire grows in *realms* as outlined below and illustrated in Figure 9-1:

Realm 1   Fuel, oxidizer, and heat brought together to cause ignition

Realm 2   Growth to a flame height of about 25 centimeters (10 inches), called *established burning*

Realm 3   Growth to a flame height of about 1.4 meters (4.5 feet), called *sustained burning*

Realm 4   Growth of flame to ceiling height so that top of flame begins to spread out

Realm 5   Growth in all directions to involve essentially all of the room or compartment (flashover)

Realm 6   Any spread of the fire beyond the room of origin.

Fires usually follow a so-called *t*-squared growth rate. A set of specific *t*-squared fires, labeled slow, medium, and fast, with fire intensity coefficients (α) such that the fires reached 1055 Kw (1000 BTU/sec) in 600, 300, and 150 seconds is found in NFPA 72. An ultra-fast fire reaches 1055 Kw in 75 seconds.

$$Q \text{ kW} = \alpha \times t^n \text{ seconds}$$

where

Q is the rate of heat release and $n = 1, 2, 3$.

Fire dynamics describe the characteristics of fire. There are now a number of computer programs, such as FPE Tool and FDS, among the 62 or more models used to predict the growth of fire. Fire is a dynamic process of interacting physics and chemistry. It is best to use a model that applies proper heat transfer to the given space configuration.

Several factors in this breakdown of fire growth are important. Below a flame height of about 25 centimeters, enough heat is conducted through the material to keep it burning, but very little heat is radiated. If a fire can be kept below this size, it will cause very little damage and will be relatively easy to extinguish. Until the flame reaches a height of about 1.4 meters, the oxygen in the air is sufficient to keep it going. Beyond that, ventilation contributes significantly to the fire. The shape of the room and openings in the walls and ceilings begin to become important at this point. When the flame reaches the ceiling, it is forced to spread sideways. Heat radiates in all directions, and smoke begins to form a blanket at the ceiling. All fuel content in the room very quickly becomes involved in the fire. Up to this point, barriers have had little effect. Now, the characteristics of the walls, the ceiling, and the floor are important in determining if and how the fire will spread to adjacent rooms or areas.

The fire plume develops as air is entrained into the plume. The fire-plume jet is the dominant air mover and produces a ceiling jet as the fire becomes

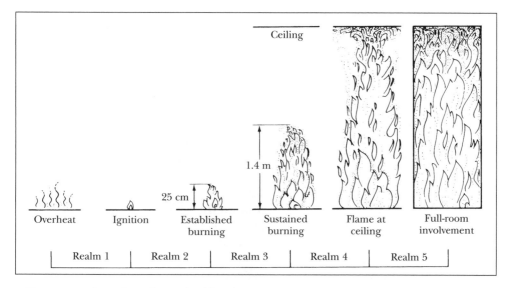

Figure 9-1. The realms of growth of flame.

established and smoke travels across the ceiling. A condition called flameover or rollover can occur as the flames move through or across unburned gases as the fire progresses. Flameover is different from flashover as only the fire gases are involved and not the surface contents of the room. Thermal layering of the gases is where the gases form into layers across the ceiling as to their temperature, sometimes known as heat stratification and thermal balance. As the fire grows in a compartment, large volumes of hot, unburned fire gases can collect. It is a condition of low oxygen, high heat, smoldering fire, and high fuel-vapor concentrations. When oxygen is introduced by venting a window, for example, a large, explosive force called backdraft is created.

Fires in unconfined areas also grow as in realms 1, 2, and 3. Beyond that point, the fire will grow in any direction where fuel is available, as long as a sufficiently high temperature can be maintained.

The time required for this flame growth varies, depending primarily on the rate of evaporation of the fuel. The American Society for Testing Materials has developed a test (ASTM E-84) for measuring and classifying the flame-spread rate of different materials. Due to varying rates of evaporation and oxidation of the fuel, fires can be very rapid, average, or smoldering.

In designing for protection against fire, there are several points in the pattern of flame growth that need special attention. First, it must be realized that a flame cannot grow at all if there is no more fuel to be consumed. It must be remembered that the fuel for a fire in a building usually includes not only the contents but also the building itself—wood in the structure, and especially paint and other finish materials. Even if there is a great deal of fuel in the room, if it is not close to the flame, it will not burn until the ambient temperature is high enough to ignite it. That condition seldom occurs until the whole room is involved, unless the fuel is in line with a forced flow of air, or a draft, from the flame. At a point at which the flame is still relatively small, then, lack of continuity of fuel will cause self-extermination of the fire. This factor is important in designing for protection against fire.

If a fire is not extinguished by the time it has reached a flame height of 1.4 meters (the realm of sustained burning), it will be difficult to extinguish it without a major fire-fighting effort.

The next major point in the development of a fire in a building is when the flame reaches the ceiling and starts to spread sideways. By this time, the ceiling is already hot and is radiating heat down into the room. At this point, the dimensions of the room—not only the height of the ceiling but also the width of the room as well—become important factors. As the fire spreads sideways, heat is radiated to walls and major objects in the room. As they become heated, they in turn radiate and/or conduct heat to other fuels and back to the fire itself. This phenomenon is often referred to as *feedback* and is responsible for a rapid increase in room temperature. Therefore, growth to full-room involvement occurs very rapidly if the room is narrow or if there are large objects that act like walls to radiate heat.

At full-room involvement, or even when the flame reaches ceiling height, attention should be given to containment of the fire within the room of origin. This involves barriers. At this point, too, attention must be given to the structural integrity of the wall and/or support structure of the building. These subjects will be discussed more fully later in the chapter, but fuels and fuel load will be analyzed first.

Fuel-Load Analysis

The fuel load, also called the fire load, is the total amount of fuel that could contribute to a fire in a given area or room. In designing for protection against fire, one should consider several characteristics of this fuel load. For the sake of convenience, the more common combustible materials are divided into two fuel families—cellulose materials constituting one family, and petrochemicals constituting the other. The total amount of combustible material is expressed as the equivalent mass of wood. For estimating purposes, it is assumed that all cellulose materials have the same heat potential per pound as wood. This is not exactly correct, but for estimating fuel load, it is close enough. Petrochemicals are assumed to have twice the heat potential of cellulose materials.

In an analysis of the potential for fire, the density of the fuel load is important. The greater the mass in a given area, the greater will be the heat generated in a fire. The fuel-load density is merely the equivalent mass of wood per given area in kilograms per square meter or pounds (mass) per square foot. In some literature, the term *fuel load* or *fire load* refers to the density of the fuel rather than the mass of the fuel.

The next step of the analysis is to determine or estimate the heat potential of this fuel load. Since all the fuel has been converted to the equivalent of wood, the heat potential available is that of the equivalent mass of wood. For lighter fuel loads (up to 200 kilograms per square meter), this value is about 18.6×10^6 kilojoules per kilogram. For medium loads (200 to 250 kilograms per square meter), it is about 17.7×10^6 kilojoules per kilogram, and for heavy loads (over 250 kilograms per square meter), it is about 16.7×10^6 kilojoules per kilogram. Since so many variables affect these data, they can be used only as a rough estimate. Fire-load estimates are used in the design of sprinkler systems, as will be discussed later in the chapter.

Another important factor is continuity of fuel. A fire cannot grow if there is no more fuel available. This statement is not as simple as it seems, however. Fire can progress more deeply into an object already burning on the surface. Since all fuels must be in the form of a gas or vapor, solid and liquid fuels can only burn on the surface. As they burn, new surfaces are exposed to evaporation. An object with a large surface-to-mass ratio, then, will burn faster than the same material in a form with a lower surface-to-mass ratio. One hundred separate sheets of paper will burn much faster than a tightly closed 100-page book. Refuse left lying around will ignite and burn faster than the same material compressed or even stacked together.

Continuity of fuel also involves other objects in the path of the spread of flame or heat, horizontally, vertically, or in any direction. If a small fire can be prevented from reaching new sources of fuel, it will be easy to extinguish. Because hot air is lighter than cooler air, it will rise quite rapidly, carrying heat and flame to fuels overhead. If there is a draft, no matter what its source, it will carry the heat and flame in the direction of the flow of air. Fires in return-air ducts and in ventilation ducts are usually started in this way. Potential fuels should be kept away from likely sources of ignition or flame-spread. Usually these patterns of airflow can be predicted quite easily.

Figure 9-2 is a diagram of the sources of fuel in a building. It is important that the fuel contributions of the building itself be separated from the

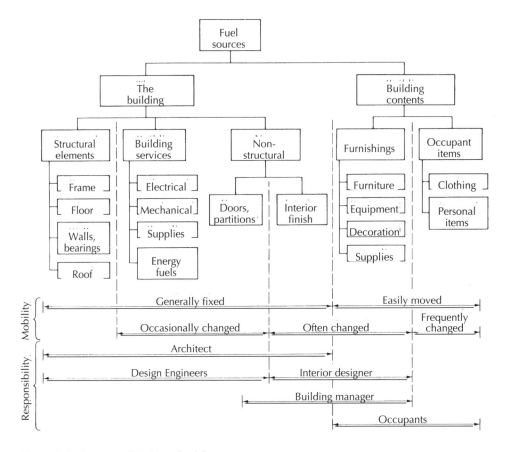

Figure 9-2. Sources of fuel in a building.

contributions of the contents, for several reasons. Certain items are fixed, yet can contribute to a fire in two ways. They can supply fuel to a fire and, because of geometry and arrangement, influence the growth of the fire. These fixed items must be designed properly the first time, since they are not easily changed. On the other hand, fires more often start in the contents of the building. These items are frequently moved and are controlled by the management and the occupants of the building. It is important that all responsible parties—architects, engineers, and managers—be made aware of the possible consequences of arranging potential fuels in such a way that they constitute a high density of fuel load or provide continuity of fuel to feed a fire.

A popular concept in building design, now used by the military and others, is the Whole Building Design (WBD) approach, which is used to create a successful high-performance and safe building by applying an integrated design and team approach. Another concept that is becoming recognized by Authorities Having Jurisdiction (AIDs) is Performance-Based Fire Safety Design, an option of NFPA 101 to the established prescriptive design methods adhering to the adopted codes. See the first section of this chapter, "Setting Goals." Fire-safety designs are created to meet the established set goals. A good reference is the Society of Fire Protection Engineers' "Introduction to Performance-Based Fire Safety."

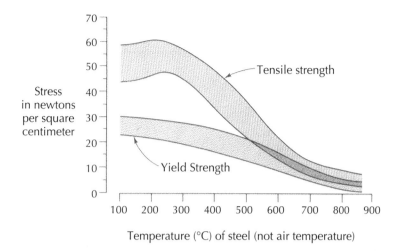

Figure 9-3. Range of yield strengths and tensile strengths of ASTM A7 and A36 structural steel at elevated temperatures. (Adapted from American Iron and Steel Institute, *Fire Resistant Steel Frame Construction*, 2nd ed. [Washington, DC: AISI, 1973], p. 19.)

Building Design

Since the structural components of a building are the most permanent and unchangeable, they should be the first to be considered in designing for fire safety. Fire-resistive and noncombustible construction are preferred. NFPA 220 or 101 (2009), Table A8.2.1.2, describes the various types of construction.

Practically all new industrial buildings, as well as the larger commercial, institutional, and residential buildings, are constructed with a steel and/or concrete support structure. Steel and concrete provide strength and rigidity without requiring the massive frame necessary in a wood structure. And, of course, steel and concrete are noncombustible.

This is not to say that steel and concrete are not susceptible to fire. They do not burn, but they do lose their strength and collapse at high temperatures. In fact, in some situations, steel and concrete lose their structural integrity sooner than wood when exposed to fire. A massive wood beam burns very slowly because of its low surface-to-mass ratio, and therefore is capable of supporting a load for quite a long time after a fire starts.

Several types of steel are available for use as structural members, but the most common types are low-carbon steels, referred to as ASTM A7 and A36 structural steels. Figure 9-3 illustrates the range of yield strengths and ultimate tensile strengths for these steels at elevated temperatures. Note that there is a great loss of strength as the steel heats to a temperature of 500°C or more. This rise in temperature is much slower in the steel itself than in the air around it, but a large fire will almost always burn long enough so that unprotected steel support members buckle and collapse.

Steel can and should be protected against this rise in temperature, and several methods exist to do this. In some cases, where space and mass permit it, the steel is encased in masonry, either brick, hollow tile, or cinder blocks. More often the steel is covered with a plaster. The plaster may be gypsum plate, gypsum mixed with perlite or vermiculite, or mineral fibers. Portland cement is sometimes used.

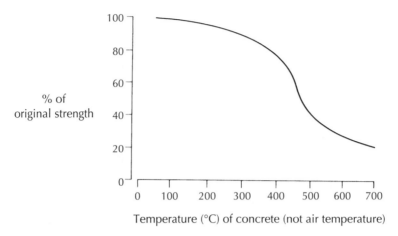

Figure 9-4. Effect of temperature on typical concrete structure.

Protective coatings, such as mineral fiber and cementitious and intumescent materials, can be sprayed onto the steel. These materials swell at higher temperatures and form a layer of insulation over the steel. All of these protective methods merely slow down the transfer of heat to the steel. If a fire were of sufficient magnitude, the steel would eventually reach temperatures high enough to damage it.

Concrete also deteriorates in strength at high temperatures. The loss in strength varies greatly due to varying aggregates and cement mixtures, but Figure 9-4 shows a typical loss in compressive strength at elevated temperatures. Prestressed concrete is usually stronger than normal concrete, but the reinforcing bars are low-carbon (structural) steel and add only slightly to the strength at elevated temperatures.

As mentioned earlier, wood beams retain their strength for quite a long time in a fire, even though they burn and add fuel to the fire. The charcoal formed as the surface of the wood burns tends to act as an insulator and delays burning. Wood laminates are often used because of their relatively high strength-to-weight ratio, low cost, and good appearance. Depending on the glue used, the laminations tend to separate at higher temperatures, thus exposing greater surface area to the fire.

Ordinary wallboard crumbles rather quickly, and, therefore, little fire resistance is realized. Specially formulated gypsum board, identified as Type X or Type XX, is commonly used to increase the fire-resisting capabilities of wood and steel. One layer of ⅝-inch-thick Type XX board can provide a one-hour fire rating in many assemblies. These assemblies are UL listed as to their fire-resistance ratings.

An increasing number of buildings being constructed today depend entirely on columns and cross members for support. The walls are not load-bearing. This type of structure requires a sturdy frame but allows much greater flexibility in the design of other components of the building, especially walls.

The roof of a building is important in planning for fire safety. Among the characteristics that should be considered are the roof support, the cover material, the roof shape, and roof openings. A building with a fairly horizontal roof will usually have a metal decking laid across girders or roof trusses.

Trusses are made of lightweight steel members joined together to provide a high strength-to-weight ratio. This type of roof is very popular and desirable in many respects, but it is difficult to protect from fire without destroying its other advantages. Since the trusses are located just under the roof, they are in the area of highest air temperatures in most fires, and they collapse more quickly than heavier girders. The roof deck is also exposed to high temperatures and fails quickly if it is made of thin material and is not insulated.

One of the most common roof materials is asphalt or a combination of asphalt and gravel. Since asphalt has an ignition temperature of only 485°C (905°F) and a melting point below 370°C (700°F), it melts and burns quickly if exposed to the heat attained in most large fires. Furthermore, if the roof deck buckles, the asphalt roofing is exposed directly to the fire and quickly melts and drips through the deck.

The shape of the roof is significant in buildings in which dust or smoke is generated in normal operations. Large quantities of smoke and dust should be vented rather than allowed to accumulate. A high roof with an inverted V-shape that has a monitor section aids in venting. A roof monitor is the raised portion of the roof with panels or louvers that can be opened and closed. These same roof features can aid in venting smoke from a fire. The roof of any industrial building should have some way to vent smoke and heat. Today roofs are usually vented with smaller, automatic-unit vents.

Floors are a major item in designing for fire safety. It is important that a nonabsorbing material be used for the floor, especially in a building where oil and other flammable liquids are used. Wood plank, wood block, and even porous concrete will absorb these liquids and act as a reservoir of fuel for a fire. Concrete with a sealed surface or a hard tiled surface is the most satisfactory material and is commonly used in industrial plants. Steel plate is used in certain areas where removable sections are needed or where a high strength-to-weight ratio is desired. Steel is good for fire safety, but becomes very slippery when worn smooth, and especially when wet, as mentioned in Chapter 7. An abrasive surface on steel reduces slipperiness and does not allow any appreciable accumulation of liquid fuel.

The floor support (floor joists) is important because its failure in a fire can cause considerable damage to equipment resting on the floor. In some cases, a failure of floor joists can cause the spilling of liquid fuels, adding to the fire.

Columns and girders are, of course, the major support members of a building. These members can and should be more massive than other structural components. They can be made of protected steel, concrete, or, in smaller buildings, heavy wood timbers and beams.

Walls and partitions can be a help or a hindrance in fire safety. Room size and shape influence the spread and the intensity of a fire. The smaller the room, the easier it is to confine a fire to a small space. However, a small room provides greater feedback, as discussed earlier, which tends to make a fire more severe. Also, a building with many small rooms usually hinders fire-fighting efforts.

Walls and partitions made of combustible materials will add fuel to a fire when the fire or heat gets close enough. Since heat and flame travel vertically faster than horizontally, walls and partitions burn very rapidly and carry flames to ceiling height quickly. Conversely, if walls and partitions, including any finish or coverings, are noncombustible, they will retard the growth of flame.

Noncombustible wall construction is often used for the purpose of preventing the spread of fire; these walls are called *fire walls*. To be effective fire barriers, such fire walls must have the following characteristics:

1. They must extend all the way from a nonflammable floor up to a nonflammable ceiling or roof.
2. They must have no unprotected openings that will allow flame or heat to pass through.
3. They must resist the transfer of heat to the back side of the wall for an acceptable length of time.

As mentioned in Chapter 8, fire walls are rated on the basis of the length of time they will retain their intended strength and resist the passage of heat and flame. Ratings are specified as one-half hour, one hour, two hours, three hours, and four hours. A two-hour fire wall will meet the standard test without failure for at least two hours.

Obviously, internal walls will have openings cut into them to allow people, materials, pipes, wires, and ducts to pass through. In a fire wall, all penetrations for pipes, wires, or ducts are *fire-stopped* to resist the passage of heat and flame for the same period of time as the wall. Doors must be rated fire doors and close tightly enough so that they, too, serve as a barrier.

Doors are rated as fire doors the same way that fire walls are. Fire doors may be either swinging or sliding doors, but must be mounted in such a way that they close automatically when the temperature on either side of the door reaches a predetermined point. A fire door must not be blocked open or in any way made inoperable, unless a responsible person is present to supervise its operation.

Fire doors are often used in places where people frequently travel, as in stairwells and other types of exits. Any door that is part of an exit must be a hinged door, and it is usually desirable to have a window in the door. The Life Safety Code permits 1- and 1-1/2-hour-labeled fire doors to have a wired-glass vision panel that does not exceed 645 square centimeters (100 square inches). Three-quarter-hour-labeled doors can have panels up to 1296 square inches.

Smoke doors, or *smoke stops*, are distinguished from fire doors by the fact that they merely resist the passage of smoke and gases. Remember that the products of fire are considered to fall into two categories: heat and flame, which damage property; and smoke and gases, which injure people. Smoke doors do very little to stop heat and flame. Large buildings with many occupants and hospitals should have smoke partitions to divide the building into smaller areas and to contain in one area the smoke and gases generated from a fire.

Ceilings also affect the spread of a fire once the flame or sufficient heat reaches ceiling height. When the flame reaches the ceiling, it can only grow sideways. Noncombustible ceilings are desirable to limit the continuity of fuel.

Many buildings now have hung ceilings with space overhead, called a *plenum*, in which most of the building services are located. If a fire breaks through a ceiling or is carried into the plenum by way of a duct, flame and hot gases easily spread throughout the plenum area unless stopped by a fire wall. Although heat and flame generally travel upward, when they get into a plenum area, they can also ignite ceiling material in adjacent rooms.

When fire fighters are called to help extinguish a fire, they are sometimes frustrated because they cannot get into the plant or building with hoses and sometimes cannot even get fire trucks close to it. Keeping access areas open at all times is a management problem, but part of the problem may be design. The building may have too few windows or other openings through which to bring hoses. The building shown in Figure 9-5 illustrates this problem. Windows serve other purposes as well, but they are essential for fire fighters outside the building.

There are several other features of a building that, though they may not be an integral part of the building, should nevertheless be planned along with the structure of the building. These include drains, the water supply, auxiliary power, storage areas, and countermeasures for potential explosions.

Drains in industrial plants and laboratories often carry flammable and even explosive materials. Vents and traps must be provided, and safe procedures for handling chemicals must be established.

Water must be made available for sprinklers, standpipes, and hoses, as well as other plant needs. Fittings compatible with municipal fire department lines must be provided.

Auxiliary power is needed during a fire for several purposes. Electric power is usually one of the first systems to fail in a fire, but exhaust fans and emergency lighting are still needed. Many industrial companies have some process operations involving chemicals that must be controlled continually, and auxiliary power is needed for these processes, as well as for an electric fire pump and a fire-alarm system. Multistory buildings in which elderly or handicapped persons are likely to be on upper levels require elevators to evacuate the building during an emergency. Elevators should not normally be used during a fire for two reasons. The first is that electrical power is likely to fail while people are in the elevator, and they will be trapped. The second is that many elevators are heat-sensitive and automatically direct the elevator to a floor where heat is being generated—that is, the floor where the fire is located. An independent power source and nonheat-sensitive controls are now required for new high-rise buildings.

Storage areas are particularly susceptible to fire, and precautions should be taken in planning and using them. People usually try to make the best use of space in a storage area, with materials stacked as compactly and as high as possible. This immediately creates several problems. Vertical openings between stacks create a chimney effect, bringing more air to a fire in that area and allowing heat to be built up very rapidly. They also hamper efforts to extinguish a fire. If the stored material is flammable or combustible, the compactness creates a very heavy fire load.

Different types of materials should be segregated in a storage area so that leaking, venting, or an accident cannot bring together two materials that react with each other to produce fire or explosion. Strong acids are kept apart from flammable materials because the acids act as oxidizers. Other oxidizers often found in storage include nitrates, nitrites, peroxide, fluorine, and chlorine. Refuse that ignites easily should never be kept in a storage area containing a heavy fire load.

Flammable liquids must be stored in a protected area. In fact, only a limited quantity may be stored in internal storage areas. NFPA and OSHA standards are very specific concerning the storage of flammable and combustible liquids. OSHA 1910.106 and NFPA 30 define these categories of liquids, specify the design of tanks and vents, and restrict the quantities that may be stored inside

Table 9-3. Maximum Allowable Size of Containers and Portable Tanks

Container Type	Flammable Liquids			Combustible Liquids	
	Class IA	Class IB	Class IC	Class II	Class III
Glass or approved plastic	1 pt.	1 qt.	1 gal.	1 gal.	1 gal.
Metal (other than DOT[a] drums)	1 gal.	5 gal.	5 gal.	5 gal.	5 gal.
Safety cans	2 gal.	5 gal.	5 gal.	5 gal.	5 gal.
Metal drums (DOT[a] specification)	60 gal.	60 gal.	60 gal.	60 gal.	60 gal.
Approved portable tanks	660 gal.	660 gal.	660 gal.	660 gal.	660 gal.

[a]Department of Transportation

Source: OSHA 1910.106, Table H-12.

a building in various types of containers. Table 9-3 is reproduced from Table H-12 found in the 1910.106 standard. There are a few exceptions to these requirements, such as bulk plants, service stations, refineries, chemical plants, and distilleries. Medicines, beverages, foodstuffs, and other consumer items are also exempt from this restriction. There are several other restrictions on larger tanks stored under various circumstances, and the NFPA and OSHA standards should be consulted when planning such facilities.

When large quantities of flammable liquids are to be stored in small containers, they should be stored in cabinets designed and approved for that purpose, called *flammable-liquid cabinets*. An alternative is to construct a room of concrete with a trough around the outer edge, a venting system, and a tight-fitting door. A trough to collect spilled liquids is desirable in any storage area where appreciable quantities of liquids are stored. Another desirable feature, if a trough does not extend all the way around the storage area, is a partial trough or curb at all doorways to the storage room to prevent spilled liquids from running on the floor to other areas. Buildings and rooms used for storage of flammable solvents should conform to NFPA 30.

A few new buildings are including a system that helps contain a fire in the room of origin by creating a slightly positive pressure in all adjacent areas. A fire increases the air pressure in a room by heating the air and by releasing gases during the chemical reaction. This increased pressure tends to force open doors, collapse thin or weak walls, and force hot air and gases into the adjacent areas. Increasing the pressure in adjacent areas helps to balance this higher pressure in the room of origin, and thus helps to keep the fire from spreading. Operation of such a system is usually activated by a smoke detector on the floor of origin.

Stairwells are continually pressurized for this same reason. Stairwell doors should be *positive-latching* to resist the fire pressures; otherwise, the doors could be forced open by the fire gases. In high rises, the fire floor is sometimes exhausted to create a negative pressure, and the floors above and below the fire floor are pressurized to keep smoke out.

Fire-Safety Practices

No matter how much effort may be put into designing a fire-safe building, it will not prevent a fire if people in the building fail to work in a fire-safe manner. Good safety practice requires continual supervision and training. The best practice recognizes the four priorities mentioned in an earlier chapter: (1) eliminate the hazard, (2) isolate the hazard, (3) train, educate, and supervise the workers, and (4) use personal protective equipment.

How can the potential for fire be eliminated? The only absolute way is to eliminate fuels, sources of ignition, and/or oxidizers. The potential for fire can be reduced to an acceptable level by eliminating all unnecessary sources of fuel, oxidizers, and sources of ignition. Beyond that, isolation can be very effective, but it involves continual care on the part of everyone.

First, people must learn to recognize what constitutes a fuel. Any hydro-carbon is a potential fuel, including any vegetable or animal fiber (wood, paper, most fabrics), petroleum-based liquids, alcohols, most of the plastic materials, and many other chemicals. Other materials may become fuels under certain conditions. Most metals, especially magnesium, titanium, sodium, potassium, and zirconium, will burn furiously under certain conditions.

Solid fuels in the form of dust burn very rapidly due to their high surface-to-mass ratio. Two practices can be followed to reduce fires in dust. First, the formation of dust can be lessened by using sharp cutting tools, thus reducing the friction that produces dust. Second, dust can be collected with a vacuum or liquid-wash system as close to the point of generation as possible. If dust can be collected and compacted into a larger solid object, it will not ignite as easily.

Scrap materials, refuse, and packing materials are common sources of fuel. Good housekeeping—a place for everything, and everything in its place—is the only practical way to prevent fire among these materials. They should not be allowed to lie around in unprotected areas. Furthermore, smoking and all other sources of ignition must be prohibited where these materials are generated or used. Nonmetal dusts usually can be kept moist.

If it is necessary to have large piles of bulk materials, such as coal, grain, or sawdust, they should be carefully guarded. Spontaneous combustion often occurs in these piles because the pressure and vibration of the particles in the pile cause the temperature at the bottom to increase. There is always enough oxygen available within the pile, and when the temperature reaches the ignition temperature of the material, combustion takes place without an outside source of ignition. Unfortunately, the fire can smolder for a long time before it is noticed.

To reduce the probability of spontaneous combustion, piles of bulk materials should be kept small and should be moved or turned over frequently. In some cases, such as with piles of coal, the pile can be kept cool by spraying it with water.

Liquid fuels should be clearly labeled regarding their flammability. Some difficulty exists in identifying and classifying the flammability of many commercial products, especially those whose ingredients and composition are considered trade secrets. This problem has been greatly reduced by the Hazard Communication Standard, often referred to as the Right-to-Know Law, OSHA 1910.1200. Manufacturers must identify any hazardous chemicals in their products and clearly indicate what hazards they may present. The Right-to-Know Standard is further discussed in the chapter on hazardous materials.

A common problem regarding labeling occurs at the work station when a worker pours a quantity of a liquid fuel from a larger, labeled container to a smaller, unlabeled container. The worker then forgets that it is flammable, or someone else comes along who is not aware that it is flammable.

Flammable liquids should be handled and stored in containers designed for that use, generally called *safety cans*. A major feature of a safety can is a flame arrester inside the pouring spout, consisting of two concentric perforated cylinders. Flame cannot pass through these perforated barriers, but liquid can be poured through them. The principle of the flame barrier can be demonstrated with screens and a small Bunsen burner. Another important feature of a safety can is a spring-loaded cap over the spout. This cap has a sealing gasket, and the spring allows the release of pressure if it develops inside the can. Some cans, designated as type-I cans, have one spout used for both filling and emptying. Figure 9-5 shows a type-II can with separate openings for filling and emptying. The cans have a squat shape to reduce spilling and a minimum of joints and crevices on the inside to reduce corrosion; they are made of heavy-gauge, corrosive-resistant material. If an extended pouring spout is used, it should be made of a nonsparking material. Larger containers, especially, should be electrically grounded to prevent static electricity from providing a spark that would ignite the fuel.

Flammable liquids are frequently found in less obvious but just as hazardous places as solid fuels. Flammable oils, solvents, and finishes are commonly used and are often splashed or dripped onto the floor, absorbed in rags and clothing, and carried along on chips, boxes, and workpieces. Oily rags and oil-soaked floors, boxes, and chips are excellent reservoirs for flammable liquids and are frequently a major source of fuel for a fire. Leaking fluid from hydraulic systems has been a major source of fuel in many industrial fires. ULIFM oily waste cans should be used to contain used oil saturated rags. A hazardous spill program should be in place to initiate cleanup and disposal of spills using absorbent materials and approved, labeled 55 gallon drums for containment and disposal.

Many solvents, sprays, and finishes that are nonflammable, or less flammable than others, have been developed and should be used in preference to the more highly flammable materials. Wherever dipping or spraying operations involving flammable liquids are being done, there should be some means of

Figure 9-5. A type-II safety can for flammable liquids with separate openings for filling and emptying. A type-I can has one combination pouring and filling spout. Courtesy Justrite Manufacturing Company.

collecting excess or spilled liquids. Spray-painting booths with a waterfall collector at the rear of the booth are quite satisfactory. Electrostatic painting greatly reduces the excess paint. Such operations should always be well vented to carry away flammable fumes and mists.

When large quantities of flammable liquids must be stored, they should be stored outdoors or in a structure isolated from occupied buildings. Such a structure should have a drain suitable for handling the liquids stored there, and a sloping floor to carry spilled liquids to the drain. The structure, as well as each container in it, should be vented. It should have explosion-proof lighting if lights are needed at all. And there should be a grounding circuit to be attached to containers that are being filled or emptied.

Flammable gases such as hydrogen, acetylene, propane, and so on, should be stored only in cylinders designed specifically for each one of them. These cylinders should be strap-secured to the wall to prevent them from falling. Gases can be distributed through pipelines, but great care must be taken in doing so. Pipes, fittings, and connections must be free of any foreign matter and must not be lubricated with any material that will react with the gas involved. Since these gases will almost always be stored under pressure, extra precautions must be taken; this matter will be discussed further in Chapter 14.

Care must be taken not only to restrict the availability of fuel for a fire, but also to restrict sources of ignition. Some sources of ignition are obvious and others are not. Open flames and sparks are usually quite noticeable. Many fires are attributed to welding and cutting sparks. Burning cigarettes should be an obvious source of ignition, yet they are often ignored and are a common cause of fires. Smoking must never be permitted in areas where there are highly flammable materials.

Other sources of ignition are so common in our daily environment that they are overlooked. Any kind of heater, whether an electric heating element, a gas burner, or a hot plate, is a source of ignition. A visible spark or flame is not necessary to raise the temperature of a fuel to the flash point or ignition point. Hot chips and workpieces, hot water or steam pipes, machine parts heated by friction—all are capable of producing temperatures sufficient to start a fire. The chemical reaction involved when two-part adhesives are used has been enough to start a fire. The sun shining through a lens or even a bottle can produce high temperatures.

The sources of ignition that must be present should be shielded from possible sources of fuel. Any unnecessary sources of heat should be eliminated in any area where potential fuels exist.

It must be assumed that there is always a sufficient quantity of oxygen available in the air to support a fire. In addition to the air, oxidizers may be available from other sources, as mentioned earlier. Water can be broken down into its elements, hydrogen and oxygen. It is not likely that water could ever be dissociated to produce free atoms of both hydrogen and oxygen, but hydrogen can be extracted and combined with other elements to leave free oxygen. Care must also be taken with other materials that act as oxidizers. Fluorine is a particularly active oxidizer. Whenever fluorine, chlorine, concentrated acids, nitrates, nitrites, or peroxides are used, a responsible person in the company should consult a chemist to learn how these materials may react with other materials. Only UL-listed perchloric acid fume hoods should be used when heating perchloric acid.

College campuses, government complexes, and corporate buildings, among others, are now requiring Fire Emergency Communications Systems or Mass Notifications Systems to be integrated into the Fire Alarm system to alert and notify building occupants to safety should there be a terrorist attack or impending natural disaster in addition to a fire.

Fire Detection and Suppression

People are the most versatile detectors of fire. People can see, hear, smell, and feel the various characteristics of fire. They also have the ability to decide what to do about a fire when they detect it. Unfortunately, people are not very dependable as fire detectors. They are often not present when and where a fire starts. Sometimes, too, they fail to identify fire accurately, and when they do identify it, they often panic or react irrationally. Many sad stories are told of incidents in which people detected a fire and failed to do anything to put it out or to call for help.

Of the several types of automatic fire detectors, the ones that sense smoke are generally the best, because smoke is usually generated in detectable quantities sooner than heat. However, in some less common types of fire, usually with nonhydrocarbon fuels, very little smoke is generated. Heat sensors are also preferable where smoke is normally produced by processes in the area. The installation of fire detectors and detector systems is covered by NFPA 72. Fire alarm systems must now conform to ADA (Americans with Disabilities Act) guidelines.

There are two types of heat detectors. One type senses a fixed temperature that is considered to be indicative of a fire. The other senses a rate of rise in temperature. Smoke detectors are also available in several forms. One common type uses an ionized beam to generate a small electric current. It detects smoke when carbon particles interrupt this beam and the flow of current is reduced. Other devices use photoelectric cells, infrared or ultraviolet light, monochromatic light beams, or laser beams to detect products of combustion, visible smoke, or flames. Each of them has some advantages and disadvantages.

Carbon monoxide detectors are used to detect unsafe levels of CO and hence the fire that produces it.

In areas where there are other high-pitched sounds, detectors with flashing strobe lights may be used. Flashing strobe lights are preferable to audio signals where deaf and hard-of-hearing persons are working.

Detectors may be used to do several things. They may merely actuate an alarm. They can actuate a fire suppression system, notify a central station, call a fire department, start exhaust fans and open vents, and operate pressurized systems. Some buildings may have several detection systems to accomplish several things.

Fire-suppression systems may be manual or automatic. The manual systems consist primarily of portable fire extinguishers and hoses. Fixed systems could conceivably be operated manually, but this is rarely done. A wide variety of portable extinguishers is available, although many types formerly available, such as soda-acid and carbon-tetrachloride types, have been discontinued for a number of reasons. Some introduced hazards, such as toxicity or the possibility of explosion. Some were just too confusing to operate.

In selecting a fire-extinguishing system, one must have some knowledge of the several extinguishing agents, as well as their advantages and their limitations. Each of the four ways by which any fire can be extinguished—removing the fuel, removing the oxidizer, lowering the temperature to a point below ignition temperature by cooling (endothermic), and lowering the temperature to a point below ignition temperature by repeatedly interrupting the chemical reaction—involves breaking the fire pyramid. All extinguishing agents terminate a fire by one or more of these four means. However, the means by which an extinguishing agent causes a fire to cease burning is not as important, in most cases, as are other factors. One of the major factors is how the extinguishing agents behave with different types of fuel.

For convenience in identifying different types of fires, five classes of fire, based on fuel, have been established, as follows:

Class A   Fires involving ordinary solid fuels such as wood, paper, and so on.

Class B   Fires involving liquid and gaseous fuels.

Class C   Fires in energized electric circuits and equipment.

Class D   Fires involving combustible metals.

Class K   Fires involving combustible cooking media such as vegetable oils, animal oils, and fats.

In Europe there are five classes, liquid and gaseous fuels being considered separately. Hence Class A involves solids fuels; B, liquid fuels; C, gaseous fuels; D, electrical circuits; and E, metal fuels.

Water extinguishes a fire by cooling. It is inexpensive and is very effective on Class A fires when applied as a continuous stream or a spray. Water damages many items in a building, however, and is messy to clean up afterwards. It should not be used on anything but Class A fires. A fine spray is sometimes used to reduce water damage, and a spray can be used with caution on some fires involving electrical circuits.

Aqueous foams with additives to improve adhesive and cohesive properties act as a blanket to reduce the supply of oxygen to a fire; they also provide cooling. A foam will not penetrate into rubbish and, therefore, is not very effective for most Class A fires. It is most effective for Class B fires. Some foams contain an additive with a very high rate of expansion, thus providing a better penetrating ability. These foams are effective for Class A as well as Class B fires. No water-based material should be used on water-sensitive fuels, such as sodium, potassium, and calcium. Class A foams, developed for fire department pumpers, have proven to be very effective on combustible structure fires.

Carbon dioxide is a more versatile extinguishing agent than water. It terminates a fire by cooling, by reducing available oxygen, and by retarding the chemical reaction. No residue is left by CO_2. It is most effective on Class B and C fires, and less effective on Class A fires. A major disadvantage of carbon dioxide is that since it reduces the supply of oxygen to the fire, it also reduces the oxygen available for people to breathe. It also creates a mist that greatly reduces visibility. Therefore, people have to be evacuated before a large quantity of CO_2 is used.

Dry chemicals are also quite popular. They tend to adhere to the fuel better than the gaseous CO_2 and are, therefore, more effective outdoors or in any location where there is a draft. Some dry chemicals do not penetrate into a Class A fire very well and are thus most effective on Class B and C fires. One type, ammonium phosphate, is used on Class A, B, and C fires. Dry chemicals, like carbon dioxide, deplete the supply of oxygen and create a mist, so people must leave the area before a large quantity is used. Dry chemicals also leave some residue to be cleaned up. Dry chemicals, such as potassium bicarbonate and sodium bicarbonate, and "wet" chemical agents are effective on cooking fires; they react with grease in a process called *saponification* to extinguish the fire.

Halogenated hydrocarbons, better known by the trade name Halon, were phased out by the year 2000 due to their adverse effect on the ozone layer. These agents were replaced with *clean agents* that are friendly to the atmosphere.

Figure 9-6 provides a summary of these extinguishing agents and their effectiveness. Class D fires have been omitted from this list because they cannot be extinguished by any of the extinguishing agents listed. Several of the listed agents become hazards when applied to the extremely high temperatures produced in metal fires. Special materials have been developed for each type of metal, and these must be used with great care. Graphite and sodium chloride are used on burning magnesium; a blanket of these materials is laid on the metal to smother the fire. Another material, trimetoxyboroxine, or TMB, is also used on burning magnesium. Sand may be effective in many metal fires. To be effective, these materials must be spread quickly and completely over the burning materials.

In an attempt to make extinguishers more easily identifiable, all extinguishers made in recent years have carried a label with the class or classes of fire they will extinguish clearly marked with the letter A, B, or C, with a color code, and with a shape code. Unfortunately, most people have not learned to associate these codes with a particular type of fire. Now extinguisher manufacturers are using a picture code to identify them. These are illustrated in Figure 9-7. The background color is blue for the types of fire for which that particular extinguisher is to be used, and black for those types for which it is not to be used. Note, too, that a diagonal slash means "Do not use for these types of fire."

Extinguishant	Class of fire		
	Green **A**	Red **B**	Blue **C**
Water	Good	Very poor	Hazardous
Carbon dioxide	Poor	Good	Good
Foam	Fair	Good	Good
Dry chemical, ordinary forms	Poor	Good	Good
Dry chemical, multipurpose	Good	Good	Good
Halogenated hydrocarbons	Good	Good	Good

Figure 9-6. Summary of fire extinguishing agents and their effectiveness.

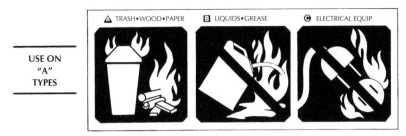

Figure 9-7. Example of the picture symbols for fire extinguishers. (Developed by the National Association of Fire Equipment Distributors and adopted, with permission of NAFED, by NFPA in 1978.)

Portable fire extinguishers are available with each of the extinguishing agents generally used, as are automatic systems. Automatic systems have a detector that senses a fire and actuates the extinguishing system. In the case of carbon dioxide and dry chemical systems, the detecting unit must sound an alarm to evacuate people before actuating the extinguishing system, usually with a 30-second interval. Automatic wet-chemical fixed systems are used to protect UL-300 kitchen hoods.

Other materials have been used effectively as extinguishing agents but are not recommended because of severe limitations. Carbon tetrachloride is very effective but is also extremely toxic and has been banned as a fire-extinguishing agent. It breaks down to produce phosgene gas.

It is very important that people know how to use certain extinguishing agents and extinguishers properly. Most extinguishers that have to be inverted for use have been discontinued because of the confusion they caused. The soda-acid type of fire extinguisher has been banned, not because of the necessity to turn it upside-down, but because the acid corrodes the container. The container is under pressure, and when corrosion has reduced the thickness enough, it explodes. Occupants should be properly trained in the use of portable fire extinguishers. Fire extinguishers should be properly inspected and maintained in accordance with NFPA 10 and 25 standards.

Design of Sprinkler Systems

Sprinklers provide the first line of defense in most buildings. They have proven to be successful in extinguishing fires in over 95 percent of all cases reported. They are usually effective even on Class B and C fires, for several reasons. One reason is that many Class B and C fires start out as Class A fires. Another reason is that a single stream of water, as from a portable extinguisher, will usually spread the fuel in a Class B fire, but a spray, as from a sprinkler, will not. A third reason involves the fact that in a Class C fire, electric current can travel a meter or two in a stream of water and slightly farther in other extinguishing agents. Since this current could conceivably cause injury to a person using a portable extinguisher, such an extinguishing method is not used on an electrical fire. However, an electric current will do no harm in a spray of water discharging from an automatic sprinkler.

NFPA Standards 13, 13D, and 13R govern the design and installation of automatic sprinkler systems. NFPA Standard 25 covers inspection and maintenance practices for water-based systems. These standards are meant to ensure dependable operation of the sprinkler system. Dependable operation requires the presence of the following components:

1. An adequate source of water with sufficient pressure to provide continuous flow,
2. A piping system independent of any other water-distribution system in the plant or building, and
3. Sprinkler heads, simply called sprinklers, maintained continuously in operable condition.

Sprinkler systems may be either wet-pipe systems or dry-pipe systems. In a wet-pipe system, the pipes contain water at all times. When heat opens a sprinkler, the water discharges immediately. In a dry-pipe system, there is no water in the distribution system until a valve is opened. This delays a flow of water, but has an advantage in that the system can be used in areas that are subject to freezing temperatures. Short lengths of a wet-pipe system can be used in subfreezing temperatures by filling that portion of the pipes with a water-soluble antifreeze.

A pre-action sprinkler system is basically a dry-pipe system in which the main valve is actuated by a separate detector system. The detectors are usually very sensitive, and they often set off an alarm first, which allows time for manual fire suppression before the water is discharged. This is very useful in areas where valuable materials could be damaged by a shower of water.

A deluge system is a dry-pipe system in which the sprinkler heads are open. Deluge valves may be actuated pneumatically, hydraulically, or electrically when a sensor detects sufficient heat. Furthermore, the detectors, which are separate from the sprinklers, control sprinklers over a wider area around the fire, so a larger flow of water results. This type of sprinkler is used where the potential for a fast-spreading fire is greater than normal. Foam is often used in this system instead of water.

Wet-pipe systems are by far the most common. The discussion of sprinkler-systems design that follows will assume a wet-pipe system, although many of the factors apply equally well to dry-pipe systems.

Sprinkler heads are available in a wide range of styles and operating temperatures. Today there are many different kinds of sprinkler heads to choose from. The correct head must be chosen for the correct application. The four basic types of heads are "upright-spray," "pendant-spray," "sidewall-spray" and "large-drop." They also vary as to their response time (RTI), discharge pattern, coverage, flow, and orifice size. Some various kinds are "Extended Coverage," "Quick Response," "Residential," "Large Orifice," "Rack Sprinklers," and "ESFR" (Early Suppression–Fast Response).

It is desirable that a sprinkler operate soon enough to prevent the fire from growing too much. At the same time, it should not respond to high temperatures that result from normal processes. Therefore, the fusible links and glass bulbs are rated and selected according to expected ambient tempera-

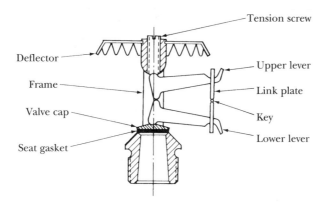

Figure 9-8. Typical sprinkler-head mechanism.

tures in a given area. Figure 9-8 shows the operating mechanism of a typical sprinkler head. Table 9-4 lists typical ranges of operating temperatures and expected room temperatures for which those operating temperatures would be appropriate. NFPA has established a standard for such operating temperature classifications; however, manufacturers use slightly different temperature ranges and classes for different styles of sprinklers.

Once the necessary temperatures have been selected, the next step in designing a sprinkler system is determining the *demand performance* characteristics required for sprinklers in each area of the building. Earlier in this chapter, fire load and heat potential were discussed. Figure 9-9 is a blank chart that can be used to estimate the fire load in any given area.

As an example, let us assume an assembly area of 100 square meters and a total equivalent mass of 8000 kilograms, giving a fire load of 80 kg/m^2. The heat of combustion of cellulosic materials is about 18.6×10^6 kJ per kg, so there are about $8000 \times 18.6 \times 10^6 = 148.8 \times 10^9$ kJ of fuel available. Ordinary combustibles burn at a rate of 50 to 60 kg/m^2 per hour. Therefore, if the fire lasts 90 minutes, the rate of burning would be $148.8 \times 10^9/90 = 1.65 \times 10^9$ kJ/minute. Cold water absorbs energy at a rate of about 2.26×10^6 kJ/kg per minute. Then $1.65 \times 10^9/2.26 \times 10^6 = 730$ kg (1608 pounds or 193 gallons) of water per minute are needed to control it.

Table 9-4. Typical Specifications for Sprinkler Heads*

Maximum Ceiling Temperature		Temperature Rating		Temperature Classification	Color Code	Glass Bulb Colors
°F	°C	°F	°C			
100	38	135 to 170	57 to 77	Ordinary	Uncolored or black	Orange or red
150	66	175 to 225	79 to 107	Intermediate	White	Yellow or green
225	107	250 to 300	121 to 149	High	Blue	Blue
300	149	325 to 375	163 to 191	Extra high	Red	Purple
375	191	400 to 475	204 to 246	Very extra high	Green	Black
475	246	500 to 575	260 to 302	Ultra high	Orange	Black
625	329	650	343	Ultra high	Orange	Black

*Temperature Ratings, Classifications, and Color Codings. NFPA 13-1996.

Sources of fuel	Fuel load in kilograms	
	Cellulosic	Petroleum/Chemical
Building		

Structural fuels
  frame
  floor
  load-bearing walls
  roof		

Building services
  electrical
  mechanical
  supplies
  energy fuels		

Nonstructural
  doors, partitions
  interior finish		

Building contents		

Furnishings
  furniture
  equipment
  decoration
  supplies		

Occupant-related
  clothing
  personal items		

Subtotals	kg	kg
Multiplying factor	× 1	× 2
"Wood equivalent" mass	kg	kg
Total equivalent mass		kg
Area		m²
Fire load		kg/m²

Figure 9-9. Chart for calculating fire load.

Sprinkler systems are classified as Light, Ordinary –1 and –2, or Extra Hazard –1 and –2. These groups vary as to the density per a given remote design area. Light Hazard design, for example, is 0.10 gallons per minute over 1500 remote square feet.

Let's calculate the demand for sprinklers and hose for a 75,000 sq. ft. grocery store (mercantile occupancy). The commodity is classified as Ordinary Hazard 2, which gives a density of 0.19 gal. per min./sq. ft. over 2000 sq. ft., using NFPA 13-2010. Maximum coverage per head is 130 sq. ft.

Demand = 0.19 × 2000 + 250 = 630 gal. per min.

The maximum area per system is 52,000 sq. ft., so the store needs two sprinkler systems.

Using a large orifice head with k = 11.4, the required end head pressure is

$Q = 11.4 \sqrt{Pressure} = 130 \times 0.19 = 24.7$ gal. per min.

Therefore, the end head pressure = 4.7 pounds per square inch (use the minimum of 7 psi).

The available supply of water is determined by conducting a water-flow test, usually from fire hydrants on the street. The sprinkler-system pipe is then sized through hydraulic calculations so that the supply exceeds the demand by a factor of safety of 5–10 psi.

Systems for large discount stores would be designed to NFPA 13 (chapters 14–20), since stock is usually up to 20 feet high and racks are used. These sometimes require three wet systems with system demands of 1700 gal. per min. for the ceiling and 2000 gal. per min. for the rack sprinklers.

Large orifice sprinklers with k-factors of 25.2 are often used in the sales area of "big box" stores, 100,000–150,000 square feet with Class I-IV commodities 12 feet high, to eliminate the need for in-rack sprinklers, requiring sprinkler and hose demands of around 1600 gpm.

Sprinklers are available with several different orifice sizes and pipehead sizes, and to make further calculations, one must refer to data provided by the manufacturer of the sprinklers to be used, or by several manufacturers if a choice is to be made. At present, sprinklers are not readily available in metric sizes. Furthermore, constants used in calculations are furnished by the manufacturer of the sprinkler heads, and these constants are given only for calculation in English units. At this point, then, only English units will be used.

There are several types of sprinkler piping arrangements. Before further calculations can be made, however, an overall piping arrangement should be designed. There are "traditional" piping arrangements, with several variations, and loop systems. The *tree* type of system is like a tree. Some other types are the *looped* and *gridded* systems. Each system has a main control valve and an alarm-check valve or dry-pipe valve or deluge valve, depending on the type of system.

Pipe-schedule design was used primarily until the early 70s in sizing the pipe diameter. Looped and gridded systems became popular because smaller pipe sizes could be used, reducing cost. Computer programs were developed to size the pipe, taking friction and elevation losses into account. Today most sprinkler systems are hydraulically calculated.

NFPA 13, Installation of Sprinkler Systems, is used in the design of most building systems. NFPA 13 Chapters 12–20 cover General Storage, Miscellaneous Storage, Class I-IV Commodities, Plastic and Rubber Commodities, Rack Storage, Rubber Tire Storage, and Roll Paper Storage.

Four basic traditional systems are illustrated in Figure 9-10. Figure 9-11 illustrates a looped piping arrangement. Traditional arrangements include a main which is usually underground or in the basement, a double-check back flow preventer, control valves, alarm-check valves, one or more vertical risers, cross mains, and branch lines. The sprinkler heads are in the branch lines. Looped systems include a main, vertical riser(s), loops, and branch lines. Again, the heads are in the branch lines. A major objective is to reach all points in the system with the shortest water travel. The farther the water travels, the more pressure is lost due to friction. Pressure is also lost to elevation. Loss (psi) = .433 × feet above the base of the sprinkler riser.

Spacing of sprinklers must be close enough to ensure that all areas not only are covered but also have a sufficient quantity of water delivered to them. This will be determined by the type of sprinkler, obstructions (both horizontal and vertical), and the pressure at each head. To establish a guide, NFPA 13 has

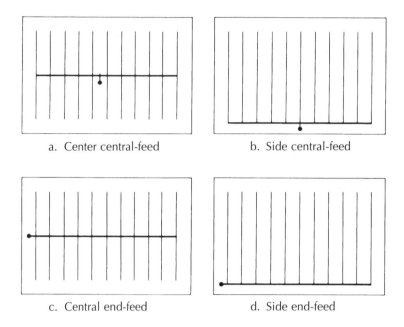

a. Center central-feed   b. Side central-feed

c. Central end-feed   d. Side end-feed

Figure 9-10. Traditional sprinkler piping arrangements.

set a maximum of 15 feet between sprinkler heads on a branch line for light- and ordinary-hazard occupancy. If solid materials below the sprinkler are stacked more than 12 feet high, the maximum spacing is 12 feet. The maximum spacing between heads for extra-hazard occupancy is 12 feet.

Placing sprinklers with too little clearance above other materials creates two problems. For one thing, the sprinkler is more susceptible to being hit. Secondly, the flow of water needs space to spread out over the area covered. As the clearance diminishes, the horizontal spread is reduced. To provide adequate protection, sprinklers would have to be placed closer together. There should be a minimum of 18 inches clearance between the top of any storage and the sprinkler head.

The maximum spacing of sprinklers for light-hazard occupancy is 225 square feet for noncombustible, smooth ceilings or beam and girder construction, 130 square feet for ordinary-hazard, and 100 square feet for extra-hazard and high-piled storage.

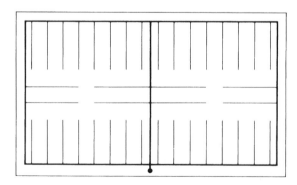

Figure 9-11. Example of a looped piping arrangement.

The pressure at any sprinkler shall in no case be less than 7 pounds per square inch (psi).

To determine the pressure needed at the sprinkler, the following formula is used.

$$P = (Q/K)^2$$

The value of K varies from 1.3 to 25.2 depending on the size of the openings in the sprinkler, where

P is the pressure in psi
Q is the water flow in gpm
K is the sprinkler constant.

Thus, if a K value of 5.6 is used, $P = (20/5.6)^2 = 12.7$ psi, which is well over the 7 psi minimum.

As water flows through piping, there are pressure losses due to friction from several sources, such as contact with the surface of the pipe and restrictions at fittings and at changes of direction. There is also a loss of pressure due to change in elevation. There are tables available, based on empirical data, giving coefficients for these losses in straight pipe, elbows, and tees, in different pipe materials, and in various sizes of pipe. Although calculation is possible, tabulated data are also used to determine pressure losses due to friction. Some of these tables are based on the velocity of flow, while others are based on the quantity of flow.

The method of calculation involves a detailed layout of the most remote section of the sprinkler system. The pressure needed at the last sprinkler is taken as the starting point. It is assumed that many sprinklers will be activated and will discharge simultaneously, although that is not likely unless the fire cannot be controlled by one or a few. The worst case should be the basis for calculations. Therefore, the quantity discharged and the resultant pressure losses are added in succession, along with other pressure losses, for the entire section to determine the pressure and quantity of flow in the cross mains, the risers, and the main. This becomes a long and tedious task even with the extensive tables of data.

Several computer programs have been developed to reduce the long calculation time. Two commonly used programs are referred to as the SprinklerCalc by Walsh, and the Haas program. These are general purpose programs and can be used for designing practically all sprinkler systems.

There are several other types of automatic fire suppression systems, including foam systems, carbon dioxide systems, dry chemical systems, and "clean agent" systems.

A major component of any automatic fire suppression system is the detector, which must not only detect a fire but also actuate the system and, in most cases, sound an alarm as well. A particular system is chosen on the basis of the advantages and limitations of the extinguishing agent, as discussed earlier. Cost of installation also becomes a factor when a system is being designed.

A predischarge alarm must be sounded before any automatic system, other than sprinklers, is activated. As previously mentioned, foam creates a fog that lowers visibility and reduces available oxygen. Carbon dioxide, dry chemical, and Halon dilute the oxygen supply when large quantities are used, as is the case in automatic systems. Sprinklers do not require this type of alarm, but a flow alarm is usually desirable to alert people that the system has been activated, which is to say that a fire has been detected.

Sprinkler systems will normally continue to discharge water until someone manually closes a valve. Each of the other types of systems has a fixed amount of the extinguishing agent and will discharge the entire amount. These can be *local application* or *total flooding* systems. The effectiveness of carbon dioxide, dry chemical, and *clean agent* systems is dependent on the concentration of the agent in the air. The recommended concentrations, however, vary greatly, depending on the volume of the space and the type of hazard for which it is designed. Automatic sprinkler protection should always be secondary or backup to other types of fire extinguishing systems, such as a clean-agent system in a computer room.

Venting of Heat and Smoke

It was pointed out earlier that smoke and toxic gases cause most deaths and injuries in a fire, while heat causes the most damage to the building and equipment. Injury to personnel, damage to the building and equipment, and even growth of the fire can all be reduced by controlling the movement of heat, smoke, and toxic gases. This control can be achieved quite effectively by venting the heat and smoke to an area outside the building. There are many factors to be considered, however, in the design of a venting system. First, one should weigh the advantages of venting against the disadvantages. The advantages include the following:

1. Release of heat reduces spread of fire.
2. Release of heat prevents activation of sprinklers outside the area of the fire, thus conserving needed water and reducing water damage.
3. Release of heat, if properly done, may reduce exposure of structural members to high temperatures.
4. Release of smoke and gases improves visibility and makes it easier for fire fighters.
5. Release of heat and gases reduces the possibility of explosion of accumulated flammable gases.

The disadvantages include the following:

1. Hot gases may become more concentrated at structural members near the roof. These may include carbon monoxide, which is combustible and may explode if ignited.
2. Venting tends to create a draft, which draws more air, and hence oxygen, toward the fire to fan it.
3. Venting of hot gases may endanger property outside the building.
4. Venting may reduce the effectiveness of high-challenge sprinkler heads.

These advantages can be maximized and the disadvantages minimized when designing the building and the layout of facilities, and by practicing good housekeeping. One of the first steps toward good venting is the separation of areas within a building. An area with a heavy fire load should be separated and kept small. The flammable materials within any area should be separated

Figure 9-12. A unit-type roof vent in operation. Courtesy the Bilco Company.

to avoid a continuous path of fuel. Areas in which a great deal of heat could develop from a fire should have a minimum of obstructions overhead to block the movement of heat and smoke. This area should also be free of utilities that would be easily damaged by heat or smoke. Appropriate roof vents should then be designed.

There are several types of roof vents. Raised portions of the roof in a variety of shapes, often called *monitor roofs*, are among the most common. They can be arranged in a sawtooth shape, or as narrow rectangular sections running the length of the building, or as many individual square or rectangular sections. Monitor roofs normally have windows. If monitor windows are designed only to let natural light into the building, as many of them are, they are usually fixed. Such windows must be broken in order to allow venting. It is far more effective if such windows, or even panels, are hinged to allow opening, either manually or, preferably, automatically.

Industrial plants today depend less on natural lighting, and monitor roofs for that purpose are becoming less common. Smaller, unit-type vents are replacing monitor roofs in popularity. Since these units are smaller, a large number of them are required, but they gain versatility from their size and from the fact that they are units. They can be placed almost anywhere and can be operated automatically quite easily. Figure 9-12 shows such a unit in operation.

Automatic vents should not open immediately when a fire starts, but only after some quantity of heat and smoke has developed. They are usually

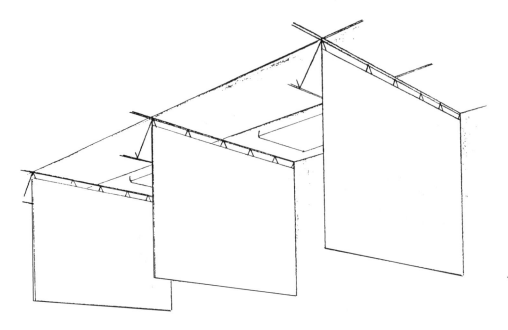

Figure 9-13. Curtain boards used with venting system.

actuated by fusible links that melt when a certain temperature, about 85°C, has been reached at roof level.

These smaller, automatic-unit vents, 16–100 sq. ft., are usually fuse-link activated or drop-out plastic types. NFPA 204, *Guide for Smoke and Heat Venting*, dictates the design criteria for curtain boards used to collect heat under the roof and vents. Manual venting is sometimes recommended when using high-challenge sprinkler heads in high-piled stock warehouses.

The depth of the curtain boards should be not less than 20 percent of the ceiling height, and the distance between them should be at least 2 times the ceiling height but not exceed 8 times the ceiling height (see Figure 9-13).

The area of the vent(s) should not exceed 2 times the smoke layer or curtain board depth squared. The required area of the vent is derived from equations in fire dynamics and plume flow (see NFPA 204).

Mechanical ventilators can also be used in venting.

Multistory buildings present problems in venting heat and smoke. Except on the top floor, venting must be done horizontally. This almost always means that hot gases and smoke will move across the ceiling in order to reach a window or other opening. Since hot gases and smoke rise, it is necessary to provide some means of moving them. It is important that they be moved before the layer becomes so thick that it extends down to the level at which people are breathing.

High-rise buildings can be equipped with pressurized stairwells with roof-top fans or fans at every other landing. These are designed in accordance with NFPA 92A, *Smoke Control Systems*, and ASHRAE standards. Some high rises will exhaust on the fire floor, creating a negative pressure, and pressurize on one or two floors below and above the fire floor.

NFPA 92B, *Smoke Management Systems in Malls, Atria, and Large Areas*, is the design guide for these areas. An engineered smoke-control or smoke-removal system that is designed to keep the smoke layers interfaced 6 feet above the highest floor level of exit access is required for atriums in accordance with NFPA 101 section 8.6.7 and NFPA 92 B.

The Georgia Dome, used for exhibits with possible heavy fire loading in addition to football games, utilizes a smoke-control system. Two fires, a 171 MBTU/hour (50 MW) and a smaller 17 MBTU/hour (5 MW) fire, were modeled using BRI-2, a Japanese Building Research Institute fire model. Twenty-six 1997 cubic meters/minute (67,000 CFM) side-wall smoke exhaust fans located 49 m (162 ft.) above the floor level, and six 566 cubic meters/minute (20,000 CFM) roof-top fans located about 70 m (230 ft.) above the field floor level were calculated to exhaust or purge a smoke layer to a level of 3.1 m (10 ft.) above the top deck seats in 15 minutes.

The most important criterion for the design of any venting system is that a life-sustaining atmosphere be maintained for as long as people are in the area. This, of course, includes fire fighters. When people must work, or at least exist, in an area that is unventilated, it may be necessary to provide suitable atmosphere by supplying oxygen or air by other means. While engaged in normal activity, an adult uses about 0.03 cubic meters of oxygen per hour and generates about 0.026 cubic meters of carbon dioxide.

Case Studies

There have been many fires in industrial plants, but one in particular makes an interesting case study. It occurred on August 12, 1953, and involved many infractions of good fire-safety practice.

General Motors Corporation had a new manufacturing plant in Livonia, Michigan. Most of the building was single-story, but there was a two-story office building attached to the manufacturing building at one end and separated from it by a concrete block wall. Except for a few small, specialized areas, the manufacturing area was undivided by partitions or fire walls. It had a heavy asphalt roof over a lightweight metal deck, and a creosoted wood-block floor that had become saturated with oil. There was a sprinkler system, but it covered only a small portion of the area. There were no roof vents and no fire detectors except for the fusible links in the sprinkler system. The area measured 264 meters (866 feet) wide by 366 meters (1200 feet) to 488 meters (1600 feet) long. Figure 9-14 shows the burned-out plant.

The fire started when a repair crew was working on an overhead pipeline. There was an overhead conveyor nearby carrying parts that had just come from a dipping operation. The excess oil from this operation was dripping from the parts into a trough or drip pan, which carried it back to the dipping tank. The flash point of this oil was reported to be 36.5°C (97.7°F). Sparks from a welding torch ignited the oil, and the fire spread along the drip pan and the conveyor quite rapidly. Enough heat soon developed to cause the drip pan to buckle and spill the burning oil onto the floor; it also buckled the roof decking and melted some of the asphalt roofing. As the fire spread across the floor, oil and hydraulic fluids from machines added to the fuel. The roof-support structure weakened and buckled, allowing the asphalt roofing to fall and add to the fuel as well. Oil condensate all over the inside of the building also added fuel.

Figure 9-14. Aerial photo of the burned-out automobile-parts-manufacturing building. From the *Detroit News*.

The fire started in an unsprinklered area, and by the time it had reached the sprinklered area, it had grown too large to be controlled by the sprinklers. The drip pan was located more than 4 meters above the floor, which was too high for effective use of portable extinguishers. When the fire department arrived, their hoses could throw a stream only about 25 meters (82 feet), which was not very effective. Since there were no roof vents, except those created when the roof buckled, the area quickly filled with smoke.

The intense heat caused practically every structural member to collapse. However, the office building, separated by only a concrete block wall, suffered very little damage. No drawings or records were lost. Fortunately, too, all employees, even in the immediate area of the original fire, were evacuated.

The tremendous heat load took 14 hours to burn. It was estimated that over 1.8 million kilograms of asphalt roofing burned, constituting a major part of the total fuel load, although many types of flammable liquids were also major contributors.

As with so many disasters, this fire brought about a review of fire-prevention standards and practices and resulted in the tightening of existing regulations. Many of the hazardous conditions that existed in this plant were common practice at that time, but would not be considered so today. In fact, several of these conditions would not be permitted today.

In November of 1990, a fire in jet-fuel storage tanks at Denver's Stapleton International Airport burned 1.6 million gallons of aircraft jet fuel and caused about $30 million in damage. In designing this storage facility, care had been taken to prevent fire and to contain a fire if one got started. The filling of the tanks and the unloading of the fuel were both done by remote controls. It is believed that a faulty pump began leaking fuel into an overflow pit and that the same pump somehow ignited the leaking fuel.

Automatic controls responded to a reduced pressure and stopped the pump that was filling one of the tanks. For some reason, that pump started up again a few minutes later. By this time, personnel in the control tower saw smoke and sent an alarm to the airport's fire department. A second alarm was sounded, and many pieces of fire equipment and fire fighters responded in less than four minutes. They had little hope of extinguishing the initial fire and concentrated on protecting the rest of the tanks and equipment. However, the heat was so intense that other pipes and support structures began to buckle and leak more fuel. Several types of fire suppressants were used. Foam was effective in controlling the fire in the pools, but not against the fuel leaking under pressure. The department exhausted their supply of suppressants and called on the city fire department for help. Some of the tanks were saved, but the fire could not be extinguished. It burned itself out after 55 hours.

In analyzing the fire, it was determined that much of the damage could have been prevented if there had been a different way of draining fuel from the tanks, especially in an emergency situation. Greater distance between tanks probably would have helped in containing the fire to a localized area.

Hindsight is always better than foresight, and much can be learned by studying each fire to determine what caused it, how it spread, and what other factors affected it. All fires have some unique features, but all share some common characteristics, and much can be learned from each one.

Large industrial plants, factories, warehouses, mills, and petroleum facilities are typically insured by HPR (Highly Protected Risk) insurers like F.M. Global, Zurich HPR, and GE-GAP, which adhere to strict underwriting principles and fire-protection features including sprinkler protection with higher densities, greater fire-hose demands, fire-wall compartmentation, sustainable construction, dual water supplies, fire pumps, security, fire detection/alarm and central station, and heat/smoke venting.

Another historical fire disaster case is the MGM Grand Hotel and Casino fire of November 21, 1980, at the Las Vegas Strip, Paradise, Nevada. This was the third worst hotel fire in modern history. The hotel, a 23-story luxury resort of over 2000 rooms, had about 5000 occupants at the time of the fire. Eighty-five people were killed, mostly due to smoke inhalation and carbon monoxide poisoning.

The fire was discovered around 7 a.m. in the restaurant. The cause was determined to be an electrical ground short in a wall soffit. The vibration of a refrigeration unit caused two wires to short together and arc. The fire spread quickly to the lobby, due to plastics in the casino, resulting in a massive fire ball that blew out the main entrance, killing seven people. Toxic fumes and smoke spread throughout the high-rise hotel via the ventilation system, stairwells, and other vertical shafts. Most of the victims died in their sleep of smoke inhalation. Hundreds were rescued from the roof by helicopters.

The casino and restaurant were not protected by automatic sprinklers, which were exempted by a county building inspector, and the owner thought it was too expensive. The entire building is now protected by automatic sprinklers.

An investigation revealed 83 building code violations, design flaws, and installation errors. This disaster resulted in stricter fire regulations, plan reviews by fire officials, inspections, and code enforcement. If automatic sprinklers had been installed, one or two sprinklers would have extinguished the fire and activated the fire alarm, with no loss of life and very little property damage.

The Sofa Super Store fire in Charleston, South Carolina, is another example where automatic sprinklers could have saved lives and property. The building was a one-story, 42,000 square-foot retail furniture store that connected to a 17,000 square-foot warehouse. It was of steel frame construction. There was no automatic sprinkler system.

The fire occurred June 18, 2007, around 7 p.m. near the side loading dock, where employees were smoking. The store was open for business at this time. Flashover occurred at 7:41 p.m. with 16 fire fighters inside. The roof collapsed shortly after flashover, sending a fireball and heavy smoke out the front of the building and shooting 30 feet into the air. Nine fire fighters were killed.

A panel was created to study the fire, resulting in a number of recommendations, including a number of safety-related issues concerning fire-fighter protection and the method of attacking and fighting the fire. The building was a total loss.

The Park Haven Nursing Home fire in Ashtabula, Ohio, on March 4, 2012, resulted in one death and six injured. Fire safety had been lacking at the home, which had been cited for 18 violations. The home is a two-story attic building with 31 rooms, built in 1950 of ordinary construction. It had a sprinkler and smoke detector system, which were factors in containing the fire. The house did not have a fire safety evacuation plan.

The fire started in a secret methamphetamine lab in one of the rooms. Meth is a drug that is a growing problem because it is intensely addictive. Cooking meth is a highly flammable and dangerous process that can result in a violent reaction and ensuing fire.

Nursing home and assisted living care residents are unable to care for themselves; thus the need for a written and practiced fire emergency evacuation plan is imperative for the safety of the residents.

Some other causes of nursing care home fires include careless smoking, cooking and grease fires, as well as fires that were purposely set. Sprinkler and fire alarm systems, automatically closing room doors, and stairwell fire doors should always be maintained in good working order.

The Providence College Dormitory Fire in Providence, Rhode Island, is yet another example of a deadly college residence fire. This was a 38-year-old, four-story dormitory for women. The fire occurred around 2–3 a.m. Tuesday morning on December 13, 1977, resulting in seven deaths. Most died in their rooms; two jumped to their deaths. The cause was hair dryers that were left on in a closet while drying clothes. The dorm was not sprinklered.

College dorm fires can be deadly. The major causes of dorm fires include alcohol, cooking, careless smoking, portable heaters, overloaded electrical outlets, unsafe extension cords, candles, incense, pranks, and purposely set fires. They occur on and off campus. The NFPA reports there were 39 deaths and nearly 400 injuries in campus housing between 2000 and 2005. The university installed smoke detectors in each room and implemented fire safety improvements for all their dorms.

On January 19, 2000, at 4:30 a.m., another deadly dorm fire occurred in Seton Hall, New Jersey, killing 3 and injuring 62 others. There was no automatic sprinkler system. Two students were charged with starting the fire. An NFPA 13R Automatic Sprinkler System would have extinguished the fire with one or two sprinklers, activated the building fire alarm, and reported the fire, saving lives.

QUESTIONS AND PROBLEMS

1. How can a company design for the extent of injuries and damage that will be caused by fire in its plant?

2. Explain the fire triangle and the fire pyramid.

3. Isopropyl alcohol was being used to clean saw blades in a sawmill while the temperature was 50°F. Without thinking, the worker lit a cigarette directly over the pan of alcohol. Did the alcohol ignite?

4. Name four major priorities for good fire-safety practices.

5. How does a cementitious coating protect a steel column from fire?

6. Name three characteristics of a fire wall, and how it is different from a fire-barrier wall.

7. Study a laboratory building, or a building with several laboratories in it, and try to determine if fire walls or smoke doors would be beneficial.

8. What is the largest size glass or plastic container that can be used to store the following materials inside a building where people work: (a) kerosene, (b) formaldehyde, (c) methyl alcohol, (d) turpentine?

9. Name several substances that are subject to spontaneous combustion.

10. Why is a smoke detector usually preferable to a heat detector as a means of detecting fire?

11. Examine the fire extinguishers in one or more buildings on campus to determine the extinguishing agent. Considering the most likely causes of fire in those areas, are the extinguishing agents good choices?

12. How would you protect a restaurant from a cooking fire?

13. What is the rate of water discharge from a sprinkler in a typical grocery store?

14. Why is a wet sprinkler system preferred over a dry-pipe sprinkler system?

15. What advantage is offered by a looped sprinkler system as compared with a traditional arrangement?

16. From your calculations in question 13 above, what pressure is needed at the sprinkler assuming a K factor of 5.6?

17. A CO_2 system is being designed for an electrical room with dry transformers. If the room measures 6.2 m × 8.0 m × 4.0 m, what is the minimum volume of CO_2 that should be supplied?

18. Name several methods of smoke control in a large public high-rise building.

19. List the factors contributing to the intense fire discussed in the first case study, and provide a way by which each of these factors could have been avoided or improved.

BIBLIOGRAPHY*

Ballast, David K., 1987. *Fire Protection in Office Buildings*. Monticello, IL: Vance Bibliographies.

Cote, Arthur and Percy Bugbee, 1988. *Principles of Fire Protection*. Quincy, MA: National Fire Protection Association.

Drysdale, D. D., 1983. *Technical Report TR83-5*, "Ignition: The Material, The Source and Subsequent Fire." Bethesda, MD: Society of Fire Protection Engineers.

———, 1998. *An Introduction to Fire Dynamics*, 2nd ed. New York: John Wiley & Sons.

Hoover, S.R., 1991. *Fire Protection for Industry*. Quincy, MA: National Fire Protection Association.

James, Derek, 1986. *Fire Protection Handbook*. Stoneham, MA: Butterworth.

Ladwig, Thomas, 1990. *Fire Prevention and Protection*. New York: Van Nostrand Reinhold.

National Fire Protection Association, 2008. *Fire Protection Handbook*, 20th Edition, Volumes 1 and 2. Quincy, MA: NFPA.

_____. *Fire Protection Guide to Hazardous Materials*.

_____. *Industrial Fire Hazards Handbook*.

_____. *National Electrical Code Handbook*.

_____. *NFPA Inspection Manual*.

_____. 2010 *NFPA 10, Portable Fire Extinguishers*.

_____. *NFPA 11, Low-Expansion Foam Systems*.

_____. *NFPA 11, Low-, Medium-, and High-Expansion Foam Systems*.

_____. *NFPA 12, Carbon Dioxide Extinguishing Systems*.

_____. *NFPA 12A, Halon 1301 Extinguishing Systems*.

_____. *NFPA 13, Sprinkler Systems*.

_____. *NFPA 17, Dry Chemical Extinguishing Systems*.

_____. *NFPA 17A, Wet Chemical Extinguishing Systems*.

_____. *NFPA 25, Inspection, Testing, and Maintenance of Water-Based Systems*.

_____. *NFPA 30, Flammable and Combustible Liquids Code*.

_____. *NFPA 72, National Fire Alarm Code*.

_____. *NFPA 80, Fire Doors and Windows*.

_____. *NFPA 92A, Smoke Control Systems*.

_____. *NFPA 92B, Smoke Management Systems*.

_____. *NFPA 204, Smoke and Heat Venting*.

_____. *NFPA 2001, Clean Agent Extinguishing Systems*.

National Technical Information Service, 1989. *NTIS PB89-853281*, "Intumescent Coatings and Paints." Springfield, VA: NTIS.

Robertson, James C., 1989. *Introduction to Fire Prevention*. 3rd ed. New York: Macmillan.

Society of Fire Protection Engineers *SFPE Handbook of Fire Protection Engineering*. 4th ed. Bethesda, MD: NFPA, 2008.

U.S. Coast Guard. *A Manual for the Safe Handling of Flammable and Combustible Liquids and Other Hazardous Materials*. Houma, LA: Marine Education Textbooks.

---

*Codes, training aids, and pamphlets on all areas of fire prevention and suppression are published by the National Fire Protection Association and by Factory Mutual System.

10

Noise and Noise Control

Richard T. Beohm, PE, CSP, ARM

Scott E. Brueck, MS, CIH

Noise that is loud (85–90 dBA) and prolonged can permanently destroy the tiny hair cells inside our ears, causing hearing loss. Today over 20 million Americans are exposed to hazardous noise on the job as well as off the job.

Studies show that, in general, the ear sustains damage after about an eight-hour exposure to noise of 80 decibels. OSHA and most U.S. standards set the permissible exposure limit (PEL) at 90 dBA of noise exposure over an eight-hour time period. Worldwide, the majority of PELs for noise are set at 85 dBA over eight hours. NIOSH estimates that 92 percent of workers should be safe from noise-induced hearing loss over their lifetime at this PEL. OSHA reports approximately 90 percent of U.S. workers have time-weighted average (TWA) noise exposure of 95 dBA or less.

Noise and ototoxic substances in the workplace can have a detrimental effect on a worker's hearing and could even cause tinnitus, or ringing in the ears, and psychological stress in older workers.

The Physics of Sound

The term "noise" is distinguished from the term "sound" by the fact that noise is unwanted. Otherwise it has all the characteristics of any other kind of sound, and the word "sound" is used in describing these characteristics. Only when sound is undesirable is it called noise.

Sound is a form of energy that radiates from a source in all directions until it is absorbed or reflected. The source is considered to be a point, at least in discussing basic characteristics. Sound energy is produced by some mechanical action at a given level of power. The term "sound power level" is used frequently in literature on this subject, and its meaning is usually restricted to the initial power at the source.

Sound has both frequency and intensity. It travels about 1,130 feet per second in air, 4,700 feet per second in water, and 16,500 feet per second in steel.

As sound energy travels in an unobstructed path, it moves radially to form spherical wave fronts. As the sphere gets larger, the pressure of the energy wave is spread over an ever-increasing imaginary surface. This pressure, measured in N/m^2, diminishes as the pressure wave moves farther from the source.

The pressures associated with sound range from $0.00002 \ N/m^2$ to about $20.0 \ N/m^2$. This range is much too wide for a linear scale if any reasonable degree of accuracy is desired. Furthermore, a small increment of change is more significant at the low end of the scale than at the high end. For these reasons, a logarithmic scale is used. Rather than an absolute scale, a relative scale is used based on the arbitrary reference pressure of $0.00002 \ N/m^2$, which is considered to be the threshold of hearing. Units on this logarithmic scale are called bels, and the more common unit is a tenth of a bel, or decibel, dB. Decibels are a measure of the sound pressure level with respect to the base or reference pressure.

The sound pressure level is inversely proportional to the square of the distance from the source. This is expressed as

$$L_1 = 10 \log (P_1^2/P_0^2)$$

where L_1 is the sound pressure level in dB resulting from a pressure P_1 in N/m^2 with respect to the reference pressure P_0 which is $0.00002 \ N/m^2$.

This is more easily handled by reducing it to

$$L_1 = 20 \log (P_1/P_0).$$

Since the sound pressure level is reduced so rapidly as distance from the source increases, a design using this fact is often an easy way to reduce exposure to noise. Then it is handy to be able to calculate the sound pressure level at a given point with respect to that at another point. This can be done by using the relationship

$$L_2 = L_1 - 20 \log (d_2/d_1)$$

where

L_2 is the sound pressure level in dB at a point d_2 meters from the source,

and

L_1 is the sound pressure level in dB at a point d_1 meters from the source, d_2 being greater than d_1.

Another characteristic of sound that is often significant in designing controls is the velocity at which it travels through a given material. The velocity of sound is dependent on the density and the modulus of elasticity of the material through which it is traveling. It is also dependent on temperature, especially in air. Table 10-1 shows the commonly accepted values for the velocity of sound traveling through various materials. The velocity of sound in air can be calculated by the expression

$$V = 20.05 \sqrt{273 + C}$$

where

V is the velocity in m/s
C is the temperature in °C.

Table 10-1. Velocity of Sound through Various Materials

Material	Velocity in m/s
Aluminum	5106
Steel	5005
Glass	4728
Wood	3050–4575
Water	1462
Lead	1228
Air at 20°C	344
Air at 0°C	332

Sound moves at some frequency that remains constant throughout the life of the energy wave. If the sound has only one frequency, it is said to be a pure tone. Usually, sounds consist of many frequencies superimposed on one another, making up what is called a complex sound. The frequency of any sound may seem to change if the hearer moves toward or away from the source of the sound, but relative to the point of its source the frequency remains constant.

The frequencies of interest in noise control are those in the audible range, which is from about 20 hertz (Hz) to about 20,000 hertz (see Figure 5-1). Studies indicate that the human body may be affected by frequencies above or below this range, but since information on this phenomenon is limited, little importance has so far been attached to those frequencies. The range from 20 to 20,000 Hz has been divided into ten parts called octave bands, or simply bands. The size of the bands increases in a geometric progression, each one approximately double the size of the one before. They are identified by frequencies approximately in the center of each band that are called "center frequencies." When this breakdown is too coarse for a particular analysis, a finer breakdown of one-third octave bands may be used. Normally a full octave band is implied, unless otherwise stated. These octave bands and one-third octave bands are shown in Table 10-2.

The Phenomenon of Hearing

The process of hearing was outlined in Chapter 5, and Figure 10-1 illustrates the main features of the auditory system. However, a few points bear repeating. The pressure wave of sound is transmitted to the nerve endings in the inner ear—the cochlea—primarily by way of the outer ear or ear canal. Some sound is transmitted through the skeletal structure of the body. Each nerve ending vibrates when hit by a pressure wave of a particular frequency. Nerve endings deteriorate as they are subjected to continued vibration, and this deterioration is hastened if the intensity of the pressure wave is great. A short exposure to a high intensity results in only a temporary impairment, but continued exposure results in permanent damage. Such impairment is usually gradual; that is, only a slight loss of hearing is evident at first, but gradually a higher and higher intensity is required to vibrate the nerve ending. Eventually, such a high intensity is required that the person is considered completely deaf at that frequency.

Table 10-2. The Standard Octave Bands and One-third Octave Bands

Frequency in Hz					
Octave			One-third Octave		
Lower Band Limit	Center	Upper Band Limit	Lower Band Limit	Center	Upper Band Limit
11	16	22	11.2	12.5	14.1
			14.1	16	17.8
			17.8	20	22.4
22	31.5	44	22.4	25	28.2
			28.2	31.5	35.5
			35.5	40	44.7
44	63	88	44.7	50	56.2
			56.2	63	70.8
			70.8	80	89.1
88	125	177	89.1	100	112
			112	125	141
			141	160	178
177	250	354	178	200	224
			224	250	282
			282	315	354
354	500	707	354	400	447
			447	500	562
			562	630	707
707	1,000	1,414	707	800	891
			891	1,000	1,122
			1,122	1,250	1,414
1,414	2,000	2,828	1,414	1,600	1,778
			1,778	2,000	2,239
			2,239	2,500	2,828
2,828	4,000	5,656	2,828	3,150	3,548
			3,548	4,000	4,467
			4,467	5,000	5,656
5,656	8,000	11,312	5,656	6,300	7,079
			7,079	8,000	8,913
			8,913	10,000	11,220
11,312	16,000	22,624	11,220	12,500	14,130
			14,130	16,000	17,780
			17,780	20,000	22,390

Since practically all noise is complex sound, it is quite rare that a person becomes deaf to only one frequency. The nerve endings that respond to the higher frequencies are more sensitive to damage, and, therefore, most people suffer hearing losses at the higher frequencies first. If a person is subjected to high intensities of sound at lower frequencies, hearing impairment will, of course, occur at those frequencies. When older workers suffer hearing impairment, it is very difficult to determine if the impairment was caused by exposure to excessive intensities of sound or by the normal, gradual deterioration called "presbycusis." The excessive intensities to which people are exposed in industrial atmospheres are commonly the same frequencies as those in other situations.

In referring to a person's normal hearing, we should not use the term "intensity" unless mention is made of specific frequencies, because people do

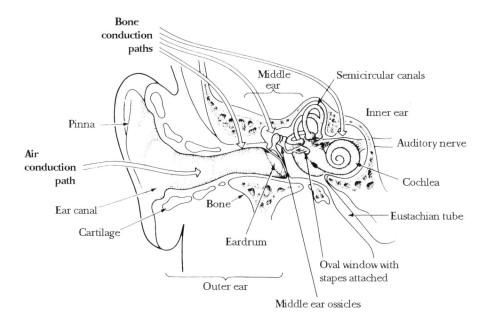

Figure 10-1. Cross-section of the ear.

not interpret intensities the same at different frequencies. Human perception of intensity is called "loudness." Several scales have been developed in an attempt to correlate loudness and intensity. Perhaps the most meaningful of these scales involves a series of equal-loudness contours. Units of these contours are called "phons." The scale of phons was made to coincide with decibels of intensity at a reference frequency of 1000 Hz. At lower frequencies, intensities must be higher in order for people to consider them equally loud. At frequencies above 6000 to 8000 Hz, intensities must also be higher, but the difference between decibels of intensity and phons of loudness is not nearly as great at the higher frequencies as it is at the lower frequencies, as illustrated in Figure 10-2.

Research has shown that as intensities increase, this difference between decibels and phons diminishes. Researchers have also found fluctuations in the scale of phons in the range of frequencies between 2000 Hz and 16,000 Hz. In this range, the perception and concept of loudness change as the person grows older. The relationship between intensity and loudness, then, is very complex.

Another problem is the aging workforce. About one-third of Americans between 65 and 75 have hearing problems, and about half the people 85 and older have hearing loss. The threshold of hearing loss rises progressively with age, and loss of hearing is greatest in the higher ranges of frequency and more pronounced in men than in women. Taking a frequency of 3000 Hz as a standard, the loss of hearing to be expected at various ages is as follows (Grandjean 1988):

50 years: 10 dB
60 years: 25 dB
70 years: 35 dB

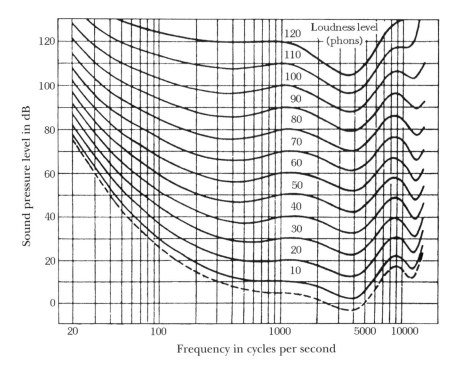

Figure 10-2. Equal-loudness contours (phons) compared with intensity in dB. From ISO Recommendation 226, "Normal Equal-Loudness Contours for Pure Tones and Normal Threshold of Hearing under Free Field Listening Conditions."

Everyday communications, accuracy of verbal instructions, and effectiveness of auditory warning systems are all subjects of concern with workers who may have noise deafness, or hearing loss due to aging.

A question raised by the discrepancy between intensity and loudness is whether noise measurements should be made in decibels or phons. It has been reasoned that since the purpose of noise control measures is to protect people's hearing, units should be used that correspond to human hearing—that is, phons. But it would be extremely difficult, if not impossible, to duplicate the scale of phons, especially considering that it changes with the hearer's age and to some extent with different degrees of loudness.

Noise Measurement

Different instruments are used to measure the intensity of sound pressure waves. Several types of detectors are used, but in each case a small cylinder called a microphone detects pressure against a diaphragm or capacitor and converts it to an electric impulse. This signal is amplified and fed to a rectifier which, in turn, feeds the root-mean-square value to some device which records and/or displays this RMS value. The RMS value must be taken over a certain length of time to be consistent. The length of time is designed by the manufacturer to be either a "fast response" or a "slow response." Fast response involves an RMS value over a period of approximately 200 milliseconds, while a slow response takes an RMS value over a longer time, approximately 500 milliseconds.

Another type of device is used for sounds that last less than 200 milliseconds. This type involves a storage oscilloscope which holds the peak amplitude and records or displays this value. Such short sounds are referred to as impact noise.

Sound-level meters, as they are called, are divided into the following three categories depending on their accuracy:

- Type 1 for precision measurement
- Type 2 for general purpose use
- Type 3 for survey measurements only

A sound-level meter must meet the specifications of ANSI S1.4 (1983). The user must periodically compare the sound-level meter being used with a calibrating device to assure that the meter is accurate. A calibrater is a device that emits a fixed level of sound pressure at each of the several frequencies. The sound-level meter can be adjusted if necessary.

Sound-level meters can be used in several ways and are calibrated with two different scales, referred to as the "A Scale" and the "Flat Scale." Early models also had B and C scales. The B Scale was intended to approximate the ear's response at mid-level pressures (from 55 to 85 dB); and the C Scale was intended to approximate the ear's response above that range. The A Scale is the closest to the human ear's response over the whole range. The B and C Scales are no longer used.

The Flat Scale measures intensity, not loudness, in dB. To measure loudness reasonably well, the A Scale is corrected, or modified, at each of the frequencies as shown in Table 10-3. Units of the A scale are expressed as dBA.

When a sound-level meter is used, several factors must be taken into consideration. It was mentioned earlier that sound radiates uniformly in all directions until it is absorbed or reflected. In order for sound pressure waves to be detected, it is necessary to place an obstruction, the microphone, in their path. Some of the sound is absorbed; that is what is measured. Some of it is reflected and, therefore, not measured. The microphone should be as small as possible, and it should be held in the proper orientation with respect to the line of travel of the sound from the source. The manufacturer of a sound-level meter provides instructions indicating the proper angles at which a particular microphone should be held and the correction values for other angles.

Table 10-3. Modifications Needed to Achieve the "A Scale" on a Sound-Level Meter

Frequency in Hz	A Scale Modifications in dB
16,000	−6.6
8,000	−1.1
4,000	+1.0
2,000	+1.2
1,000	0
500	−3.2
250	−8.6
125	−16.1
63	−26.2
31.5	−39.4

Although it might seem that sound should ideally be measured in a free field—that is, one in which no objects exist to absorb and reflect it—such a free field is really desirable only when a particular source of noise is being analyzed. It is also a very difficult condition to achieve because walls, equipment, and people always get in the way. In any event, people are subjected to sounds from many sources and from many directions. It is often more meaningful to measure the total noise or sound pressure level at various locations. There may not be any single direction in which it is best to orient the microphone. Trial and error may be the best way to get the measurement with the least loss. The most useful locations will be discussed later.

Very often it is helpful to measure sound intensities at one or several frequencies. This provides information often needed in identifying the source or sources of a complex noise. An instrument called an octave band analyzer can be used for this purpose. With this instrument, each of the octave center frequencies listed in Table 10-2 can be selected and the intensity at this frequency will be displayed or recorded. Some octave band analyzers also provide for the selection of one-third octave band center frequencies, and some models of sound-level meters also incorporate an octave band analyzer.

Another type of instrument is the impact sound-level meter. It has been argued that a measurement of the very peak impact intensity may not be as useful as one of a level somewhat below that peak. Therefore, some manufacturers have included an "impulse" intensity reading along with, or instead of, the peak or impact intensity reading. Some models available today include many or all of the features described above. Figure 10-3a shows an instrument with all of these features except the one-third octave band center frequency selector. Calibrators are also available, as shown in Figure 10-3b.

Sometimes it is inconvenient or even impossible to make a satisfactory analysis of noise in the space or under the conditions available. For this reason, tape recorders with exceptionally high fidelity have been developed. Recordings of noise can be made in several locations and then replayed in a more convenient atmosphere. Sound-level measurements can be made from such recordings with nearly the same accuracy as could be made at the actual location.

Sound-level measurements must, for several reasons, be made during a relatively short span of time. A worker, however, is subjected to the noise throughout the work period. Samples may be taken during this work period, but even samples do not provide a very good measurement of a worker's total exposure. This is especially true if the worker moves about quite a bit or if the noise is intermittent. As mentioned in Chapter 5, dosimeters have been developed that are worn by the worker as close to the ear as possible during the entire work period. The dosimeter is basically a sound-level meter with an integrating device that produces an overall sound-level exposure for any period of time.

Noise Exposure Standards

The noise exposure standards first adopted by OSHA were essentially obsolete by the time they were adopted in 1971. The American Conference of Governmental Industrial Hygienists (ACGIH) had already established more stringent standards. Many changes have been made in the OSHA standards, however, and research continues in this field.

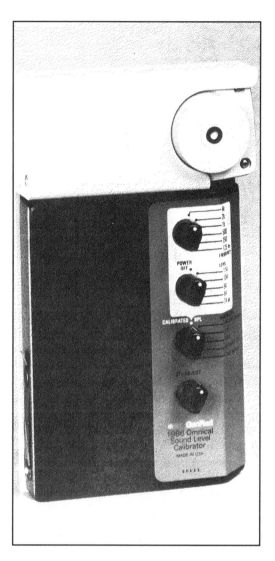

Figure 10-3a. Sound-level meter incorporating a frequency band analyzer. Courtesy GenRad, Inc.

Figure 10-3b. A sound-level calibrator for calibrating a sound-level meter. Courtesy GenRad, Inc.

Table 10-4 shows the basic permissible noise exposure limits, which are the same as earlier OSHA limits. However, there are many factors that must be considered in using these data, and there are other more detailed sets of data.

These exposures are time weighted averages (TWA), that is, the average of all noise exposures during the whole day. OSHA 25 CFR 1910.95 establishes the maximum permissible exposure levels (PELs) for harmful noise. If there is evidence that a worker may be exposed to a TWA of 85 dBA or more, a hearing conservation program must be administered and exposure to noise must be monitored. All exposures from 80 dBA to 130 dBA must be included. A more complete set of data, shown in Table 10-5, is used for these

Table 10-4. Permissible Noise Exposures (Table G-16 in the OSHA Standards)

Duration per day, in hours	Sound level in dBA, slow response
8	90
6	92
4	95
3	97
2	100
1$^1/_2$	102
1	105
$^1/_2$	110
$^1/_4$ or less	115

Table 10-5. Reference Data for Calculating Permissible Exposures to Noise (Table G-16a in OSHA Standards)

TWA sound level L, in dB	Reference duration in hours	TWA sound level L, in dB	Reference duration in hours
80	32	106	0.87
81	27.9	107	0.76
82	24.3	108	0.66
83	21.1	109	0.57
84	16.4	110	0.5
85	16	111	0.44
86	13.9	112	0.38
87	12.1	113	0.33
88	10.6	114	0.29
89	9.2	115	0.25
90	8	116	0.22
91	7.0	117	0.19
92	6.1	118	0.16
93	5.3	119	0.14
94	4.6	120	0.125
95	4	121	0.11
96	3.5	122	0.095
97	3.0	123	0.082
98	2.6	124	0.072
99	2.3	125	0.063
100	2	126	0.054
101	1.7	127	0.047
102	1.5	128	0.041
103	1.3	129	0.036
104	1.1	130	0.031
105	1		

calculations. In this table of data, T is the permissible time in hours a worker may be exposed to a given sound level, L. The value of T is determined by the formula

$$T = 8/2^{(L-90)/5}$$

For example, the permissible time exposure to 95 dBA is calculated to be

$$T = 8/2^{(95-90)/5} = 8/2^{(5)/5} = 4 \text{ hours}$$

In many cases, a worker is exposed to several different levels of noise for varying periods of time during the day. The overall exposure is referred to as the "dose" the worker receives and is expressed as a percent. This is calculated by the expression

$$D = 100(C_1/T_1 + C_2/T_2 + \text{ --- } C_n/T_n)$$

where C is the time in hours the worker is exposed to a given sound level, and T is the permissible exposure time to that same sound level.

For example, if a worker is exposed to a sound level of 85 dBA for 3 hours, to 92 dBA for 3 hours, and to 95 dBA for 2 hours, the equivalent dose is $D = 100(3/16 + 3/6.1 + 2/4) = 100(.187 + .491 + .500) = 100(1.178) = 117.8\%$. Any dose up to 100 percent is acceptable. Note that, even though each exposure was within its time limit, the sum of them is not an acceptable exposure. Even though 85 dBA is below the 90 dBA allowed for an 8-hour day, it must be included in this calculation. Exposures below 80 dBA or above 130 dBA are not included. Permissible time limits for them are not even listed.

OSHA standards make no mention of noise above 130 dBA or of impact noise. Table 10-5 indicates that a noise level of 130 dBA is permissible for up to .031 hours, or 1.86 minutes a day. It is not likely that noise of 130 dBA will be encountered for any appreciable time. Nevertheless, this problem needs to be addressed.

Impact noise is a high-intensity noise of very short duration, such as that produced by a hammer blow or a small explosion. The ACGIH long ago recommended exposure limits for these high-intensity noises in terms of the number of impacts that are permissible per day, as follows:

Sound level, dB	Number of impacts per day
120	10,000
130	1,000
140	100

Exposure to impulsive or impact noise should not exceed 140 dB peak sound-pressure level.

The safety culture should include a hearing program that addresses hearing baseline testing, the use of PPE such as earmuffs and earplugs, audits to reduce noise through engineering methods, compliance, and rules regarding the wearing of hearing PPE. This will not only protect the worker, but increase workplace safety and reduce workers' compensation losses.

Of major interest, though, is the OSHA requirement of hearing conservation programs and the monitoring of noise within such a program. The requirements are quite complex. Basically, any worker who is in danger of suffering loss of hearing because of exposure to high-intensity noise must be given audiometric tests, and the noise exposure of that worker must be monitored. This is to be accomplished by a program administered by the employer. The program must identify any worker who is subjected to an 8-hour time weighted average of 85 dB or more, monitor those exposures, and conduct periodic audiometric testing. In monitoring exposures, all continuous, intermittent, and impulsive sound levels from 80 dB to 130 dB shall be integrated into the noise measurements.

OSHA standards make it clear that engineering controls must be used to reduce noise levels if at all feasible. Then, as a last resort, personal protective equipment must be used to protect a worker's hearing. Employees exposed to an 8-hour time weighted average (TWA) of 85 dBA noise dosage must be fitted with proper hearing protectors. These protectors have a Noise Reduction Rating of about 20–29 dBA. Some approximate time weighted averages are: a conversational voice, 60–65 dBA; a shout, 90 dBA; a noisy office, 80 dBA; a factory, 90 dBA; a punch press, 110 dBA.

EN 458 (2004) Hearing Protectors has recommendations for selection, use, care, and maintenance of hearing protectors.

If the TWA is greater than 85 dBA, a Hearing Conservation Program is required. If the TWA is greater than 90 dBA, the worker is required to wear hearing protection to reduce the sound level below 90 dBA.

An example of using an earmuff with an NRR of 25 dBA and earplugs with an NRR of 29 (NRR highest) is:

TWA(A-weighted) − [NRR(highest) − 7] × .5 = the noise exposure level
Example: TWA(A) = 90 dBA, the OSHA 8-hour/day standard exposure level
7 is a conversion factor for A-weighted sound levels from C-weighted levels

Assume a PPE hearing protection of 29 dBA
The noise exposure level with PPE = 90 [(29 − 7) ×.5]
$$= 79 \text{ dBA}$$

Noise Surveys

Before an attempt is made to control noise, an analysis should be made of the locations of objectionable noise, noise levels in particular areas, especially where people are working, and the sources of this noise. This information can be obtained from a survey, which can be made in any of several different ways.

One of the simplest ways to survey a work area is to have two people with normal hearing ability move about areas where workers are likely to be and converse with each other. A very rough guide to the noise level is provided by having the two people stand 1 meter apart, measured from the mouth of the speaker to the ear of the hearer. If speech at a normal speaking level can just be heard and understood, the background noise is about 60 dBA. If the speaker has to shout to be heard at 1 meter, the background noise is about 80 dBA. If the speaker must move to about 0.5 meters while shouting, the background noise level is about 90 dBA. A more complete guide is shown in Figure 10-4. This method of making a survey should be used only to locate problem areas, never to establish noise levels.

If this cursory survey has revealed problem areas in which it seems obvious that the excessive noise has only one source, an analysis can be made

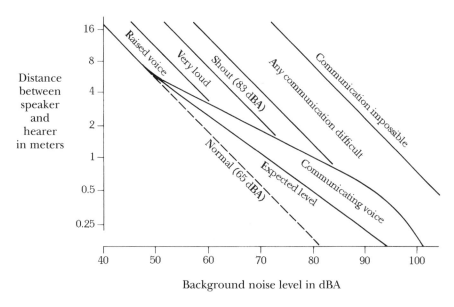

Figure 10-4. Guide for noise survey by voice communication.

quite easily with a sound-level meter and a band analyzer. The microphone should be held at each point where a worker's ear is likely to be for an appreciable period of time. It should also be held at points around the source to locate the path that the sound of greatest intensity is traveling. This should help to zero in on the point source of the noise. The band analyzer should now be used to determine the frequencies at which the greatest intensities are being produced. Special attention should be given to harmonics of the frequencies that correspond to the rpm of shafts, gears, and couplings. All of these readings and their locations should be carefully recorded, preferably on a layout drawing.

A more complete survey is often needed when there are several sources of noise. This involves the same steps as described above plus additional measurements to determine sound-level contours, such as those shown in Figure 10-5. It is very helpful to plot several contours, one for each of several conditions, with different combinations of sources.

An octave band analysis should be made for each of the combinations for which sound-level contours are plotted. Figure 10-6 shows one band analysis made to accompany the sound-level contours shown in Figure 10-5. The several sources of noise can be quite adequately identified and evaluated with the contours and band analyses. A further study of the rotational speeds of various machine components will often help to identify causes of excessive noise. If rotational speeds and/or their harmonics do not correspond with frequencies at which high sound levels occur, other components must be investigated.

In the process of making a noise survey, there are often situations in which noise comes from two or more sources. It may be necessary, then, to add two or more sound levels mathematically or to convert intensities in dB to loudness in dBA. Several techniques have been developed to do these operations. However, there are often situations in which it is necessary to add intensity levels at two or more frequencies. They cannot be added mathematically,

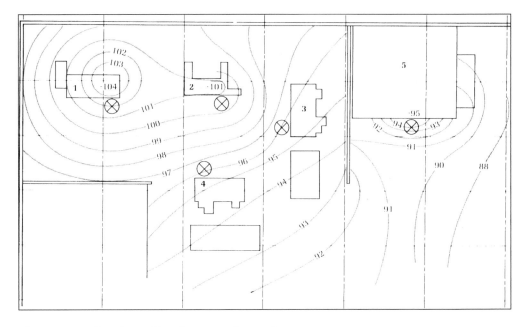

Figure 10-5. Plot of sound-level contours from noise survey.

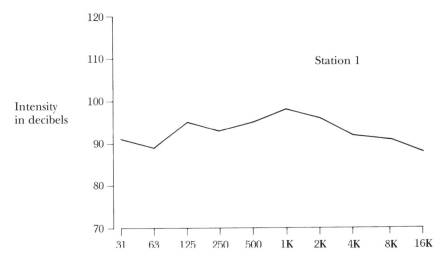

Figure 10-6. Octave band analysis made at operator's position in noise survey.

because they are logarithmic relationships. First, an octave band analysis must be made to determine intensities at each of the octave band centers. The sum of any two intensities can be calculated using the equation

$$I = 10 \times \log^{-1} dB_1/10 + \log^{-1} dB_2/10$$

where I is the sum of the two intensities and dB_1 and dB_2 are the intensities at two frequencies.

This calculation is cumbersome. Several tables and charts have been developed to provide the results when the intensities at the two frequencies differ by amounts ranging from zero up to as much as 20 dB. Figure 10-7 is

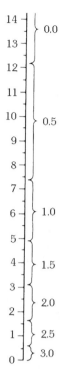

14 ┤
  │   0.0
13 ┤
  │
12 ┤
  │
11 ┤
  │
10 ┤
  │   0.5
 9 ┤
  │
 8 ┤
  │
If the difference between   7 ┤   Add this amount to the
two sound pressure levels   │   larger of the two to get
in decibels is: 6 ┤   1.0   total sound pressure level
                                  │   in decibels.
 5 ┤
  │
 4 ┤   1.5
  │
 3 ┤
  │   2.0
 2 ┤
  │   2.5
 1 ┤
  │   3.0
 0 ┤

Figure 10-7. Chart for adding two sound intensities.

one such chart. This chart uses numbers that have been rounded off to the nearest 0.5 dB and, thus, some error may occur by using the chart. However, that error will be very small.

An examination of Figure 10-7 leads one to conclude that two sources of sound together can never produce a sound intensity greater than 3 dB above the higher of the two sources alone. If the higher of the two intensities is taken as a starting point, lower intensities will add less and less as the difference becomes greater. As an example, let us suppose that two machines side by side produce sound pressure levels of 93 dBA and 86 dBA respectively, measured at a point in front of the two machines. The difference between the two is 7 dBA. From Figure 10-7 we find that if the difference is 7, we should add 1.0 to the higher one to get the total, so the sound pressure level resulting from the two machines will be 94 dBA. Three or more sound pressure levels can be added by adding any two, using the resulting sum as a single sound pressure level, and adding another one to it. This process can be done repeatedly.

Evaluation of Acoustical Materials

As mentioned at the beginning of this chapter, sound pressure waves travel radially from a source until the energy is either absorbed or reflected. As sound pressure waves travel through any material, even air, energy is absorbed by the molecules of that material. So even in air the sound pressure level is reduced with distance. When sound pressure waves hit a surface, a portion of the energy is also reflected.

If a room were very large and there were no objects in it, the sound pressure waves would eventually be completely dissipated and no sound would be reflected. As previously mentioned, this is called a free field, and it never happens in the real world. The other extreme is a very small room with hard, smooth surfaces in which sound pressure waves are reflected time and time again. This is called a reverberant field or a diffuse field. The energy is diffused as it is reflected around the room. Each point at which the pressure wave is reflected can be thought of as a new source.

Neither of these extremes is found in actual conditions. Objects always exist to absorb and reflect pressure waves. This condition is sometimes referred to as a semi-reverberant field. The pressure waves behave much as they would in a free field while they are close to the source, but more as they would in a reverberant field as they get farther away.

Within a few meters of the source, the sound pressure level decreases about 6 dB for every doubling of distance. A level of 93 dB at a distance of 1 meter drops to about 87 dB at a distance of 2 meters. As distance increases, reflected noise has a greater effect and the reduction in dB cannot be estimated on the basis of distance alone.

In noise control, materials are divided into two major categories: absorbing materials and reflecting materials. All materials have some ability to reflect, to absorb, and to transmit sound energy. The two categories, therefore, indicate relative absorbing and reflecting ability. Each of these qualities is useful in certain applications, as will be seen later.

Sound energy can be absorbed in three ways. The most common way is by conversion to heat by creating internal friction in a material. Two types of material are used for this purpose. One contains fibrous materials in which the fibers are arranged randomly, touching each other but not packed tightly. Glass fiber is a good example of this type. The other type is a viscous material that allows vibration among the molecules. In each case, the movement of the fibers or molecules converts the sound energy to heat.

Sound energy can also be absorbed by conversion to mechanical energy, or motion, by having it move or vibrate a diaphragm. This requires the design of a relatively thin member with the proper mass and stiffness to absorb the energy incident upon it. The conditions in which it is effective are limited and it is, therefore, not a common method.

A third method of absorbing sound energy involves the use of a cavity in which the energy is reflected back and forth, some being absorbed at each incident, until it is completely dissipated. There has to be a channel guiding the sound energy into the cavity, and this channel must be small enough to prevent a major portion from being reflected out again. The most common application of this principle, called a resonant cavity, is a muffler. Another less common example is a hollow concrete block designed for this purpose. Resonant cavities must be designed for a relatively narrow range of frequencies. The area of the opening is usually about 5 percent or less of the cross-sectional area of the cavity. This is a very effective method of absorbing sound energy when frequencies are known and fixed. Piping systems, including engine exhaust systems, are common applications. These cavities are sometimes referred to as Helmholtz resonators.

The ability of a material to absorb sound energy is often expressed by its absorption coefficient. The absorption coefficient is defined as the ratio of the total energy incident on the surface, minus the energy reflected from the surface, to the energy incident upon the surface, and is designated by the Greek letter alpha.

$$\alpha = \frac{E_i - E_r}{E_i}$$

A perfect reflecting material, or surface, has a coefficient of 0, and a perfect absorbing material has a coefficient of 1. Due to the manner in which this coefficient is measured, it is possible, but not likely, to arrive at a value slightly over 1.

The absorption coefficient is dependent on frequency to a great extent, and coefficients are, therefore, given for each of six frequency band centers: 125, 250, 500, 1000, 2000, 4000 Hz. Absorbing materials typically have better absorbing capability for the middle and upper frequencies within this range. Table 10-6 gives the absorption coefficients for several common materials. It is important to remember that the coefficients given here and by manufacturers of other materials are taken from specific tests under controlled conditions. In reality, these controlled conditions do not exist and variations in reverberant noise, method of mounting, and so on, may change the actual performance of these materials considerably. Nevertheless, these coefficients will help in selecting material for a given situation.

Once the noise survey is finished, a band analysis should be made. This information can then be used to select the material that has the highest coefficient at the frequencies at which noise levels need to be reduced. Sometimes it becomes a tedious task and relatively little is gained from much effort. A single index may be more useful than the six separate coefficients. For this reason, an index called the noise reduction coefficient (NRC) may be used.

Table 10-6. Absorption Coefficients for Common Construction Materials

Material	Absorption Coefficients at Frequencies (Hz)					
	125	250	500	1000	2000	4000
Celotex 5/8″ thick	0.46	0.38	0.55	0.80	0.82	0.78
Concrete block	0.36	0.44	0.31	0.29	0.39	0.25
Fiberglass,						
1″ thick	0.12	0.28	0.73	0.89	0.92	0.93
2″ thick	0.24	0.77	0.99	0.99	0.99	0.99
4″ thick	0.73	0.99	0.99	0.99	0.99	0.97
Glass (window)	0.35	0.25	0.18	0.12	0.07	0.04
Hairfelt,						
1″ thick	0.06	0.31	0.80	0.88	0.87	0.87
Plywood,						
3/8″ thick	0.28	0.22	0.17	0.09	0.10	0.11
Polyurethane foam,						
1″ thick	0.14	0.30	0.63	0.91	0.98	0.91
2″ thick	0.35	0.51	0.82	0.98	0.97	0.95

The NRC is the average of the four absorption coefficients at 250, 500, 1000, and 2000 Hz. Referring to Table 10-6, we can calculate the NRC of each of the materials shown. The NRC of 1-inch-thick fiberglass, for example, is

$$\text{NRC} = \frac{0.28 + 0.73 + 0.89 + 0.92}{4} = 0.705$$

Some manufacturers prefer to provide NRC values instead of absorption coefficients. Seldom are both values given. Like absorption coefficients, effective NRC values depend a great deal on various methods of mounting, angles of incidence of the sound pressure wave, and reverberant noise. Referring again to Table 10-6, one should note that some materials are better sound absorbers at lower frequencies while others are better at higher frequencies; the NRC does not distinguish between these conditions.

There are two types of material that are sometimes referred to as barriers and damping materials rather than as absorbing materials. The major property of a good barrier is mass, or, more specifically, mass per unit area. Limpness, or lack of stiffness, is also required. Lead is one of the best materials for barriers and is commonly used in the form of thin sheets or foil, often combined with foam material. Many other materials make good barriers as well.

Barrier materials are usually evaluated by their ability to absorb and reflect sound pressure waves. The term *transmission loss* (TL) refers to the sound power loss from the side of incidence of the sound pressure wave to the opposite side, as determined in a specified test. Mathematically it is expressed as

$$\text{TL} = 10 \log (1/r)$$

where

r is the ratio of transmitted power to incident power.

The ratio, r, is often called the transmission coefficient and is the value determined in the test mentioned above. Occasionally this transmission coefficient is specified for a given material rather than the TL. The transmission coefficient is often designated by the lowercase Greek letter tau, τ.

Transmission loss can also be calculated by the equation

$$\text{TL} = 20 \log f + 20 \log W - 33$$

where

f is a particular frequency
W is the mass, in pounds, per square foot surface area, called surface density.

A major advantage in the use of transmission loss as compared with the absorption coefficient is that the TL is expressed in dB, which frequently has more meaning. The TL values, like absorption coefficients, are given for each of several frequencies. It is often useful to give a single value; the American Society for Testing Materials has developed a method of doing this (Standard ASTM E413). A series of contours was made by plotting typical

TL values in dB against one-third octave band frequency centers, as shown in Figure 10-8.

Each contour represents a sound transmission class (STC), which is identified by the value in dB of that contour at 500 Hz. Tests of materials are fitted to these contours and thus given an STC number, which is in dB. Most manufacturers of barrier materials will give them either the TL values or the STC.

Damping materials are viscous and absorb sound energy by converting the sound energy to heat via internal friction. This principle was mentioned earlier, but damping materials, as referred to here, are thin layers applied as sheets to a surface or sprayed onto it. Applications of damping materials will be mentioned again later.

Noise Control at the Source

As was mentioned in Chapter 3, there are four types of corrective measures that can be taken once a hazard has been identified. Engineering controls include eliminating or isolating the hazard, and management controls include training and supervising workers and providing personal protective equipment. The first approach to controlling the hazard of excessive noise should be to try to eliminate it. This does not necessarily mean that the sound must be eliminated, but rather the excessive part of it, so that the overall sound is reduced to an acceptable level.

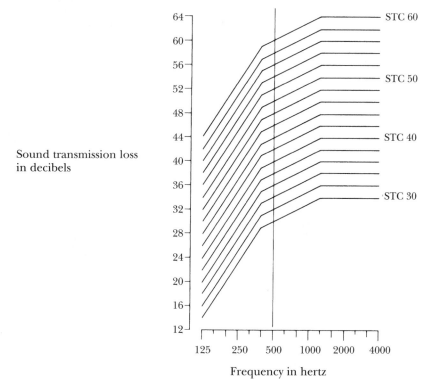

Figure 10-8. Standard contours for determining the sound transmission class (STC) of materials.
Adapted from ASTM E413.

When one is attempting to reduce noise, there are three potential areas of treatment. The first and most preferable is the source of the noise. The second is the path of transmission between the source and the hearer. The third is the receiver—that is, each person who is subjected to the excessive noise.

To reduce the noise at the source, it is necessary to identify specifically the cause of the noise. If a band analysis has been made, the frequencies at which high intensities were recorded should provide a clue to the sources. If equipment is involved, rotating shafts, gears, and bearings should be prime suspects. The noise may have one or more of several sources.

Since noise is a series of pressure waves of a particular frequency, any vibration will produce sound, provided the frequency involved is within the audible range. Therefore, any vibration of rotating members should be examined. This vibration can be caused by looseness between a shaft and a bearing that allows the shaft to move radially during each rotation. A worn bearing or bearing surface on the shaft or a lack of lubrication may be the cause. Proper lubrication is often a simple way to reduce noise.

If a rotating member has teeth, vanes, or any other uniformly spaced elements, these interruptions in rotation essentially constitute vibration. The frequency of this vibration will be the rotational speed of the member in revolutions per second times the number of interruptions, or $f = \text{rps} \times \text{N}$. For example, a spur gear with 24 teeth rotating at a speed of 1200 rpm will produce pressure waves with a frequency of $f = 1200/60 \times 24 = 480$ hertz.

A shaft or other rotating component that is bent, out-of-round, or otherwise unbalanced, will cause vibration with a frequency equal to the speed of rotation or multiples of that speed. If this member moves back and forth from a central position, the resulting frequency will normally be twice the speed of rotation. Other forms of unbalance may produce other multiples of the rotational speed.

Methods of reducing noise from rotating members include the following:

1. Proper lubrication
2. Replacement or repair of worn or damaged parts
3. Changing rotational speed to move the frequency to a less bothersome range
4. Changing the spacing of interruptions to spoil the repetition of vibrations

The intensity of noise generated by powered shafts is related to the power being transmitted by that vibrating member. The relationship is not directly proportional but can be estimated by an empirical relationship:

$$L_2 = L_1 + 17 \log (W_2/W_1)$$

where

L_2 is the increased intensity in dB

L_1 is the initial intensity in dB

W_2 is the increased power in watts or horsepower

W_1 is the initial power in same units as W_2.

Intensity is also related to the speed of a rotating member if the cause of the vibration is unbalance. Obviously, the amplitude of the unbalance is a factor, but intensity of noise is proportional to the square of the speed. Thus, the higher the speed, the greater the influence of speed on the intensity of the noise generated.

Gears, pulleys, and other members which rotate in essentially one plane can be balanced statically by adding or subtracting mass at the appropriate places. A shaft or other rotating member supported in such a way that the distance between supports is greater than the largest radius on the member should be dynamically balanced; that is, it must be balanced lengthwise as well as radially. Care should be taken to assure that masses added or removed do not complement other interruptions and add to the noise being generated.

Components that consist of flat or nearly flat parts often deflect and vibrate at either the frequency of the driving force or at the natural frequency of the part. Machine members are almost always restrained so that they do not respond as free-vibrating masses. Vibrating masses are modelled mathematically as moving masses attached to a nonmoving support by a spring and a damping device. There may be friction with the supporting member or some other member, as well as internal friction within the mass. Machine components can rarely be modelled as free-vibrating masses. Thus, the mathematical analysis of such systems becomes complex, and, although it's possible, it will not be attempted here.

Reducing noise generated from such a vibrating system may involve one or more of the following steps.

1. Remove the driving force by structurally separating the flat member from the member forcing the vibration.
2. Change the speed of the driving member to avoid the natural frequency of the flat member. The ratio of the driving frequency to the natural frequency should be greater than 1.4 to 1.
3. Change the natural frequency of the flat member by
   a. increasing or reducing its mass, or
   b. adding stiffening ribs or webs.
4. Add a damping member to the flat member.

Commonly, the frames and housing of machines and their subassemblies are made of steel plate, aluminum plate, or castings. Castings have been widely used for machine frames. The internal friction in cast iron provides excellent damping of vibration, particularly at low frequencies. Castings are heavy and occupy a great deal of space, so many machine manufacturers are now using welded steel plate. The addition of diagonal ribs, welded to steel plate or cast integrally into a casting, not only changes the natural frequency but adds strength as well. This is the most common and most effective method of reducing noise from flat panels.

In lighter equipment, sheet metal is often used for enclosures, chassis, and covers. The vibration of these members can be reduced by using a thicker metal or by creasing large flat areas to increase stiffness.

In some cases, a thin layer of an elastomer material attached to a flat surface is sufficient to damp vibrations. Damping materials may either be applied as a sheet, cut to the proper size and attached with an adhesive, or sprayed onto

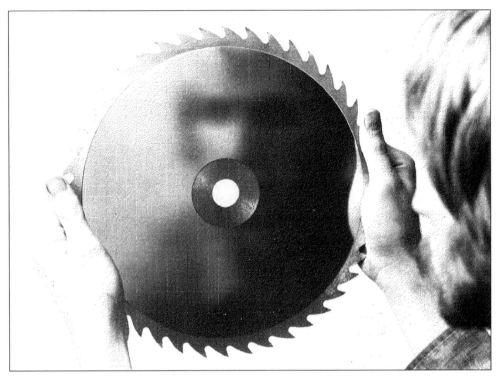

Figure 10-9. Saw blade with viscous damping material applied. Courtesy 3M Structural Products Department.

the surface as a liquid. Liquid spray is easier to apply in areas that are not flat or that have odd contours. Figure 10-9 shows a circular saw blade with a thin sheet of damping material cemented to it. It was reported that the application of this material on a 16-inch steel cut-off saw reduced the noise level from 108.6 dBA to 95.5 dBA when 1-inch blocks of hardwood were being cut.

Other machine parts generate noise when one part strikes another. The magnitude of the noise depends on several factors, including the material in the two parts, the distance between them, and the force pushing one against the other. This condition often results from wear of parts that should be contiguous or from inaccuracies in manufacturing. Lubrication will help to reduce the space between parts and will also act as an energy absorber.

This problem is particularly bothersome when one part is steel and the other is cast iron. Quite often one part can be replaced by a part of a different material, such as brass, aluminum, or even plastic. In many applications, steel or cast-iron gears can be replaced by nylon gears.

Parts that hit one another repeatedly, either by rolling action or by reciprocating action, can sometimes be designed so that the parts hit one another at an angle. This often helps to reduce noise. A good example is the use of spiral or helical gears rather than spur gears. Wood and metal cutting tools are often designed in this fashion.

Moving fluids generate noise in several kinds of applications. Two phenomena are involved: the internal friction of the fluid when turbulence occurs, and the formation and collapse of bubbles within the fluid, called cavitation, caused by fluctuations in pressure.

Pressure and velocity are closely related in moving fluids; one estimate of noise generated by moving fluids is expressed by the relationship

$$I_2 - I_1 = 60 \log (V_2/V_1)$$

where

V_1 and V_2 are respective velocities.

This indicates that doubling the velocity produces an increase in noise intensity of about 18 dB.

Fluids flowing in pipes can generate a great deal of noise when the piping system changes direction. The noise is generated by impact against the pipe at the corner and by turbulence within the fluid. Large bubbles cause hammering as a liquid passes by corners, allowing the induced-pressure-wave front of the liquid to strike the wall of the pipe.

The most effective method of obtaining minimal noise in fluid flow is to design the system with the fewest possible changes in pressure, velocity, and direction, and to keep the flow continuous and uniform. However, achieving these characteristics is often so costly that it is not feasible.

Noise Control in the Path of Transmission

If attempts to eliminate excessive noise at the source have not been successful, the next step is to try to prevent its transmission to workers, who are the receivers or hearers. The three types of paths by which sound energy may be transmitted to the receiver are direct transmission, reflected transmission, and reverberated transmission. Direct transmission occurs over a straight path through the air directly from the source to the receiver with no obstructions between the two. Reflected transmission involves reflection of the sound pressure wave from a wall, the ceiling, the floor, an object, or a series of these elements, between the source and the receiver. Reverberated transmission occurs when vibrations are transmitted through solid objects between the source and the receiver, usually involving a path from the source to the floor, along the floor to the feet of the receiver, and through the skeletal structure of the receiver.

Direct and reflected transmission can be discussed together because one is rarely found without the other, and because the methods of treatment are the same. In brief, the object of treatment is to absorb the energy or reflect it away from all receivers. This involves the use of the acoustical materials discussed earlier. It is usually desirable to interrupt a path of transmission as close to the source as feasible, so one should start at the source.

Since noise radiates in all directions from the source, the most effective means of control of transmission is to enclose the source with absorbing and reflecting material. It has been proven that it is more effective to place the absorbing material toward the source and the reflecting material outside the absorbing material, as shown in Figure 10-10. A complete enclosure can often be placed around sources of noise such as pumps, air compressors, and other noisy equipment that operate without the attention of a person. Even large equipment that requires occasional attention can be enclosed, and access can be provided by means of a door. Several machines have been modified to place controls outside an enclosure. Figure 10-11 illustrates an enclosure around a large piece of equipment.

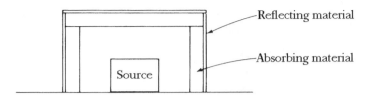

Figure 10-10. Use of absorbing material and reflecting material in an acoustical enclosure around a noise source.

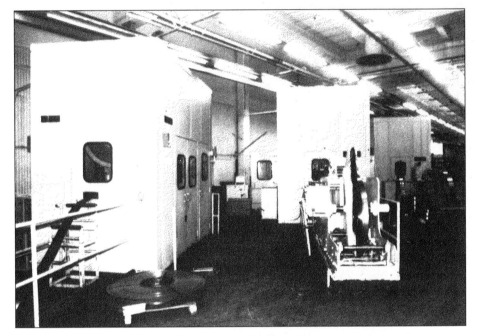

Figure 10-11. Noise control enclosures around punch presses. Note coiled, flat-steel stock being fed through the enclosure. Courtesy Eckels Industries, Inc., Eckoustic® Division.

When complete enclosures are not feasible, the use of baffles often helps. They are never as effective as a complete enclosure because some reflected sound pressure waves always escape. Different types of baffles can be used, depending upon paths of sound transmission and available space for them. Two common designs are partial partitions supported from the floor and curtains hanging from above. When a smooth ceiling reflects sound, several forms of baffles can be used on the ceiling. Figures 10-12 and 10-13 illustrate some of the various types of baffles. As with enclosures, the most effective baffles are those with an absorbing material facing the source of the sound and a reflecting material on the back side of the absorbing material.

In treating cabinets, machine surfaces, some ceilings, and other objects, it is helpful to create a rough surface, either by attaching a layer of material with adhesive or by applying a crackle-finish paint to the surface. When the rough surface reflects the sound in diverse directions, the sound becomes diffused.

The treatment of noise following a reverberant path of transmission should also occur as close to the source as feasible. This noise is most often generated

Figure 10-12.   Vinyl-glass-fabric noise barriers. Two styles
are available with NRCs of .70 and .75.
Courtesy Noise Control Associates, Inc.

by machines or equipment mounted on the floor or suspended from the roof-
or wall-support structure, such as heavy machines, air compressors, fans, pipes,
and pumps.

Vibrating floor-mounted equipment transmits its vibration to the floor,
which in turn transmits it directly to a receiver standing on the floor, or, in
many cases, to other structural members that carry it to various parts of the
building. Vibration-isolation mountings are often placed under such equipment to absorb the vibration. A wide variety of mountings is available. Their
selection or design depends on the mass they must support and the frequencies forced upon them. In many cases a wide range of frequencies is involved
and "general use" absorbers are selected. When only one frequency is involved
and the vibration is a simple harmonic oscillation, the absorber or vibration
damper is a spring, the natural frequency of which is very different from that
of the vibrating equipment. It is recommended that the natural frequency of
the isolator be no more than 0.16 times the disturbing frequency (in hertz)
and preferably 0.10 times that frequency.

Figure 10-13. This installation of transparent acoustical barrier material reduced the noise at the operator location from 94 dB to 86 dB. Courtesy The W.B. McGuire Co., Inc.

Vibration isolation is never complete, but it can reduce transmission to an acceptable level. Vibrating equipment also transmits some sound energy through the air, and isolation mountings also reduce that transmission by reducing the vibration. Figure 10-14 shows typical isolation mountings. Those isolation mountings are most often used with floor-mounted equipment, but they can also be used with equipment that is supported from overhead.

When isolation mountings are not feasible, it may be satisfactory to isolate segments of the floor. It is very difficult to do this with a steel support structure or even with a monolithic, reinforced-concrete structure, but when floors are made of concrete slabs, it can be done quite easily. The concrete slabs can be set upon vibration-absorbing pads and separated from one another by strips of the same material. This greatly reduces the transmission of vibration.

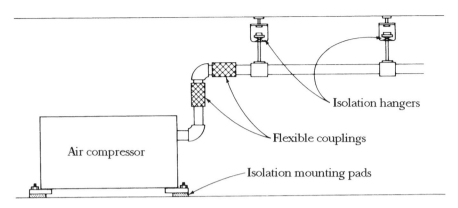

Figure 10-14. Typical application of vibration-isolation mounting devices.

Noise Control at Receiver

When none of the above control measures are successful, there are still a few steps that can be taken at the receiving end. One of the most effective measures at this point is to increase the distance between the hearer and the source of the noise. It was mentioned earlier in this chapter that the sound pressure level is inversely proportional to the square of the distance from the source. The reduction in sound pressure level is approximately 6 dB for every doubling of distance. It becomes apparent that increasing this distance is very worthwhile close to the source but less so farther away.

The nature of some operations makes it impossible or very impractical to place an enclosure around the equipment or any sort of noise barrier in front of it. However, some operations can be remotely controlled so that the worker can perform the required functions inside an enclosure; that is, the hearer is enclosed instead of the noise source. This is commonly done with sandblasting operations, steel rolling mills, and traveling equipment such as cranes, as well as rail and highway vehicles.

The same types of materials used for enclosures for noise sources are used for enclosures for hearers. Absorbing materials should be placed toward the source of noise as before. Absorbing materials should be on the outside and reflecting materials on the inside when the enclosure is for the hearer.

Barriers or baffles arranged around the hearer may be used instead of a full enclosure if the worker needs a wider range of motion. As before, however, some noise will always be reflected around barriers, so they are not as effective as a full enclosure.

The last resort is the use of personal protective equipment: earplugs or earmuffs. As noted in Chapter 5, personal protective equipment is never a desirable alternative if the hazard can be eliminated or made acceptable by any other means.

To illustrate the use of several noise control measures, Figure 10-15 shows a series of applications and the resulting octave band analysis. Isolation mountings are most effective for low frequencies, but they have very little, if any, effect at high frequencies. Absorbing materials and reflecting materials are most effective at the upper frequencies. Also, it is interesting to note that a complete enclosure of reflecting material alone is more effective than a complete enclosure of absorbing material alone.

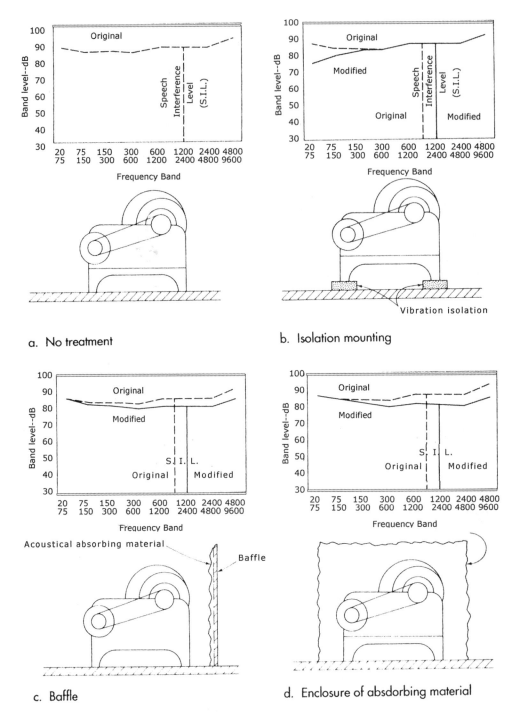

Figure 10-15. Typical noise reductions with several types of acoustical treatment.

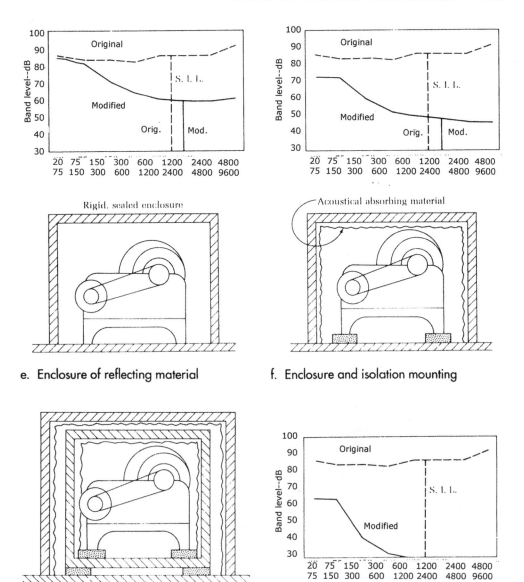

e. Enclosure of reflecting material

f. Enclosure and isolation mounting

g. Double enclosure and double isolation mounting

Figure 10-15. Cont'd.

QUESTIONS AND PROBLEMS

1. What is the difference between "sound power level" and the "sound pressure level"?
2. Calculate the sound pressure level 20 meters from a 90 dB source 10 meters away.
3. Calculate the velocity of sound through water at 20 degrees C.
4. Name several types of meters to measure sound.
5. Calculate the permissible time exposure for a worker at 110 dBA.
6. Describe a rough way to tell what the noise level is in a noisy shop.
7. What is the sum of 90 dB and 35 dB?
8. Name three ways in which sound energy can be absorbed.
9. What physical property is attributed to a good sound barrier?
10. Name several types of sound barriers.
11. Describe the difference in hearing loss caused by presbycusis and that caused by loud noise exposure.

BIBLIOGRAPHY

Bell, L. H., 1982. *Industrial Noise Control.* New York: Marcel Dekker.

Beranek, Leo L., 1989. *Noise and Vibration Control.* Poughkeepsie, NY: Institute of Noise Control Engineers.

Bies, David Q. and Colin H. Hansen, 1988. *Engineering Noise Control: Theory and Practice.* Winchester, MA: Unwin Hyman, Inc.

Feldman, Alan and Charles T. Grimes, 1985. *Hearing Conversation in Industry.* Baltimore: Williams and Wilkins.

Haight, Joel M., ed. 2012. *The Safety Professionals Handbook*, 2nd ed. 2 vols. Des Plaines, IL: ASSE.

Knowles, Emory. ed. 2003. *Noise Control: A Guide for Workers and Employer*, 3rd ed. Des Plaines, IL: ASSE.

Kryter, Karl D., 1985. *The Effects of Noise on Man.* 2nd ed. San Diego: Academic Press.

Loeb, Michel, 1986. *Noise and Human Efficiency.* New York: Wiley.

Miller, Maurice H. and Carol A. Silverman, eds., 1984. *Occupational Hearing Conservation.* New York: Prentice-Hall.

Miller, Richard K. and Albert Thuman, 1986. *Fundamentals of Noise Control Engineering.* Englewood Cliffs, NJ: Fairmont.

OSHA Compliance Manual, 2009. Neenah, WI: J.J. Keller & Associates, Inc.

OSHA General Industry Regulations. 2012. Davenport, IA: Mancomm.

Tempest, W., ed., 1985. *The Noise Handbook.* San Diego: Academic Press.

U.S. Department of Labor. *OSHA Safety and Health Standards for General Industry.* 29 CFR 1910. Washington, D.C.: Government Printing Office (updated annually).

11

Explosion

Richard T. Beohm, PE, CSP, ARM

Richard W. Stickle, PE, CSP

Characteristics of an Explosion

Explosion has been defined in several ways. An explosion most commonly begins with the ignition of a fuel that burns very rapidly, producing a large and sudden release of gas. By the broadest definition, however, an explosion need not involve a fire. When a container bursts from increased internal pressure, no matter what the cause, the sudden release of pressure is often referred to as an explosion. By some definitions, this would be considered merely a mechanical failure rather than an explosion. It is important for companies that handle, store, and process chemicals and materials that are explosive to develop a strong cultural awareness in safety. There should be a thorough plan review process and an effective safety program with frequent audits and process reviews. Any serious accident could result in multiple casualties, loss of property and business, poor public relations, environmental damage, and unfavorable litigation. Management should incorporate safety into the corporate culture of doing business and compliance with industrial standards and government regulations.

The National Fire Protection Association's (NFPA) Standard on Explosion Prevention Systems, NFPA 69, defines an explosion as "the bursting or rupture of an enclosure or container due to the development of internal pressure from a deflagration."[1] *The NFPA Fire Protection Handbook*, 20th edition, defines it as "a rapid release of high-pressure gas into the environment." In either definition, the key word is "pressure" and its effects on the surrounding environment.

The definitions are not specific to the means by which the high-pressure gas is produced. Therefore, an explosion might result from a chemical reaction (combustion of a flammable gas mixture), from the over-pressurization of a structure or an enclosed container/vessel, by physical means (bursting of a tank), or by a physical/chemical means (boiler explosion).

It is the first condition that will be discussed most thoroughly here. The spread of burning, referred to as the rate of flame spread, varies from less than 1 centimeter per minute to above the speed of sound. A flame-spread rate of less than the speed of sound produces what is called a deflagration, while a flame-spread rate that is above the speed of sound produces what is called a detonation.

When a fire burns at a high rate of flame spread, either above or below the speed of sound, it is accompanied by a pressure wave, which also travels at a very high speed. A pressure above atmospheric pressure is called overpressure. An overpressure of as little as 3.5 to 5 kilopascals (0.5 to 0.7 pound per square inch) is enough to break windows. An overpressure of 2.75 kilopascals (0.4 pound per square inch) is considered the maximum at which there will be no damage to personnel. A charge of 0.6 kilogram (1.0 pound) of TNT will produce up to 2,200 kilopascals overpressure at a distance of 0.6 meter from the point of initiation (320 pounds per square inch at 2 feet).

The terms *explosive material* or *explosive substance* should be used to describe materials that are capable of causing an explosion that, to a large extent, is influenced by confinement. A material or device that has been designed to release large overpressures when detonated is called an explosive. The terms are often used too loosely in the literature. In this discussion, the terms will be used as defined.

There is a wide range of materials that will explode, given the right conditions. Many materials that are flammable and burn slowly under some conditions will explode under other conditions. The flammability limits described in the previous chapter have also been called explosive limits, which is unfortunate. Some materials will explode at limits well within their flammability limits. An increase in temperature or pressure is often enough to change a flammable material into an explosive material. Furthermore, a mixture of the fuel with pure oxygen instead of air, which has only 21 percent oxygen, will cause some materials to explode rather than burn.

Reactivity of any hazardous substance is one of the characteristics indicated in an identification symbol developed by the NFPA. This will be discussed more completely in the chapter on hazardous materials. The symbol is shown in Figure 11-1. The square on the right side of this symbol is reserved for the reactivity index. An index of zero (0) indicates no reactivity, whereas an index of 4 indicates the most serious condition of reactivity.

Suppression of an explosion is difficult due to the rapidity of the reaction. Detonation occurs in a matter of milliseconds after the initial ignition. Automatic detectors, in conjunction with a fire-extinguishing system, have been used successfully. The detectors generally consist of pressure sensors or ultraviolet radiation detectors which initiate fire-extinguishing systems consisting of clean agents and dry chemical agents. A better way to deal with explosions is to

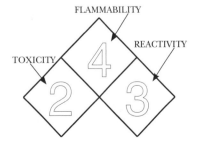

0—Stable, nonreactive with water
1—Unstable at high temperatures and pressures
2—Reacts vigorously but will not detonate
3—Will detonate with high heat or strong ignition
4—Very sensitive to shock, explosion

Figure 11-1. NFPA symbol showing the ratings of reactivity.

prevent a reaction and thus avoid explosion. This, of course, involves the same principles as fire prevention, plus a few more precautions.

Many of the substances prone to detonate are sensitive not only to increased temperature but to vibration, friction, static electricity, and increased pressure as well. Such substances must be handled and stored in well-ventilated atmospheres, as free from vibration and jarring as possible. Those materials that are particularly sensitive to vibration and jarring should be kept in small, sturdy containers and should be cushioned when being transported.

It is extremely important that explosive materials be identified and that containers be clearly labeled. The Department of Transportation (DOT) administers regulations adopted to provide safe transportation of a wide range of hazardous materials.

The six classes of explosive materials designated by the Department of Transportation are as follows (See Figure 11-2):

- New classification system (Effective 10/2001). Class 1 is explosives, which is subdivided into six hazard Divisions by the Department of Defense and the Department of Transportation.
- Division 1.1 (Mass detonation or blast hazard)
  Explosive material and explosive items containing this material are expected to mass detonate. If one of these devices detonates, all others placed close to it will detonate practically simultaneously.
- Division 1.2 (Non-mass detonating; fragmentation hazard.)
  While blast is still a hazard, the principle hazard is considered to be fragmentation. These items are not expected to mass detonate, but those in close contact will detonate one or a few at a time and high-velocity fragments will be projected.
- Division 1.3 (Mass fire hazard.)
  This explosive material, and devices in which it is loaded, will burn with intense heat and violence. Some explosions may occur, but high blast pressures are not expected.
- Division 1.4 (Moderate fire hazard.)
  Material will burn, but the fire is not expected to be sizable or intense. Small explosions may occur, but hazardous fragmentation will not be projected great distances.
- Division 1.5 (Very insensitive but will mass detonate.)
  Explosive items are assigned to this division for transport purposes only. They are so insensitive to external events that there is almost no probability of accidental initiation or transition from burning to detonation. When exposed to fire, these materials will burn or melt. In storage these items are treated as Division 1.1.
- Division 1.6 (Extremely insensitive.)
  The division includes ammunition that contains insensitive explosives. These items have shown through testing that the mass and confinement effects of the case are negligible on the probability of initiation or transition from burning to detonation while in transport or storage.

Specific labeling instructions and requirements are described in the DOT regulations. Placards must be attached to commercial transportation vehicles, including railroad cars, aircraft, water vessels, and public highway vehicles,

Figure 11-2. DOT placards identifying cargos of explosive materials.

when they are transporting explosive materials, as well as other hazardous materials. The placards for explosive materials are illustrated in Figure 11-2. The background color for explosives is orange.

Designing Facilities for Use of Explosive Materials

It should be obvious that when a material is determined to be, or is suspected of being, an explosive material, a search for a substitute should be made, unless an explosive is desired. A wide range of chemicals is being used today by many different industries for processing other materials. Many of these commercial materials contain components that in some stage of processing are explosive. Often it is not possible or feasible to substitute a less hazardous material. If this is the case, great care should be taken to ensure that unwanted explosions do not occur. In some cases a pressure rate-of-rise detector can actuate a device or system to extinguish a potential explosion before it reaches an explosive stage. However, such applications are not commonly feasible.

Materials that have the potential to become explosive should be stored and carried in containers of as small a size as possible. Storage areas as well as work stations should be well ventilated to prevent any accumulation of vapors. Temperatures and pressures should be maintained well below those required to detonate or ignite an explosive or flammable mixture. It should be remembered that processing often increases either temperature or pressure or both. All oxidizing agents such as oxygen, chlorine, and fluorine should be kept out of the area. Many of the common explosive substances act as oxidizers for other substances. Blasting caps and detonating primers must not be stored in the same storage magazine with other explosives.

Some recent notable accidental explosions include the following:

- The Imperial Sugar Refinery dust explosion in October 2009, killing 14 and injuring 36. The cause was possibly an overheated bearing that ignited sugar dust, causing a series of powerful dust explosions along the length of the facility's conveyor. The Chemical Safety and Hazard Board (CSB) cited poor maintenance, housekeeping, and equipment design as factors.
- The Salt Lake City oil refinery explosion on November 4, 2009. The CSB determined the cause to be incorrect pipe thickness.
- The Corpus Christi refinery blast in July 2009. A blocked control valve caused an explosion and fire, and a release of HF vapors.
- The Puerto Rico (near San Juan) terminal blast on October 23, 2009. The cause was most likely a tank overfill, which resulted in a massive explosion and fire involving a number of storage tanks and a release of about 30 million gallons of solvent.

Carrying and transferring liquid explosive materials, especially unstable substances, should be done very carefully. A container of such material should be opened slowly and without vibration. The container should be grounded electrically to prevent ignition by a spark from static electricity. Plastic as well as metal containers should be grounded, since static electricity can be generated on plastic surfaces also. The liquid must be poured slowly and gently.

When transferring liquid or gaseous explosive materials, two precautions may be in order. It is always possible that, even with hose and/or pipe connections, some leakage will occur. The air around possible leaks should be vented, or an inert atmosphere should be provided. Second, a collecting cup of some sort should be located under the connection to catch leaking liquid.

After the liquid or gas is evacuated from the original tank or container, some vapor will be left. This vapor is often more hazardous than the liquid or a container completely filled with gas, since it is more likely to mix with the air in explosive proportions. Therefore, this original tank should be vented to the outside or filled with an inert gas as the container is emptied. Any transfer of an explosive liquid or gaseous substance should be done in an area that is isolated from adjacent areas by a solid barrier or by pressurization. Pouring and transferring are among the most hazardous steps when explosive materials are being used. Figure 11-3 shows a transfer air-lock system designed for use when hazardous liquids are being transferred.

There are several conditions that call for purging of containers, pipes and hoses, and all other parts of a system containing an explosive material or other hazardous material. Purging is the process of completely eliminating all traces of a given material. This pertains primarily to liquid and gaseous systems, and especially to the latter.

Purging must be done when a container that has held a hazardous material is going to be used for another purpose, or if it is going to receive a new supply of the same material, or if it is suspected that foreign matter has leaked into the system. In purging, another material is forced into the system in a manner that completely replaces the previous material with the purging material. An inert gas is commonly used for this purpose, but air can sometimes

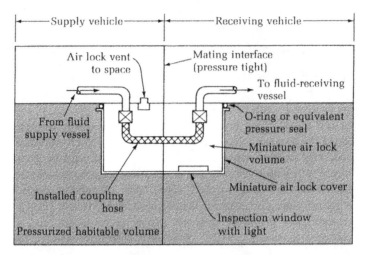

Figure 11-3. A design for transferring explosive or other hazardous liquids from one vessel to another.

be used. In a system containing flammable gases or liquids, it is important that nothing flammable be left in the system. Inert gas or air can be used to replace it, or another material may render the initial material nonflammable.

The purging process must be done with care since a flammable or other hazardous material is being forced out. This may create an uncontrolled, hazardous atmosphere. Care must be taken to see that unauthorized persons are out of the area. Preparations must be made to capture the material being purged out, if possible. Fire-fighting equipment must be ready to use. The supply of oxygen may be reduced by both the purged gas and the purging gas, unless the purging gas is air; therefore, breathing apparatus may be necessary.

The NFPA and ICC gas codes are to be revised to prohibit the purging of gas lines indoors and to require the use of combustible gas detectors following the CSB investigation of the June 9, 2009, natural gas explosion at a food processing plant in Garner, North Carolina. The lines should be vented to the outside, away from ignition sources.

Design of Buildings

Explosives must be stored in special buildings, but explosions can occur in any type of building. There are several building design features that are affected by the possibility of an explosion in a building. In designing such facilities, one must be cognizant of the hazards of an explosion. An explosion results in a pressure wave and an atmospheric overpressure. If the explosion is in a container, two hazards exist, assuming that no person is in the container. One is the flying fragments of the container, which may injure or kill personnel in the area and damage property. The other is the pressure wave transmitted from the explosion. If the container is almost strong enough to restrain the explosion, most of the energy is absorbed by the container and there is only a slight pressure wave. If

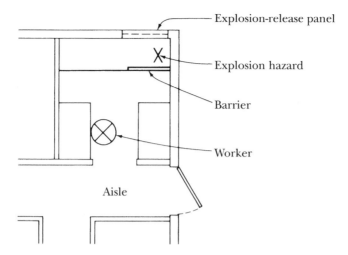

Figure 11-4. Layout of a room containing a potential explosion hazard.

not, the pressure wave may cause a great deal of property damage, and, if people are in the way, it may blow them against objects and cause injury.

If the explosion takes place in a more open room where there are people, additional hazards exist. Like fire, an explosion consumes oxygen and produces carbon dioxide, carbon monoxide, and possibly other toxic gases. The hazards for personnel are the same as those produced in a fire except that an explosion occurs much more rapidly. There is no opportunity to evacuate the area. Everything happens so fast that even suppression is rarely effective.

When a potential explosion hazard is recognized, all facilities should be located and arranged with respect to it. The hazard should be located along an outside wall if possible. The room containing the hazard should be fairly small and contain as little equipment as possible. People should be kept out of the room as much as possible. The room should not contain a large number of operations or many workers. Figure 11-4 illustrates a layout that might serve as a guide for planning such a room. There should be a direct route out of the room for personnel, but the door should not permit the shock wave to pass unobstructed into another work area. People and easily damaged equipment should be shielded from an explosion, if possible, but the shield or barrier should not in itself become a projectile if an explosion occurs.

There are two approaches that may be taken in designing walls and/or barriers around a potential explosion hazard. One approach involves the construction of a wall or barrier strong enough to absorb the energy of such an explosion. This is called an explosion-proof wall or barrier. The other approach involves a portion of an outside wall that will break away at some predetermined overpressure and allow the pressure wave to be transmitted harmlessly to the outside atmosphere. This is called an explosion-release panel and is noted in Figure 11-4.

Each of these approaches offers advantages in certain applications. For testing materials, munitions, and so on, a test cell with massive walls can be constructed. Personnel can be evacuated from the area before testing is done.

Explosion-proof walls do not need to be rebuilt, assuming they were designed and built properly in the first place.

When explosion potential can be reduced to a reasonable amount but cannot be eliminated or predicted in time, an explosion-release panel may be more practical. An explosion-release panel may be a single window frame, a wall panel, a whole wall, or even a roof section. The panel is attached to the wall or the roof with notched or necked rivets or screws. When an overpressure reaches a calculated limit, the fasteners break and allow the panel to swing out, or lift off the roof. After this happens, new rivets or screws must be installed.

Another form of release panel is a window with a scored line in the glass near the perimeter. The scored line is designed to break at a predetermined pressure. The major disadvantage is that broken glass must be picked up afterward. With any type of release panel, a clear space must be maintained outside the building.

The U.S. Department of Defense (DOD) now requires antiterrorism codes such as UFC 4-010-01 and UFC 4-021-01 for their buildings. These codes are available on the web. Some of these features are blast walls, stand-off distances, and seismic restraints for electrical and sprinkler systems.

Explosives

The word *explosive* implies a material designed to be used so as to utilize the tremendous power released when it is detonated. Included are ammunitions, dynamite, water gels, blasting caps, pyrotechnics, and smokeless propellants, as well as many other materials. Ammonium nitrate is one of the commonly used explosive materials. Aluminum and other metal powders are also very explosive. It is always best to use the least hazardous material that will adequately serve the purpose. Water gels are now commonly used instead of dynamite because they are more stable and, therefore, easier to handle and store.

A few years ago, a water gel was developed that could release a pressure wave at least as effective as the pressure wave of dynamite and that was much more stable. Known by the trade name Tovex, this gel can be carried and stored with much greater safety than dynamite; it also takes up less space and produces less smoke and fewer fumes.

All explosives must be stored in special buildings called magazines. Such magazines must be constructed according to specifications prescribed in OSHA Standard 1910.109. Tables H-21 and H-22 are particularly helpful. Magazines are classified as Class I (quantities in excess of 23 kilograms, 50 pounds) or Class II (quantities of 23 kilograms or less). Such structures must be located a minimum specified distance from inhabited buildings, railways, highways, and other magazines. The ground around these magazines must slope away from the structure to allow proper drainage. The amounts of explosives covered in the table of distances range from 0.9 kg (2 lbs), with a minimum distance of 1.8 meters (6 feet), to 136,000 kilograms (3 million pounds), with a minimum distance of 117 meters (385 feet). The design and construction of the magazines is specified as well as the distance to other structures.

Vehicles that transport explosives must carry placards (as described earlier). Explosives must be protected from sparks or any source of heat, including people smoking, while being stored or transported. Routes followed by vehicles transporting explosives must be planned to avoid congested areas or other hazards.

Even though they are more stable than most other types of explosives, water gels are subject to the same standards of storage and transportation.

Any engineer who designs facilities to accommodate explosives or explosive materials should seek further information and should especially consult the ANSI and NFPA codes regarding explosives. The following NFPA codes may be appropriate:

Code No.	Title
68	Venting of Deflagration
69	Explosion Prevention Systems
495	Explosive Material Code
498	Standard for Safe Havens and Interchange Lots for Vehicles Transporting Explosives
8502	Furnace Explosions/Implosions

Dust Explosions

Another special explosion hazard is the accidental explosion of grain dust; rather infrequent, it can be devastating when it occurs. The grain industry includes the movement, storage, and processing of grains such as wheat, corn, oats, soy beans, and others. Grain processing is an ever-growing industry due to a growing population. The processing and storage of grain involves the use of grain elevators, dryers, conveyors, and silos. Dust collectors are frequently used to control and collect grain dust and often present a serious fire hazard. Dust explosions are the most serious of all hazards in the grain industry. The first recorded dust explosion occurred in 1785 in Turin, Italy. Since 1976, dust explosions have continued to decrease in frequency and severity.

Three combustible dust explosions in 2003 took 14 lives and injured numerous others. Combustible dusts caused 281 incidents that killed 119 workers and injured 718 from 1980 to 2005, and an additional 16 deaths and 84 injuries occurred from 2006 to 2008.

The Sago Mine methane explosion in West Virginia in 2006 killed 13 workers and resulted in the Mine Improvement and New Emergency Response Act.

On February 7, 2008, at about 7:15 p.m., something, possibly an overheated bearing, ignited sugar dust in Imperial's refinery in Port Wentworth, Georgia, causing a series of powerful dust-fueled explosions along the length of the facility's conveyor.

The blasts and subsequent fires destroyed much of the facility; the loss was estimated at $275 million. Fourteen employees were killed and dozens injured. The U.S. Chemical Safety and Hazard Investigation Board (CSB) cited poor maintenance, housekeeping, and equipment design as factors. Today Imperial is undergoing a safety culture change, where safety standards are valued and where its success relies on the attitude of the workers, and change management is incorporated into the safety program.

NFPA codes were integral to the rebuilding process at Imperial's refinery. They were NFPA 10, 13, 61, 68, 69, 70, 72, 101, 499, 505, 654, and 780. (NFPA Journal, March/April 2010).

Dust can be defined as a solid, organic, or unoxidized metal material that is not larger than 500 microns in cross-sectional area. The smaller the particles, the greater the potential for an explosion to occur. The primary areas for a grain explosion to occur are in the bucket elevators, storage bins, hammer mills, dust collectors, and enclosed equipment. Sometimes a secondary explosion occurs when the primary explosion's pressure wave disturbs dust layers, creating a dust cloud that is ignited by the flame front, and this is often more severe than the first. Preventive maintenance and good housekeeping are of utmost importance in preventing a dust explosion.

The explosion triangle consists of FUEL (dust powder, clouds, or layers), an IGNITION SOURCE (open flames, sparks, hot surfaces, overheated bearing, etc.), and OXYGEN (air).

When the product is contained in an enclosed space, such as a silo, the potential for explosion is great due to its concentration-oxygen ratio. The explosion range for most dusts is between the limits of 50–100 g/m³ and 2–3 kg/m³. Some loss-control measures to consider would be to use open structures and/or explosion venting, as well as separating the various buildings and strengthening them to withstand explosion-strength forces.

The severity of the explosion is determined by a number of variables, such as the surface area of the dust, the distribution of the particle mixes and shapes in the dust cloud, the particle spacing within the cloud, the strength and duration of the ignition source or spark, the concentration of the oxidant, the heat of combustion of the dust, and the volume and shape of the enclosures.

NFPA 61 is the designer's guide for minimizing losses from explosions at grain-handling facilities. It recommends a separation of 100 ft. (30.5 m) between personnel-intensive areas and concrete elevator headhouses and silos. Some designs include an open tower structure and relief venting, the elimination of the conventional headhouse, explosion venting on the bucket elevators, locating the bucket elevators on the outside, and eliminating the gallery.

NFPA 68, Guide for Venting of Deflagrations, can be used to calculate the venting area. Av (vent area) sq. ft. = $C^* \times As$ (internal surface area) sq. ft./ sq. root of Pred, where Pred (psi) is the maximum internal overpressure that can be withstood by the weakest structural element.

$$C^* = .10 \text{ for grain dust}$$

NFPA 484, the standard for combustible metals, applies to operations where metal or metal alloys are subject to processing or finishing operations that produce powder for dust, such as machining, sawing, grinding, buffing, and polishing.

NFPA 654, the standard for prevention of fire and dust explosions from manufacturing, processing, and handling of combustible particulate solids also contains comprehensive guidance on the control of ignition sources to prevent explosions, and to minimize the danger and damage from an explosion.

Facilities should carefully identify the following in order to assess their potential for a dust explosion in conduction a facility dust hazard assessment:

- Materials that can be combustible when finely divided;
- Processes that use, consume, or produce combustible dust;
- Open areas where combustible dust may accumulate;
- Means by which dust might be dispersed into the air; and
- Potential ignition sources.

A process hazard analysis (PHA) should also be conducted and reviewed at least every five years to update any changes that could affect the status of the facility.

Venting, suppression, and containment methods are often used to control or minimize the effects of an explosion. Doors, windows, explosion-relief panels, and light-gauge structural coverings on steel structures are sometimes used to accomplish venting. Fire walls and fire-rated enclosures are used to separate various processes and hazardous areas.

There are a number of technologies and strategies available to prevent and/or reduce dust explosions. William J. Stevenson describes this in terms of a dust pentagon with the following components:

1. Fuel (combustible dust)
2. Suspension (agitated dust)
3. Confinement (with an explosible concentration of dust)
4. Oxidizer (usually air)
5. Ignition source (of sufficient strength and duration)

Avoidance of an explosion (these assume the pentagon can be avoided):

1. Avoid ignition sources.
2. Avoid dust concentrations in the explosion range.
3. Inert to reduce oxygen concentration.

Mitigation of an explosion (these all assume that the pentagon could occur):

4. Locate the explosion containment vessel outside so that an explosion will cause no consequential damage.
5. Locate the vessel inside, but adjacent to an outside wall so that a vent can be directed outside through a straight, short duct.
6. Locate the vessel indoors as above, but install the duct up through the roof.
7. Q-Rohr® venting devices for inside venting-containment.
8. Actively suppress.
9. Contain.

There are now many devices available to detect, warn, and control faulty operation of mechanical equipment for use in Class II, Group G, dust atmospheres. Electrical wiring should conform to NFPA 70 Articles 500, 502, and 504, if applicable. Safeguards also include electrical interlocking for shutdown of activities should a certain fault or hazardous condition occur.

QUESTIONS AND PROBLEMS

1. What are some of the industries that have experienced major explosions?
2. Name several factors that could contribute to a major dust explosion.
3. Name some of the NFPA codes that could be used in the design of buildings where a combustible dust hazard is a concern.
4. What are the four classes of explosive materials designated by the Department of Transportation?

5. What are some of the safety considerations for the storage of explosive materials?
6. Describe a method of purging a hazardous material from a container.
7. What are some of the requirements for the storage of explosives in magazines?
8. How does an explosion-release panel work?
9. Describe how to suppress an explosion in a dust collector.

NOTES

1. National Fire Protection Association, 2008. "Explosion Prevention Systems," NFPA 69. Quincy, MA: NFPA.
2. U.S. Department of Transportation, 1976. *Code of Federal Regulations,* "Transportation" (49 CFR, Parts 100 to 199). Washington, D.C.: Government Printing Office, 200.

BIBLIOGRAPHY*

Bartknecht, W., 1989. E*xplosions—Cause, Prevention, Protection.* New York: Springer-Verlag, Inc.

Eckhoff, R., 2003. *Dust Explosions in the Process Industries,* 3rd ed. Houston, TX: Gulf Professional Publishing.

FM Global. 2005. Data Sheet 7-76, "Prevention and Mitigation of Combustible Dust Explosion and Fire."

Hartwig, M., and H. Steen, eds. 2004. *Handbook of Explosion Prevention and Protection.* Hoboken, NJ: Wiley.

National Aeronautics and Space Administration, 2010. Safety Standards for Explosives, Propellants, and Pyrotechnics. (NASA-STD-8719.12). Washington, D.C.

National Fire Protection Association, 2007. "Deflagration Venting" (NFPA Code 68). Quincy, MA: NFPA.

_____, 2008. "Explosion Prevention Systems" (NFPA Code 69). Quincy, MA: NFPA.

_____, 2008. *Fire Protection Handbook.* 20th ed. Quincy, MA: NFPA.

_____. 1990. *Industrial Fire Hazards Handbook,* 3rd ed. Quincy, MA: NFPA.

U.S. Department of the Army. Ammunition and Explosives Safety Standards (DA PAM 385-64. Washington DC: 1999

U.S. Department of Labor. *OSHA Safety and Health Standards* (29 CFR 1910) Part 1910.109. Washington, D.C.: Government Printing Office.

U.S. Department of Transportation, 2007. *Code of Federal Regulations,* "Transportation" (49 CFR, Parts 100–199). Washington, D.C.: Government Printing Office.

---

*In addition to the codes listed, the NFPA also has several codes and standards for specific industries.

12

Radiation

Glenn M. Sturchio, Ph.D., CHP

Radiation and its properties and effects on humans are covered in physics, health physics, and industrial hygiene. The fields of safety engineering and industrial hygiene have somewhat ambiguous boundaries, and, frequently, elements from both are used together to solve problems. The field of radiation, both ionizing and non-ionizing, is very complex, and experts in the field should be consulted whenever any source is to be installed or used. This chapter is presented, then, not as an in-depth study of radiation and its control, but as a basic survey to provide some understanding of what it is and why and how it becomes a problem, as well as some of the more common methods of controlling the hazards of radiation.

Wavelength, Frequency and Energy

Radiation is a form of energy emitted from a wide range of man-made and some natural sources. The electromagnetic (EM) spectrum covers those forms of energy with wavelengths of approximately 10^{-14} meters to 10^5 meters. That corresponds to frequencies of approximately 10^{22} hertz to approximately 1000 hertz. All forms of radiation travel at the speed of light but at different frequencies. The wavelength of the wave is inversely proportional to the frequency, and often a particular type of radiation is described in terms of its wavelength, frequency, or energy. For example: radio waves are described by frequency (KROC 106.9 MHz), laser radiation is described by wavelength (KTP laser 532 nm), and X-rays are described by energy (100 keV). Figure 12-1 shows the pertinent segment of the electromagnetic spectrum and approximate locations of points of interest and concern. Very few of the points indicated can be taken as exact measurements on the respective scales, and note that a few areas overlap.

The radiation spectrum can also be broken into two segments, based on interaction with tissues. Ionizing radiation (e.g., X-rays) deposit sufficient energy in tissues to cause ionization and subsequent biological effects. Non-ionizing radiation (e.g., microwaves), although not energetic enough to

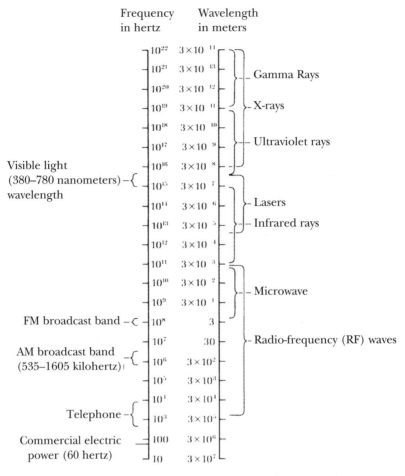

Figure 12-1. The electromagnetic spectrum, with points of interest noted.

cause ionization, may cause thermal effects that lead to subsequent biological effects. Table 12-1 lists some common sources of radiation. With machine-based radiation sources, the hazard exists only when the source is activated, such as when an RF sealer, a laser, or an X-ray machine is turned on. Radioactive materials, on the other hand, are continually emitting ionizing radiation.

Table 12-1. Common Sources of Radiation

Non-ionizing radiation	Ionizing radiation
RF heaters	X-ray machines
Microwave units	Linear accelerators
Laser systems	Nuclear reactors
The sun	Radioactive materials

Ionizing Radiation

Ionizing radiation, as discussed previously, has sufficient energy to free an electron from the atom and thereby transform the atom into an ion. These radiations may be either electromagnetic waves or particles emitted from man-made or natural sources. Any high-energy ionizing radiation, such as X-rays, gamma rays, and beta particles, can damage living cells through ionization. As a result, unnecessary exposure to any of these radiations should be avoided. The two main sources of ionizing radiation are machine sources (e.g., X-ray machines) and radioactive materials.

Radioactive materials (or radionuclides) are unstable and spontaneously emit radiation, either energetic rays or particles, to become stable. This process is known as *radioactive decay*, and each radionuclide has a characteristic set of radiations that are emitted, as well as a characteristic half-life (or the amount of time that it takes the radionuclide to decay to one-half of the original radioactivity). X-rays and gamma rays emitted are part of the electromagnetic spectrum discussed previously and may be referred to as *photon radiation*. The *particulate radiation* that may be emitted includes alpha particles (helium nuclei) and beta particles (fast-moving electrons).

Alpha particles, due to their mass and double-positive charge, deposit significantly more energy per unit-path length in tissue than beta particles or photons. Therefore, they create more ionization within the cell tissue they reach and subsequently more biological damage. Fortunately, alpha particles lack the ability to penetrate beyond the outer layer of human skin. If they are inhaled or ingested, however, they are extremely hazardous to sensitive cell tissue in the lungs, gastrointestinal tract, and other organs.

Beta particles may, depending on the energy, penetrate through the skin up to about one centimeter. Photon radiation, which is massless and chargeless, penetrates the skin and, depending on energy, can penetrate much more dense materials. The ability to create ionization within cell tissue is about the same for beta and photon radiation but much lower than alpha radiation.

A single particle or photon can disrupt a cell. The disruption can either be repaired, lead to a mutation, or cause cell death. Mutations will be passed along to future cell generations and may lead to somatic effects in the individual irradiated or genetic effects in future generations of offspring. The main somatic effect of concern is cancer, unless the exposure is to the embryo/fetus when there are developmental concerns. To date there has been no evidence of genetic effects in offspring from the radiation exposure of parental germ cells.

Radiation Units

In the United States, especially in regulations and regulatory guidance, the common English radiation units are still being used. However, in the medical field and the literature, the use of international metric units has been adopted. The amount of energy deposited by the radiation in the individual is the absorbed dose. The absorbed dose is modified by radiation and tissue-weighting factors to calculate the dose equivalent. The regulatory limits are stated in dose-equivalent units. The common and metric units and what they represent are shown in Table 12-2.

Table 12-2. English and Metric Units of Radiation

English Unit	Metric Unit	Conversion	Measure of
Curie (Ci)	Becquerel (Bq)	$1\ Ci = 3.7 \times 10^{10}\ Bq$	Radioactivity
Roentgen (R)	Coulomb per kg (C/kg)	$1\ R = 2.58 \times 10^{-4}\ C/kg$	Exposure
Rad (rad)	Gray (Gy)	$1\ rad = 0.01\ Gy$	Absorbed Dose
Rem (rem)	Sievert (Sv)	$1\ rem = 0.01\ Sv$	Dose equivalent

Radiation Dose and Dose Limits

Everyone is exposed to ionizing radiation. This includes radiation emitted from naturally occurring radionuclides found in soil and water—so-called background radiation. Even the human body itself contains radioactive elements, such as uranium, thorium, potassium, and carbon. The National Council on Radiation Protection and Measurements (NCRP) estimates that the average person in the US population receives an annual background radiation dose of about 0.311 rem, of which 0.029 rem is from internal sources.

The Nuclear Regulatory Commission (NRC) has revised the occupational-radiation-dose standards on which the OSHA standards are based, but OSHA has not revised its standards as of this writing. The dose-limitation system is designed to minimize the risk from radiation injury to be similar to the risk of injury in "safe" industries. The revised NRC standards are contained in 10 CFR 20; similar standards for most machine-produced radiation may be found in state regulations. The NRC standards change the occupational limit from 3 rem per quarter to an absolute limit of 5 rem per year and eliminate the age-based cumulative dose limit. There is a memorandum of understanding between OSHA and the NRC that states that NRC licensees may follow the NRC regulations for radioactive materials safety purposes. For comparison, the exposure limits allowed by the OSHA standards are shown in Table 12-3.

The regulations also contain dose limits for pregnant employees, minors, and members of the public. One of the main radiation-safety principles that has been written into the NRC regulations is the "as low as reasonably achievable" (ALARA) principle. ALARA suggests that each operation should be evaluated to ensure that all reasonable methods have been used to minimize

Table 12-3. Occupational Ionizing Radiation Dose Limits

Target	OSHA (rem per calendar quarter)	NRC (rem per year)
Whole body; head and trunk; active blood-forming organs; lens of eye; or gonads	1.25	5
Hands and forearms; feet and ankles	18.75	50
Skin of the whole body	7.5	50

the radiation dose to employees and the public. In the United States, routine occupational-radiation doses are well below the regulatory limits due to the successful implementation of ALARA in day-to-day practice.

If it is likely that an individual will receive greater than 10 percent of the annual occupational limit, then it will be necessary to provide personnel monitoring with radiation dosimeters. Sometimes it is prudent to provide personnel monitoring to evaluate radiation exposures during a new process, whether or not the 10 percent threshold is likely to be exceeded, to potentially identify steps where additional safety precautions (e.g., shielding) could be implemented.

Methods of Control

Only trained personnel should be allowed into areas where sources of ionizing radiation are being used; interlocks may be necessary to prevent unauthorized persons from entering such areas. The first step to minimizing exposure to radiation sources is awareness; that is why the regulations typically require the labeling of radiation sources and posting of work areas. Figure 12-2 shows the trefoil symbol used in warning of an ionizing-radiation hazard. There are three primary methods used to minimize radiation exposure from all radiation sources: time, distance, and shielding. In cases where radioactive materials are used, it is also necessary to consider contamination control.

All radioactive materials decay over time. The length of time a particular source takes to decay to 50 percent of its initial radioactivity is called its *half-life*. After another half-life period, it will decay 50 percent of the remaining radioactivity—and so on, until essentially all of its strength is dissipated. This is illustrated in Figure 12-3. Thus, just waiting until the strength of the radiation source has been reduced to an acceptable level by decay might be one way to avoid its harmful effects. That is not a practical approach to control in most cases, however, because the half-life of most sources is days to years. However, this is a standard practice for the disposal of low-level radioactive waste.

Distance may be a better method of protection from ionizing radiation. Even the use of tongs in handling radioactive materials greatly reduces the

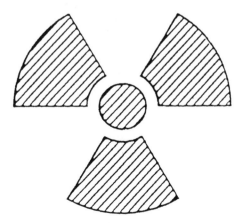

Figure 12-2. Symbol for ionizing-
radiation warning sign.

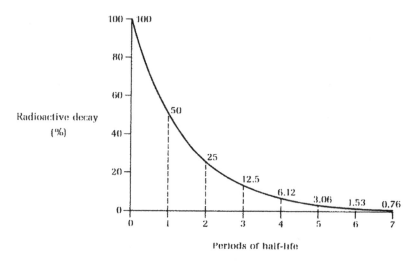

Figure 12-3. The half-life pattern of decay of ionizing radiation.

exposure. When more hazardous radioactive materials must be handled, some form of remote handling should be used. As you move away from a radiation source, the exposure is reduced by the square of the distance. This is expressed by the relationship

$$I = I_0 \times (d_0/d)^2$$

where
    I_0 is the initial radiation level;
    I is the radiation level at the new position;
    d_0 is the distance at the initial position; and
    d is the distance at the new position.

Thus, if a worker receives 100 mrem/h at a distance of 2 meters from the radiation source, at 6 meters she would receive: I = 100 mrem/h × (2 m/6 m)² = 100 mrem/h × (1/9) = 11.1 mrem/h

Shielding of ionizing radiation sources is essential, but methods of shielding vary from one type of radiation to another. Professionals in this field should be consulted to ensure the best shielding design for any given source of radiation.

Particles of radioactive materials may contaminate work surfaces and breathing air. The use of personal protective equipment may be required. Workers should not eat, drink, or in any way ingest any substances in an area where radioactive materials are present. Food and beverages should not be taken into such an area. Protective clothing worn into such an area should also be checked—with appropriate radiation-detection equipment—to ensure that contaminated materials are not brought into clean areas where other people may be exposed to them.

Non-ionizing Radiation

Non-ionizing radiation is a form of electromagnetic radiation and includes radio frequency (RF) radiation, microwaves, infrared light, visible light, ultraviolet light, and lasers. These radiations are not energetic enough to cause ionizations in cell tissue; however, they can interact with cell tissue to create biological effects. All non-ionizing radiation produced by equipment, as distinguished from that produced by the sun, can be shielded to prevent exposure to people or other objects. This is done most often by some sort of barrier around the source of the radiation. As with ionizing radiation, distance can also be used to reduce exposure from a source. It is a good policy to turn off any device that produces non-ionizing radiation when that device is not actually being used.

Radio Frequency (RF)

RF radiation emitted by these waves may have thermal effects on the body if the body is unable to dissipate more energy than it absorbs. This is dependent upon the strength of the field, the duration of exposure, and the frequency of the radiation. A review of the potential carcinogenic effect of long-term, low-level exposure to RF fields by international experts determined that it was unlikely and not a basis for limiting exposure. In addition, RF radiation also interferes with many forms of electronic instruments, including cardiac pacemakers. Figure 12-4 shows an example of a sign used in warning of a radio-frequency hazard.

Microwaves—a portion of the RF spectrum—in addition to heating cell tissue, also heat other objects sufficiently to make them sources of burn hazards. These other objects include jewelry worn by a worker and tools and work materials with which a worker may come into contact. If exposed directly to microwave radiation, cell tissue in the eye can be damaged. When antennae or other direction-controlling devices are used with a source of microwave radiation, rays should not be directed toward people. If possible, metallic objects should not be placed in the field of microwave radiation. If that is not feasible, such parts should be made inaccessible, or else gloves should be worn for protection from the heated metal.

Laser

Laser beams concentrate energy in a very narrow path and are, therefore, more hazardous than other radiation of similar wavelengths. The Food and

Figure 12-4. Radio-frequency warning sign.

Drug Administration's Center for Devices and Radiological Health (CDRH) has classified laser hazards in 21CFR1040.10 as follows:

> Class I are those that cannot under normal operating conditions emit a hazardous level of optical radiation.
>
> Class II are those visible lasers that do not have enough output power to injure a person accidentally but which may produce retinal injury when stared at for a long time.
>
> Class IIIa covers visible lasers that cannot injure the unaided eye of a person with a normal aversion response to a bright light, but may cause injury when the energy is collected and put into the eye, as with binoculars.
>
> Class IIIb consists of lasers that can produce accidental injury if viewed directly. The danger from such a laser is the direct or specularly reflected beam.
>
> Class IV includes lasers that not only produce a hazardous direct, specularly reflected, or diffusely reflected beam but also can be a fire hazard.

It is extremely important that Class IIIb and IV laser beams be shielded so that they will not reach the eyes or skin of any worker or bystander. Overexposure to laser radiation can cause permanent functional changes in the eye (e.g., blindness). This includes shielding of reflected, refracted, and diffused radiation, as well as the initial emitted radiation. Figure 12-5 illustrates such secondary laser radiation. ANSI Z-136.1, *Safe Use of Lasers*, provides guidance on the safe operation of laser systems and lists the permissible exposure limits for the eye and skin.

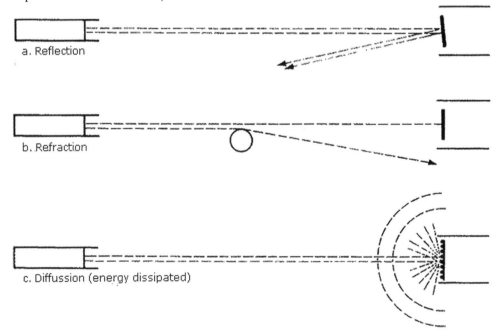

a. Reflection

b. Refraction

c. Diffussion (energy dissipated)

Figure 12-5. Illustration (exaggerated) of reflection, refraction, and diffusion of laser beams.

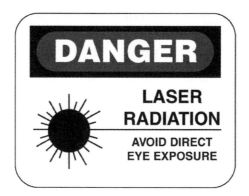

Figure 12-6. Warning sign for Class IIIb and Class IV lasers.

All laser devices should be constructed with interlocks that will stop power to a laser if any shield is opened for any reason. Unless a laser device is entirely self-contained and adequately shielded, access to the area should be restricted to those people who are trained to use that equipment.

Laser systems may also present non-radiation hazards that may need to be addressed. As noted previously, some high-powered lasers may act as an ignition source for flammable materials, so fire safety is important. The lasing of material may create a smoke plume that contains various chemical or biological hazards that should be controlled. Electrical-safety precautions may also be warranted in the general operation of the laser system and especially during any maintenance procedures.

Class IIIb and IV laser systems should be positioned to eliminate direct viewing of the laser beam and its reflections. Workers should use protective eyewear with special absorptive lenses during work with or around Class IIIb and IV laser systems. It is important that the proper optical density, laser type, and wavelength be considered in selecting the lens used, although, if possible, visible-light transmission should be maximized.

Warning signs also are required for Class IIIb and IV laser system operations. Figure 12-6 illustrates such a warning sign.

Ultraviolet Radiation (UV)

Ultraviolet (UV) radiation is produced by the sun, bulbs used for tanning, or bulbs used for germicidal purposes, as well as in welding applications and by some types of industrial equipment. UV radiation causes erythema (reddening and sunburn), premature aging, and increased risk of cancer of the skin. Long-term unprotected exposure to the sun can also increase the risk of cataracts and macular degeneration. The more harmful, shorter UVC rays are mostly absorbed by the ozone layer in the atmosphere, but UVB and UVA rays pass through the outer layers of the atmosphere. UVB rays are primarily responsible for erythema; however, both UVB and UVA rays contribute to aging and cancer risk.

Sunblock and sunscreen lotions protect outdoor workers from excessive exposure to the sun's UV rays and help to prevent sunburn, skin cancer, and skin damage. OSHA recommends a sun-protection factor (SPF) of 15 or more for effective protection against damaging solar radiation. An SPF of 15 blocks 94 percent of the UVB radiation that causes sunburn. Sunscreens with SPF values greater than 15 are available. The wearing of sunglasses can make a difference. Sunglasses that block greater than 99 percent of ultraviolet light are recommended. The ANSI Z87 Standard recommends a variety of fixed-density tinted lenses for use in specific job situations involving harmful radiation to the eyes.

Low-Frequency (LF) Radiation

Many studies concerning the effects of LF radiation on humans have been conducted. The primary driver for exposure limits is the induced electric currents in the body—leading to nerve and muscle stimulation—created by exposure to the time-varying electric and magnetic fields associated with LF radiation. Some epidemiological studies show a relationship between power-line frequency magnetic fields and childhood leukemia. Animal and laboratory experiments do not support the magnetic-field-leukemia link, and no biological mechanism has been identified to explain a causal link between exposure and carcinogenesis; therefore, it is not considered in establishing exposure limits. The best advice at this time seems to be to avoid excessive exposure and to reduce exposure as much as seems feasible. As with any form of radiation, the dose of such radiation decreases as distance from the source is increased.

QUESTIONS

1. In what ways do ionizing and non-ionizing radiation harm the human body?
2. How is radioactive material distinguished from other types of material?
3. Name one naturally occurring non-ionizing radiation source.
4. Name one man-made ionizing radiation source.
5. What is a rem?
6. A worker is assembling radioactive thorium units in a control device and uses 5-cm tweezers to handle the material. If she were to use 20-cm tongs, how much would this reduce her exposure to radiation?

BIBLIOGRAPHY

Barat, K. 2006. *Laser Safety Management.* New York: CRC Press.
Hardy, K., M. Meltz, and R. Glickman, eds. 1997. *Non-Ionizing Radiation: An Overview of the Physics and Biology.* Madison, WI: Medical Physics Publishing.
International Commission on Non-Ionizing Radiation Protection. Publications – EMF Web site (http://www.icnirp.de/PubEMF.htm)
National Council on Radiation Protection and Measurement. *Ionizing Radiation Exposure of the Population of the United State, No. 160.* Bethesda, MD: NCRP, 2009.
National Council on Radiation Protection and Measurement. *Management Techniques to Minimize Off-Site Disposal of Low-Level radioactive Waste, No. 143.* Bethesda, MD: NCRP, 2003.

National Council on Radiation Protection and Measurement. *Operational Radiation Safety Program, No. 127*. Bethesda, MD: NCRP, 1998.

Occupational Safety and Health Administration. Ionizing Radiation Web site (http://www.osha.gov/SLTC/radiationionizing/index.html)

Occupational Safety and Health Administration. Non-Ionizing Radiation Web site (http://www.osha.gov/SLTC/radiation_nonionizing/index.html#additional)

Stabin, M. 2007. *Radiation Protection and Dosimetry*. New York: Springer.

13

Hazardous Materials

Richard T. Beohm, PE, CSP, ARM

Exposure to Poisons

The characteristics that make a material hazardous may depend on who makes that judgment and what his or her particular interests are. The degree or level of the characteristics that make the material hazardous may also be subject to debate. If we took everything we read in the newspaper seriously, we might conclude that all materials are hazardous. It is important for organizations that are involved with hazardous materials to foster a cultural awareness of safety. This should involve compliance with government regulations and private standards, safety teams and audits, employee training and empowerment, emergency-response plans, good community relations, and the like, to prevent accidents, fires, explosions, casualties, loss of business and profits due to a loss, poor public relations, civil and criminal litigation, and environmental disasters.

In occupational safety and health, the term *hazardous material* will be used to indicate substances that are flammable, toxic, corrosive, or highly reactive with other materials. In most cases, it is assumed that such characteristics must be evident in a normal workplace environment to be considered hazardous. Some hazardous materials, such as flammable materials, air contaminants, and radioactive materials, have already been discussed in previous chapters.

Hazardous materials can also include materials at elevated temperature and marine pollutants. NFPA, the National Fire Protection Association, has 70 standards relating to hazardous materials.

The term *poison* is used synonymously with the term *toxic material*. A poison is a substance that causes harm to the human body when it is ingested or comes into contact with the body in small quantities. Toxicology is the study of poisons. It is difficult to be exact about the meaning of "a small quantity," but it seems to be understood sufficiently well. Specific limits of quantity can be established, but they tend to be rather arbitrary.

It is difficult, and perhaps meaningless, to try to classify poisons because there are so many variables and many poisons fit many categories, but we can name a few. The most serious are the carcinogens. Others include agents that burn or otherwise destroy sensitive tissue such as the eyes and internal organs. Asphyxiants, narcotics, and irritants make up separate classifications. But while agents that cause dermatitis might be called irritants, some forms of dermatitis are more serious than merely irritations.

Several other terms that appear frequently in literature on this subject need to be explained. In an effort to determine the harmful effects of various substances, it is customary to conduct tests on small animals and to gather data from cases of accidental contact by humans. In testing it is common to refer to the lethal dose or the LD_{50}. The LD_{50} is a dosage of a given substance which can be expected to kill 50 percent of the test animals. The probability of 50 percent is an arbitrary, but nearly universal, statistic used in such testing. The dosage is usually specified in milligrams per kilogram mass of the test animal. If the substance is ingested with the air, the measurement commonly used is the concentration in parts per million parts of air by volume. The concentration that can be expected to produce death in 50 percent of the test animals is then called the lethal concentration or the LC_{50}.

The term *ingested* means that the substance is taken inside the body, as distinguished from skin contact alone. Ingestion may be accomplished by any one of three routes: by way of the lungs (breathing), by way of the stomach (eating or drinking), or by absorption through the skin. Relatively few substances can be absorbed through the skin in large enough quantities to be very harmful. Ingestion by eating or drinking can be accidental, but the act of eating or drinking is voluntary. Breathing, on the other hand, is generally involuntary. A person must breathe more or less continuously in order to live. Furthermore, substances that one eats or drinks can be identified more easily than contaminants in the air. Therefore, the greatest emphasis is placed on poisons that are carried in the air.

In Chapter 6, air contaminants and exhaust systems were discussed, and toxic materials were mentioned briefly in relation to the design of exhaust systems. Toxic materials that are air contaminants will be discussed more fully here. These substances may be solid, liquid, or gaseous, and several names are used to describe the various forms that they can take when carried in the air. Table 13-1 defines these terms.

Table 13-1. Definition of Terms Used to Describe Forms of Air Contamination

Aerosol—A suspension of fine, solid, or liquid particles in air, as dust, fume, mist, smoke, or fog.

Colloid—A gelatinous substance made up of very small, insoluble, nondiffusible particles larger than molecules but small enough so that they remain suspended in a fluid medium without settling to the bottom.

Dust—A broad term used to describe solid particles that are predominantly larger than colloidal size and capable of being temporarily suspended in air or other gases. They do not diffuse but settle under the influence of gravity.

Fog—Suspended liquid droplets generated by condensation or by the breakup of a liquid into a dispersed state, such as by splashing, foaming, or atomization.

Fume—Fine, solid particles dispersed in air or gases and formed by condensation, sublimation, or chemical reactions, such as oxidation of metals.

Gas—A formless fluid that tends to occupy an entire space uniformly at ordinary temperatures and pressures.

Mist—Suspended liquid droplets generated by condensation from the gaseous to the liquid state or by the breakup of a liquid into a dispersed state, such as by splashing, foaming, or atomization.

Smog—A term derived from smoke and fog and applied to extensive atmospheric contamination by aerosols, which arise partly from the activities of human beings. Now sometimes used loosely for any contamination in the air.

Smoke—Carbon or soot particles less than 0.1 micron in size resulting from the incomplete combustion of carbonaceous materials such as coal or oil.

Vapor—A diffusible, gaseous form of substances normally in the solid or liquid state, which can be changed back to these states either by increasing the pressure or by decreasing the temperature.

OSHA regulates hazardous substances that are found in the workplace and represent a danger to the health and welfare of the workers. These regulations require employers to control employee exposures to air contaminants. OSHA's air contaminants standard, 1910.1000, sets permissible exposure limits (PELs) and provides other information such as ceiling values and eight-hour time-weighted averages and maximum limits (ceiling values).

One of the most complete and most widely recognized lists is that published by the American Conference of Governmental Industrial Hygienists (ACGIH). This list with recommended exposure limits is titled "Threshold Limit Values for Chemical Substances and Physical Agents." This publication is valuable also in that it provides background information and data. The OSHA standards include much of the ACGIH list of TLVs and refer manufacturers of chemicals to the ACGIH publication.

A threshold limit value (TLV) is defined as "the time-weighted average concentration for a normal 8-hour workday or 40-hour workweek, to which nearly all workers may be repeatedly exposed, day after day, without adverse effect." Concentrations are in terms of volume, unless otherwise stated. They are usually expressed in parts per million parts of air (ppm) and/or milligrams per cubic meter (mg/m³). The maximum permissible concentration of dusts may be given in million parts per cubic foot (mppcf) or so many fibers per cubic centimeter. In tables of TLVs it is assumed that the values that are listed are time-weighted average values, unless otherwise stated.

Time-weighted average values permit exposure to concentrations above the TWA, but the average exposure must not exceed the TWA value. Exposures above the TWA are also limited and are referred to as the short-term exposure limits (TLV-STEL). They indicate the maximum exposure a person may experience for a short time, which is called an excursion. The following four conditions apply to such excursions:

1. Excursions shall be for not more than 15 minutes.
2. There shall be at least 60 minutes between excursions.
3. There shall be not more than four excursions per day.
4. The total exposure must not exceed the time-weighted average value given for that substance.

In the ACGIH publication mentioned above, the STEL values are given along with the TWA concentrations. ACGIH also provides a guide explaining how the STEL concentrations are determined. For substances that do not have ceiling designations, the TWA concentration is multiplied by an "excursion factor" as follows:

If the TLV-TWA is:	the excursion factor is:
0–1	3
1–10	2
10–100	1.5
100–1000	1.25

The ACGIH also points out that some toxic substances are fast-acting and others are slow-acting. Exposure to high concentrations of a slow-acting substance for a short time would produce little harm, but such an exposure to a fast-acting substance could be quite harmful. However, distinctions between these two types of substances have not been established, and this is an issue the ACGIH intends to address in the near future.

The Mining Enforcement and Safety Administration (MESA), with the help of NIOSH, has approached this problem by defining a concentration that is "immediately dangerous to life or health." IDLH concentrations, as they are called, are identified for use in mines to allow miners to escape in case respiratory protective equipment fails.

There are several substances that are not normally in a form that can be carried in the air and breathed, but which are toxic to the skin. These substances are identified in the ACGIH and OSHA lists of hazardous substances by the word "skin" after the name or by an "X" under the heading "skin."

Table 13-2 shows excerpts from the OSHA 1910.1000 Table Z-1. This is a list of "Limits for Air Contaminants" which contains approximately 2,400 substances.

Table 13-2. Excerpts from OSHA 1910.1000 Table Z-1

Substance	PEL – TWA ppm mg/m^3
Acetone	1,000 2,400
Carbon monoxide	50 55
Dibutyl phosphate	1 5
Fluorine	0.1 0.2
Furfuryl alcohol	50 200

OSHA Table Z-2 contains the limits for special substances which are particularly dangerous and includes the following: benzene, beryllium and beryllium compounds, cadmium fumes and dusts, carbon disulfide, carbon tetrachloride, chromic acid and chromates, ethylene dibromide, ethylene dichloride, formaldehyde, hydrogen fluoride, hydrogen sulfide, fluoride as dust, mercury, methyl chloride, methylene chloride, organo (alkyl) mercury, styrene, tetrachloroethylene, toluene, and trichloroethylene.

OSHA Table Z-3 contains the limits for mineral dusts and includes crystalline silica (quartz), amorphous silica (mica, soapstone, talc, tremolite, and Portland cement), graphite, coal dust, silicon dioxide, and inert or nuisance dusts.

Exposure to more than one toxic substance is measured by adding ratios of the actual concentration to the permissible concentration for each substance. Mathematically this is expressed as

$$E = \frac{C_1}{L_1} + \frac{C_2}{L_2} + \cdots + \frac{C_n}{L_n}.$$

The sum of all of these ratios must not exceed unity. As an example, let us assume that during an eight-hour day a worker is exposed to 15 ppm of ammonia, 200 ppm of ethyl acetate, and 50 ppm of turpentine. The permissible concentrations of these substances are 50, 400, and 100 ppm respectively. The total exposure, then, is

$$E = \frac{15}{50} + \frac{200}{400} + \frac{50}{100} = 0.3 + 0.5 + 0.5 = 1.3.$$

Obviously, this exceeds unity and, therefore, exceeds permissible exposure, even though the concentration of each substance is well below the permissible level.

The duration of exposure is important, too, since the effects of toxic materials are cumulative. To calculate the total exposure for an eight-hour day, the following equation can be used. This total must not exceed the permissible TWA concentration.

$$E = (C_1 \times T_1 + C_2 \times T_2 + \cdots + C_n \times T_n)/8$$

where

> E is the total exposure
> T_1 is the time in hours of exposure to the C_1 concentration.

As an example, let us assume that a worker is exposed to a given poison at 50 ppm for four hours, 80 ppm for two hours, and 100 ppm for two hours, thus:

$$E = (50 \times 4 + 80 \times 2 + 100 \times 2)/8 = 560/8 = 70 \text{ ppm}.$$

If the permissible time-weighted average concentration for that substance is greater than 70 ppm, this exposure is permissible.

No equation has been shown to determine permissible exposure to several substances for varying portions of the day. This does not need to be calculated since exposure to any given substance is controlled by the TWA and excursion limits for that substance. Any combination within these limits is permissible if it satisfies the $C_1/L_1 + C_2/L_2 + \cdots + C_n/L_n \leq 1$ criterion.

Air Sampling

It is important to specify the limits of toxic materials to which workers may be exposed, but actual levels of contaminants must also be determined. Substances in a container can be measured, but substances in the air are not so easily measured or even identified.

Air sampling has come to be a very important activity, and many techniques have been developed to measure concentrations of a wide range of toxic substances, as well as flammable, airborne substances. Two frequently used types of sampling are referred to as "grab sampling" and "integrated sampling," or "short-term" and "long-term" sampling.

Grab sampling, or short-term sampling, involves taking samples of a given atmosphere. The kinds of toxic substances are usually known or suspected in advance, and tests are made for specific substances. A common method of

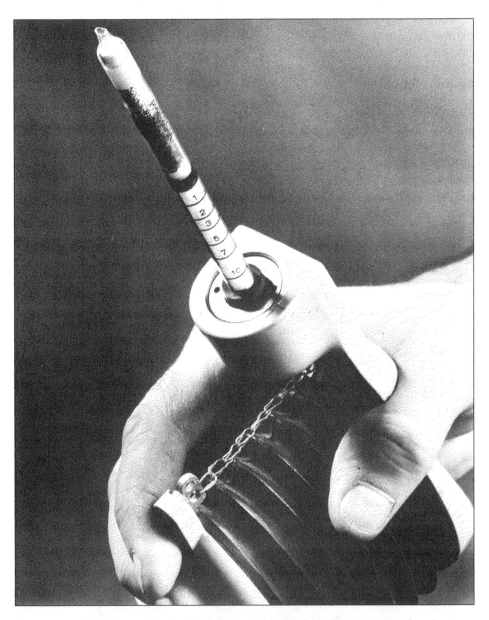

Figure 13-1. A grab-sampling device including a hand-held bellows pump and a tube for testing the concentration of vinyl chloride. Courtesy National Mine Service Co.

doing this involves a hand pump with a sample tube for a specific substance, as shown in Figure 13-1. Chemicals in the sample tube change color as the particular substance enters the tube. The tube is calibrated to indicate the concentration. The pump, if operated properly, draws a given quantity of air into the sample tube, thus giving a consistent sample-quantity for succeeding tests. This technique is sometimes referred to as the calorimetric method of sampling.

The accuracy of such a sampling procedure is always questionable, but, if done properly, it is accurate enough to indicate whether specific substances are present in significant concentrations. If this sampling suggests that a substance is present in concentrations close to or exceeding the permissible concentrations, further testing is in order.

The major advantages of grab sampling are that it is fast, inexpensive, and can be done at any location at any time by a person with relatively little training. If concentrations are likely to vary during the work period, several samples should be taken to determine peak exposures.

Integrated sampling is done continuously over a period of time in order to obtain a time-weighted average exposure. The several techniques for integrated sampling are generally more accurate than grab sampling but do not give any indication of peak concentrations. In most cases, both grab sampling and integrated sampling are required to provide needed information.

Some of the more common methods of obtaining time-weighted average samples include the use of filter-type collectors, precipitators, chromatographic analyzers, and spectrometers. Each of these techniques is suitable for only one or a few types of substances. Some form of instrumentation is needed for each method, and some of them are quite complex. As more accurate instruments are developed, the trend seems to be toward more specialized applications. Instruments used for more than one substance generally require special filters, tubes, or other units for each substance.

It is always important that samples be taken at points where workers are likely to inhale the suspected toxic substances. Samples must be representative of any variations in concentrations in order for both the peak values and the time-weighted average exposures to be determined. When very hazardous substances are involved, it is strongly recommended that continual monitoring be provided.

In determining air quality, it is also often helpful to have workers wear dosimeters, small sampling devices that can be attached to the worker's clothing near his or her face. They are integrating devices that give a time-weighted average exposure and are replaced each day. Figure 13-2 shows one type of personal dosimeter.

Air sampling is also useful in engineering efforts to reduce exposure of workers to toxic materials. If done carefully, sampling can help locate leaks and places where a substance is escaping into the atmosphere. Sampling is also useful in checking the success of attempts to contain and/or control the movement of airborne substances.

Hazardous Materials in the Plant

Research related to hazardous materials has increased dramatically during the past decade. As a result of this research, acceptable exposures to a wide range of substances have changed. Many substances once considered to be benign are

Figure 13-2. A portable gas monitor for sensing the exposure to specific toxic and harmful air contaminants and alerting the worker of any unsafe conditions. Courtesy of BW Technologies, Honeywell Gas Alert Quattro

now listed as hazardous materials. Exposure limits are being established on the basis of what we learn, and we keep learning more. Limits must be considered as being the best estimates that can be set at any given time.

Companies are now required by EPA Standard 40 CFR, Part 68, to formulate and establish a risk management plan (RMP). OSHA also has requirements for process safety management (PSM), 1910.119. The RMP involves identifying worst-case emissions and accidental releases and minimizing the potential as well as the consequences of an accidental release. Computer modeling is often used to identify and assess the potential for accidental releases.

The EPA is also continually developing new rules to supplement the Clean Air Act to control hazardous air pollutants. The Emergency Planning and Community Right-to-Know Act has a list of extremely hazardous substances. Companies also need to identify any Resource Conservation and Recovery Act (RCRA) requirements and develop an effective program to manage them.

OSHA's Hazard Communication Standard, 29 CFR 1910.1200, designates over 600 industrial chemicals as hazardous and regulates their handling and use in manufacturing. The standard includes requirements for container labeling, safety data sheets (SDSs), and employee training. The standard applies to all industries and is designed to protect approximately 32 million workers from exposure to an estimated 650,000 potentially hazardous chemicals. A written HAZCOM program and MSDS for each hazardous chemical are required. OSHA has moved to align the Hazard Communication Standard, 29 CFR 1910.1200, with the international Globally Harmonized System of

Classification and Labeling of Chemicals (GHS), which will affect labeling and SDSs as well as impact regulations from CPSC, DOT, and EPA. The HC/GHS integration will take several years to complete.

Some other examples of hazardous materials are flammable and combustible materials (see 1910.106), highly hazardous chemicals in processes (see 1910.119), hazardous chemicals in laboratories (see 1910.1450), lead and asbestos (see 1910.1001 and 1910.1025), and hazardous waste (see 1910.120).

Of major concern is the exposure of workers to hazards of which they are not aware. OSHA standards now set several requirements for the identification, labeling, and use of chemicals. Manufacturers and importers of chemicals must assess the hazards of those chemicals and communicate that information by means of warning signs, labels, and a Safety Data Sheet (SDS). The employer who acquires such chemicals must inform all employees who will handle, transport, use, or come into contact with such chemicals.

Figure 13-3 is extracted from an SDS prepared by a manufacturer on the chemical toluene. The format may vary, but the essential information must be given (see Figure 13-3).

The use of the SDS is intended to track all chemicals from the time they are manufactured or produced until they are completely disposed of. All workers who come into contact with such a substance, in or out of a container, should be made aware of its hazards and how to protect themselves and others from those hazards. Failure to communicate such information to his employees, who could not read warnings written in English, cost the owner of one company a prison sentence.

A problem sometimes encountered with liquid chemicals is that of leaking and spilling. The SDS gives instructions for dealing with leaks or spills, but measures should be taken to prevent leaks and spills from happening in the first place. Storage facilities need to be designed to segregate hazardous materials. Flammable materials must be kept away from all sources of ignition. Corrosive substances must be kept away from other materials, including containers, with which they might react.

Spills and leaks can also occur during transportation. Containers must be kept in good condition. If a container becomes damaged in any way (by rusting, puncture, cuts, or even significant dents), materials in it should be transferred to a good container. The process of transferring those materials also can be dangerous and must be done carefully.

It is common to find hazardous substances in the workplace in open and unlabeled containers. This is especially true in laboratories. When this condition is necessary in order to perform the required work, the area should be restricted to employees who are properly trained to use these materials.

Containers used in transporting hazardous materials must carry adequate identification. One of the labeling systems used in addition to the DOT is NFPA 704, which is primarily used to represent the characteristics of the material in conditions of fire. The National Fire Protection Association (NFPA) has a system of symbols, or signals, for identifying hazards. Examples of this system are shown in Figure 13-4. The types of hazards are given specific locations in the diamond form. The number on the left side (colored blue) denotes toxicity. The number on the top (colored red) denotes flammability. The number on the right side

There are sixteen new categories for the new OSHA Globally Harmonized System (GHS):

1. Identification
2. Hazards Identification
   GHS Classification:
   Health: Environmental: Physical:
   GHS Label:
   symbols
   Hazard Statements: Precautionary Statements:
3. Composition/Identification or Ingredients
   Component: CAS Number: Weight %:
4. First Aid Measures
   Eye: Skin: Inhalation: Ingestion:
5. Fire Fighting Measures
   Fire Fighting Procedures: Unusual Fire and Explosion Hazard:
   Combustion Products:
6. Accidental Release Measures
7. Handling and Storage
8. Exposure Controls/Personal Protection
   Exposure Limits: Engineering Controls: PPE:
9. Physical and Chemical Properties
10. Stability and Reactivity
    Stablity/Incompatibility: Hazardous Reactions:
11. Toxicological:Information
12. Ecological Information
13. Disposal Considerations
14. Transport Information
    USDOT: IMDG:
15. Regulatory Information
    TSCA: CWA: CAA: SARA: International Regulations:
16. Other Information
    NFPA Ratings:

Figure 13-3. GHS Safety Data Sheet (SDS) format.

(colored yellow) denotes chemical reactivity. The bottom space may be used for special markings, such as water sensitivity, radioactivity, or corrosiveness.

Background colors for this label are shown in Figure 13-4, but the numbers in the same relative locations may be used without background colors or even without the boxes. To draw attention to one predominant characteristic, one company places that number, usually a 3 or a 4, in a black box within the appropriate section.

The degree of the hazard in each category is identified by code numbers 0, 1, 2, 3, or 4, with 0 indicating no hazard and 4 indicating the most severe hazard. These degrees are explained briefly in Table 13-3. It should be pointed out that the NFPA first developed this labeling system for the benefit of fire fighters.

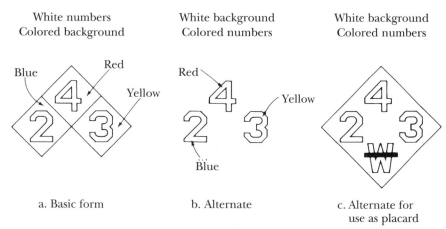

White numbers
Colored background

White background
Colored numbers

White background
Colored numbers

Blue   Red   Yellow

Red   Yellow   Blue

a. Basic form

b. Alternate

c. Alternate for
use as placard

Figure 13-4. NFPA signals for identification of hazardous materials. From NFPA No. 704 Identification System, *Fire Hazards of Materials*.

The familiar Mr. Yuk shown in Figure 13-5 is commonly used on household goods and in public places but is not satisfactory for use in industry. It gives no information about the substance to which it is attached. It merely implies that the substance is poisonous. The NFPA symbols and the DOT symbols shown in Figure 13-6 give too much information for the public in general and would probably be ignored.

Table 13-3. Hazard Classification for Use with NFPA Label

Health Hazard
0—Nonhazardous
1—Causes some irritation, but minor permanent harm will result
2—Causes temporary disability and irritation; prolonged exposure harmful
3—Causes serious harm even with short exposure
4—Can be fatal even with short exposure

Flammability Hazard
0—Nonflammable
1—Ignition point above 1472°F (800°C), flash point above 199°F (93°C)
2—Combustible solids, flash points at or above 100°F (37.8°C)
3—Dusts, flash points at or above 73°F (23°C)
4—Highly flammable, flash points below 73°F (23°C)

Reactivity
0—Stable, nonreactive with water
1—Unstable at high temperatures and pressures
2—Reacts vigorously but will not detonate
3—Will detonate with high heat or strong ignition
4—Very sensitive to shock, explosive

Corrosiveness (suggested)
0—Nonhazardous
1—Slightly irritating to skin and eyes
2—Moderately irritating even in small amounts
3—Burns skin and eyes on brief contact
4—Extremely hazardous, may cause blindness

Figure 13-5. Symbol for poisons, for use in homes and public places.

Figure 13-6. DOT symbols for hazardous materials for use on commercial carriers.

When hazardous materials are transported by a commercial carrier—railroad, highway, truck, air freight, or ship—the Department of Transportation requires that the cargo as well as the vehicle carrying it be identified by DOT symbols. Figure 13-6 illustrates some of these symbols. Others, used for explosives, were illustrated in Figure 11–2. The Code of Federal Regulations contains these symbols and the regulations governing the classification, identification, and handling of hazardous materials (49 CFR 102.1, Parts 171-179).

DOT classifies hazardous materials into various hazard classes.

Class 1 (Explosives and blasting agents)
Class 2 (Gases)
Class 3 (Flammable liquids)
Class 4 (Flammable solids and reactive liquids and solids)
Class 5 (Oxidizers and organic peroxides)

Class 6 (Poisonous materials and infectious substances)
Class 7 (Radioactive material)
Class 8 (Corrosive material)
Class 9 (Miscellaneous hazardous materials)
ORM-D (Consumer commodities)

In planning for storage of flammable, toxic, corrosive, and reactive materials, special precautions must be taken to avert serious problems in case stored materials are accidentally spilled. This problem was addressed in Chapter 9 in connection with flammable liquids, but it is equally important with storage of other types of hazardous materials.

Keeping materials that react with each other separated, not only by distance but also by physical barriers as well, is the first and most essential step. Proper ventilation is also required. When liquids are involved, a gutter, or in some cases a sloping floor with a drain opening, should be provided. However, reactive materials must never meet in these gutters or drains.

When hazardous materials are transported from a storage area to a work station, several precautions should be observed. The labels described above are often placed on a box, tub, or pallet in or on which individual containers are stored, but not on the containers themselves. Therefore, when containers are removed, these identifications do not go with them. It requires extra effort to label each container, but this effort is justified if only one accident is prevented by such identification.

Transporting these containers to the workplace may involve greater risks than transporting them from the dock into storage. Traffic, especially at intersections, and sudden stops and movements create threats to the safe handling of these materials. When materials reach the work station, they are often unloaded at a "set-down" area beside the aisle and adjacent to the work station. If such a set-down area exists, the containers of hazardous materials should be protected against mechanical damage by other vehicular traffic.

Once in the work station, materials are usually transferred to a smaller container, and identification is again lost. It is important to comply with the new OSHA Hazard Communication Standard in additional labeling, including hazard identification and target organs. If special precautions need to be observed, they should be posted. Again, this is especially important when two or more persons use the work station.

Hazardous materials that are gaseous or are likely to vaporize require an exhaust system. Exhaust systems were discussed in Chapter 6. Particular note should be made of the rule that hazardous gases must never be drawn by the worker's face. The flow of toxic fumes and gases must be exhausted to the rear of the bench or downward through a perforated bench top. If toxic substances are used in equipment rather than on a bench, care must be taken to assure that there are no leaks. Such equipment should never be allowed to exhaust toxic materials into the open air.

If materials are very toxic, some sort of monitoring device should be installed at work stations where they are used. In some industries, toxic gases and liquids are carried around the plant by a piping system. In these installations, provisions must be made for control valves at reasonable intervals. Provision must also be made for disconnect joints to allow the purging of the pipes. Monitors should also be placed at valves.

Purging with an inert gas, such as nitrogen, is done to remove any foreign matter that may have leaked into the piping system. Water or even air can be a serious contaminant to some other liquids and gases. Purging involves forcing all existing liquid or gas from the system. Even traces of the substance at crevices must be removed. Usually an inert gas is forced through the system until no trace of the hazardous substance can be detected. Obviously, the inside of such a piping system must be as smooth as possible and free of hidden recesses.

Hazardous materials that are not consumed in the process operation or exhausted must be collected in some other manner. If it is feasible, waste materials should be treated at the point where they are generated as waste in order to make them inert or at least less hazardous. Hazardous solids are usually in the form of small particles. Hazardous solids and liquids must both be treated to render them harmless before they are released to the earth as waste.

Hazardous Waste

The disposal of hazardous waste is one of industry's, and the public's, major problems. This waste material has polluted the air, waterways, and even sources of drinking water. Landfills lined on the bottom and sides with an impermeable layer of plastic now serve as a temporary solution in many areas, but this is expensive, and the space to hold all the waste that is generated is limited. The ultimate solution is to find ways to render these materials nonhazardous before they are discarded. Better still, we should find ways to use these materials in some useful way instead of treating them as waste.

Hazardous waste is solid waste that poses substantial or potential threats to public health or the environment. It is regulated under the Resource Conservation and Recovery Act (RCRA) and is defined in 40 CFR 261. The four traits of hazardous waste are: (1) ignitability; (2) reactivity; (3) corrosivity, and (4) toxicity.

The safe handling of hazardous waste should be an integral part of any hazardous waste program. All possible exposures should be identified. Exposures from splashes, vapors, and particle releases should be evaluated and minimized or eliminated. The program should include procedures for reducing the risk of fires, explosions, and chemical reactions.

The program needs to show employees how to identify, handle, and store hazardous waste at the plant. It should teach employees to follow EPA, DOT, and OSHA requirements for the preparation and shipment of hazardous waste for off-site treatment and disposal. Employees should be trained in DOT's HM-181 requirements for the shipping of hazardous waste.

Some companies have made a great effort to reduce the amount of hazardous materials they use by substituting other less hazardous materials. Some have found ways to recycle some of these materials. Others have developed more efficient ways of using them, thus reducing the amount used.

Hazardous materials that must be discarded must be carefully collected and labeled. This requires a commitment on the part of management at all levels to see that this is done in a safe and thorough manner.

Hazardous wastes should not be allowed to get into any drain system unless that system leads to a treatment facility that will render them safe. Most acids, but not all, and alkalines can be neutralized. Hazardous solids can usually be separated and removed from a drain system. Check valves and/or vents may be required in some drain systems to assure that waste and gases do not back up into other work areas.

Polyvinyl chloride (PVC) pipes are often used in drain systems and for ducts in exhaust systems that carry hazardous fumes and gases. However, vinyl chloride has been found to be a carcinogen. It should not be used where it may become rubbed to form dust particles that could be carried in the air.

Of particular interest is the problem of disposing of carcinogenic materials, including radioactive materials. We are learning that there are many more substances that can cause cancer than we knew about earlier, and the list keeps getting longer. As mentioned in Chapter 2, OSHA has adopted a "cancer policy" and has attached it to the standards for general industry.

Within the OSHA standards, there is also a section related to hazardous waste operations and emergency response (1910.120). Under this provision, any employer with employees involved with hazardous waste operations must have an acceptable written safety and health program.

To date, toxicity studies have been conducted on only a small percentage of chemicals licensed for use by U.S, industries.

Community Right-to-Know

In 1986, Congress passed a comprehensive bill aimed at protecting workers and the public from hazardous chemicals. The entire bill is called the Super-fund Amendments and Reauthorization Act of 1986 and is referred to by its acronym, SARA. Part of that act is Title III, The Emergency Planning and Community Right-to-Know Act of 1986. Emergencies may occur that involve local industries, railroads, highway trucks, and even local businesses that may sell, store, or handle hazardous chemicals. Furthermore, fire departments, rescue squads, and other services may be called upon in case of emergencies and need to know about any chemical hazard with which they may come into contact.

Title III of SARA permits states to develop more stringent rules than the federal rules but nothing less strict. Title III requires that facilities within a community must submit copies of their MSDS (SDS) or a list of MSDS (SDS) chemicals to:

a. the local emergency planning committee,
b. the state emergency response commission, and
c. the local fire department.

Trade secrets have been difficult right-to-know issues, and this legislation has included some provisions that will allow complete identification to be withheld. However, the hazards presented by the product must be made known in some way. A special section allows a company to submit specified information without divulging the trade secret. Even so, this has caused concern among the users of such chemicals.

Use of Personal Protective Equipment

Once more, the use of personal protective equipment should be the last resort after all other corrective measures have failed to control the hazard satisfactorily. If the hazardous materials could be eliminated, there would be no need for personal protective equipment. If these materials could be so well controlled that there would be no possibility of contact between them and people, there would also be no need of personal protective equipment.

In reality, there can be no guarantee that hazardous materials will be completely controlled. Therefore, workers must use personal protective equipment to ensure that any exposure will not produce injury or illness. The type of equipment will obviously vary from one case to another. The manner of exposure, concentration of the offending material, and possible duration of exposure will determine what protection is needed.

Training and supervision are important, too. A common problem is that workers cannot identify hazardous materials, especially after they are spilled or escape into the air. For example, fluoric acid has the same appearance as water, but if it comes into contact with a person's skin and is not quickly washed away, it will eat deeper and deeper into the flesh.

Workers must be informed of the hazards of materials that they may contact. It is common for workers to belittle the danger that hazardous materials present. They should not be unnecessarily fearful, but they should be made to understand the real hazards. The use of personal protective equipment will not be very effective if the workers themselves do not appreciate why it is to be used.

Hazardous Conditions Caused by Nonhazardous Materials

Many times, hazardous conditions have resulted from the lack of control of substances that in themselves are nonhazardous. Two events that actually occurred will illustrate such conditions.

A young family bought a house in a residential area and discovered that a previous owner had built a bomb shelter partially within the basement and out under the lawn. Quite a number of such shelters were built during the mid-1950s. This particular shelter was well planned and had food and water stored on shelves, an air-circulation system, a sump pump, and a tank of oxygen. Apparently all this was put into place before a small metal door was set in concrete dividing the shelter from the basement. When the new owners bought the house in 1978, they found the oxygen tank badly pitted and the valve so corroded that it could not be opened. Furthermore, they discovered that the entrance to the shelter, being angled, would not allow the tank to be removed.

Is there a hazard in keeping a tank of oxygen in the basement? Not as long as everything is intact, but the corrosion would eventually cause the tank to burst, or at least allow the oxygen to escape. Also present in the basement were a gas-burning furnace and the usual collection of combustible odds and ends. At least two potential hazards existed: the pitted tank, which could explode, and the oxygen-rich air, which could cause a fire. Eventually a local welding supply company was able to open the tank and allow the oxygen to escape outdoors through a hose.

Another incident resulted in tragedy. A man whose business involved the use of nitrogen gas kept a tank of nitrogen in his panel truck. His 18-year-old son entered the back of the truck and was later found asphyxiated. The truck was not ventilated, and the tank of nitrogen had leaked. The nitrogen had gradually displaced the oxygen in the confined space until the air could no longer support life.

QUESTIONS AND PROBLEMS

1. How has the international GHS affected U.S. regulations?
2. What is the OSHA PEL (ppm) for acetone?
3. List some of the ways in which an organization that handles hazardous materials can foster a cultural awareness of safety.
4. A researcher is exposed to 300 ppm of acetone, 400 ppm of ethyl alcohol, and 30 ppm of furfuryl alcohol. Is this exposure permissible?
5. Name four characteristics of a hazardous material.
6. Name three agencies to which Title III of SARA requires facilities in a community to send a list of their MSDS chemicals.
7. How does DOT classify hazardous materials?
8. Name some important precautions in planning for the storage of hazardous materials.
9. What are the four NFPA label hazard classifications?
10. What is the difference between OSHA PELs and ACGIH TLVs?
11. Calculate the total exposure for an eight-hour day in ppm, if the worker is exposed to a toxic chemical at 75 ppm for four hours, 80 ppm for 2 hours, and 50 ppm for two hours.
12. Name two methods of air sampling to determine concentrations of toxic substances.

BIBLIOGRAPHY

1910 OSHA Guide - 2012. Neenah, WI: J.J. Keller & Associates, Inc.

American Conference of Governmental Industrial Hygienists (updated annually). Document of Threshold Limit Values. Cincinnati, OH: ACGIH.

American Trucking Association, 1989. *Handling Hazardous Materials.* Alexandria, VA: ATA.

Best's Safety & Security Directory, 1999. Oldwick, NJ: A. M. Best Company.

Cralley, Lewis J. and Lester V. Cralley, 1985. *Patty's Industrial Hygiene and Toxicology.* 2nd ed. New York: Wiley.

Freeman, Harry M., 1989. *Standard Handbook of Hazardous Waste Treatment.* New York: McGraw-Hill.

Haight, Joel M. ed. 2012. *The Safety Professionals Handbook,* 2nd ed. (2 vols). Des Plaines, IL: ASSE.

Lewis, Richard J., Sr., 1990. *Carcinogenically Active Chemicals.* New York: Van Nostrand Reinhold.

Martin, William F. et al., 1987. *Hazardous Waste Handbook for Safety and Health.* Stoneham, MA: Butterworth.

MSDS Collection. Schenectady, NY: Genium Publishing Corp.

National Fire Protection Association, 2008. *Fire Protection Handbook,* 20th ed. Vols. 1 and 2. Quincy, MA: NFPA.

Plog, Barbara A., 1988. *Fundamentals of Industrial Hygiene.* Chicago: National Safety Council.

Sax, N. Irving and Richard J. Lewis, 1989. *Sax's Dangerous Properties of Industrial Materials.* 7th ed. New York: Van Nostrand Reinhold.

Tally, John T., ed., 1988. *Industrial Hygiene Engineering.* Chicago: National Safety Council.

Traverse, Leo, 1990. *The Generator's Guide to Hazardous Material Waste Management.* New York: Van Nostrand Reinhold.

14

Mechanical Hazards

Steven W. Hays, CSP

Classifications and Control Methods

Mechanical hazards are distinguished from other kinds of hazards by the fact that force is transferred by means of the motion of one or more objects. In some cases the object in motion is the human body or a body member. For ease in discussion, mechanical hazards have been divided into several classifications, which also help in recognizing hazards.

Table 14-1 lists and defines eight classes of mechanical hazards. The first six are usually found in or on equipment, although they could be found elsewhere. For a hazard in any of these six classes to cause an injury, a person has to place a body member close to it. The last two, however, require no action on the part of the injured person; he or she is simply located in the path of the moving object.

Squeeze points often occur when machines have tables that reciprocate slowly back and forth. The squeeze point exists between the end of the table and a wall or another machine located close to it. Another situation that falls in this category better than into any other is illustrated in Figure 14-1. In this case, the motion of the worker creates the squeezing action.

The first method that should be applied to correct such a hazard is to provide enough space to eliminate the squeeze point. If this is not feasible, a guard or barrier is usually required. In the situation illustrated in Figure 14-1, several types of guards could be installed, two of which are illustrated in Figure 14-2.

Pinch points are commonly found in machine mechanisms. There are usually several points on the vertical sliding members of a press at which a person could get a finger or hand pinched or crushed. Such presses are required to have a guard covering the whole area where these pinch points exist. Chapter 17 discusses several types of guards used with presses.

Another common type of pinch point occurs in clamping devices, actuated either manually or by pneumatic or hydraulic cylinders. Figure 14-3 illustrates a pinch point created by two pivot bars used to hold a guard in a raised position

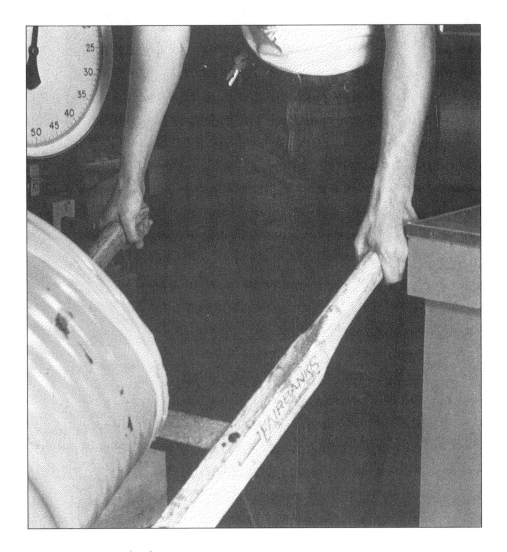

Figure 14-1. Example of a squeeze point.

so that a belt can be shifted on a step pulley. This piece of equipment violates one of the principles concerning the design of guards that will be discussed in Chapter 17: the guard itself creates a hazard.

As with any other hazard, an attempt should be made to eliminate a pinch point. In most cases, pinch points are created by a mechanism needed to provide some motion. Eliminating the pinch point by substituting a different type

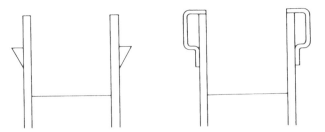

Figure 14-2. Guards on hand truck to reduce the hazard of squeeze points.

Table 14-1. Classifications and Definitions of Mechanical Hazards

Squeeze points—Squeeze points are created by two objects, one or both of which is in motion, as they move toward one another. The objects do not touch but come close enough (roughly 2 to 12 inches) so that an arm, leg, or other body member could be squeezed between them.

Pinch points—Pinch points are created by two objects that come together, or within roughly ¼ inch, in such a way that they could pinch human flesh.

Shear points—Shear points are created by two objects with relatively sharp edges, one or both of which is in motion; when they come together, the sharp edges pass one another so that a finger, hand, or other body member could be cut by this action.

In-running or nip points—In-running or nip points are created when two moving objects, one or both of which has a rotating motion, move closer and closer together until they reach the point of contact. A person's hand or other body member could be pulled into the narrowing space and crushed.

Catch points—Catch points are created by objects, either stationary or in motion, that have sharp corners or rough surfaces capable of snagging a person's skin or clothing. A catch point may itself cause injury, but, more seriously, it may pull a person into another hazard.

Sharp points and edges—Sharp points and edges are those that are sharp enough to puncture or cut a person's skin or damage another object when they hit or are hit by the other object at a reasonable speed.

Flying objects—Flying objects are those objects in motion with sufficient kinetic energy to cause injury or damage when they hit a person or another object.

Falling objects—Falling objects are objects falling by gravity with sufficient kinetic energy to cause injury or damage when they hit a person or another object.

of mechanism usually results in exchanging one hazard for another. Therefore, the most feasible corrective measure usually is to install a guard to cover the pinch point. The pinch point illustrated in Figure 14-3 could be eliminated quite easily by using a different mechanism without any significant hazard, as shown in Figure 14-4.

Shear points are not limited to edges that were designed to cut material. Paper shears and sheet metal shears as shown in Figure 14-5 have obvious shear points. Metal shears large enough to cut steel bars 15 centimeters in diameter are used in some industries. If that is the purpose of the mechanism, the shear point cannot be eliminated without substituting a completely different device; that is certainly worth considering. Otherwise, the shear point may be covered with a suitable guard.

Frequently other mechanisms not intended as shearing devices nevertheless have shear points. Pivoted arms such as those shown in Figure 14-3 may be sufficiently sharp to cause cutting in addition to pinching. Either redesign or guarding should be considered.

Cutting boards used for cutting paper are frequently found in offices, photo laboratories, and so on. Some cutting boards have been designed with a plastic strip over the paper close to the cutting edge. This encourages the user to hold the paper at a point farther away from the cutting edge, but it does not cover the shear point very well. At best it substitutes a pinch point for a shear point, which would result in a less serious injury.

In-running or nip points have caused many serious injuries. Figure 14-6 illustrates a nip point between two gears. Machines built many years ago com-

Figure 14-3. This linkage supporting a guard presents several pinch points as it is raised and lowered.

monly had exposed gears, belts, and pulleys. Such obvious nip points are almost always covered with guards on newer machines, but some less obvious ones, such as the chain in Figure 14-6, are not.

Nip points are commonly seen on a belt conveyor. Wherever a belt travels toward a rotating pulley or idler, there is a nip point. Support rolls under the belt also create nip points and are frequently exposed.

A worker in a marble shed tried to extricate a strip of metal that had become wedged between a conveyor belt and a pulley at one end of the conveyor. Wearing gloves, he reached with a stick to pry the metal strip out of the way while the conveyor was running. The nip point caught the stick and pulled it around the pulley. The stick, in turn, twisted the glove, tightening it on the worker's hand and pulling his hand into the pulley. His badly mangled hand was later amputated. Similar accidents have happened in many industrial plants and on farms.

It is usually not feasible to eliminate all nip points, although elimination should always be considered. They should otherwise be covered with a guard or moved out of the reach of workers. It is frequently very difficult to design a

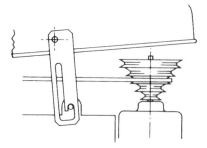

Figure 14-4. A mechanism to replace that shown in Figure 14-3, thus eliminating pinch points.

Figure 14-5. An unguarded shear point.

guard that covers the hazard sufficiently and still allows the device to be used. When designing a guard for a nip point or any other guard, the safety engineer should not attempt to do this by himself/herself. To design an effective guard that interferes with the operator or maintenance employees as little as possible or not at all, the operator and maintenance employee(s) should be involved. Depending on the equipment being guarded, the area supervisor and other engineers may also need to be involved.

Catch points are often not recognized as hazards. In some cases they are quite obvious, but often they are not. Typical catch points include burrs on the end of a rotating shaft, a key projecting from a rotating shaft, the splice on a moving belt, and even the rough surface of a moving leather belt. As mentioned in Table 14-1, a catch point may by itself cause injury, but more often it pulls a person into another hazard.

A catch point was a hazard in the tragic accident that was described in Chapter 2 and that bears repeating here. A young man working on a farm was operating a silo unloader. Because the silage was partly frozen, some areas were hard and others were still soft. The silo unloader was on top of the silage digging it out and blowing it into a chute that carried it down to the ground where it could be fed to cattle. A drive wheel continuously moved the device around the inside of the silo. The whole machine turned around a vertical shaft in the center of the silo, and a drive shaft extended from the electric motor at the center to the drive wheel at the perimeter.

Figure 14-6. An unguarded nip point.

On this occasion the drive wheel had gotten stuck in a soft spot in the silage. The farm worker went up into the silo, saw what was happening, and attempted to pull on the frame of this machine. He apparently reasoned that he could not lift it alone but might be able to do it with the drive wheel still turning. The head on a setscrew on the slowly turning drive shaft caught the worker's belt buckle and twisted his clothing around the shaft. He could not reach the switch to stop the machine, nor could he release the buckle. He was found dead with his shirt and pants twisted tightly about his chest and abdomen.

The head on the setscrew was not needed, and, in fact, headless setscrews were used in other places on the same machine. It would have been easy to eliminate the catch point.

If a catch point cannot feasibly be eliminated, a guard should be made to cover it. A major problem with most catch points is recognizing them as hazards.

Sharp points and edges can exist almost anywhere, but sometimes they are not as obvious as one might expect. Many points and edges appear to be quite dull to the touch but can easily puncture or cut human flesh when one or the other is in motion.

A woman working in an industrial plant reported to the medical dispensary with a cut on her arm. She was not aware of what had happened at the time, but recalled that she felt something scratching her arm when she passed by a particular drinking fountain. Upon investigation it was found that nickel plating on a metal strip around the fountain had left a ragged edge at the bottom. It was not sharp to the touch, but the motion of her arm was enough to cause the ragged edge to cut her.

It is common practice and a closely observed rule in many plants that all edges and corners be "broken," or filed smooth, on all jigs, fixtures, and hand tools, except, of course, the cutting edges on cutting tools. In the manufacture of metal parts, it is common to have a deburring operation to remove burrs and sharp edges prior to an operation in which the parts may be touched with bare hands. In other cases, workers wear gloves to protect their hands against sharp edges.

In some situations it would be impractical to try to remove sharp points and edges. In such cases guards or barriers should be made to cover the hazards. Good housekeeping will help by keeping objects out of aisles and work areas.

Flying objects can be anything from particles from a grinding wheel to a swinging door. In fact, the size of the object is seldom a major factor in determining the damage a flying object can do. As mentioned earlier, injury from this hazard does not require that a body member be placed close to the source. There is no limit to how far an object may be propelled to inflict injury or damage. In some cases it may be appropriate to consider a body member, such as an arm, to be the flying object. The chuck jaws in Figure 14-7 are flying objects.

With the development of robotics came a new cause of accidents. Workers are sometimes hit by moving arms of the robot. Occasionally a robot malfunctions and causes an accident, but more often a worker fails to get out of the way of the moving arm. In some cases a robotic mechanism may act as a squeeze point or a catch point. In most cases it will act as a flying object.

A robotic mechanism becomes a more serious problem when it is designed to interface with human activities. To avoid the possibility of human error, a robotic device should be designed to be as independent of human operations as possible. When parts of the mechanism must swing or move into an area

Figure 14-7. The protruding jaws indicated by the arrow on a rotating chuck are an example of flying objects.

where a human worker is likely to be, a guard should be erected (consult ANSI/RIA R15.06 – 1999 (R2009) Industrial Robots and Robot Systems – Safety Requirements for specific robotic guarding standards). In Figure 16–18 (Chapter 16), the long arm holding a hot workpiece moves in and out and also swivels to load and unload the workpiece in the machine. A railing completely encircles the robot to keep people out of the way. (The railing does not show in this photograph because the picture is a close-up within the railing.) Safeguarding devices such as light curtains and pressure-sensitive mats can be useful for guarding robots.

Falling objects are more common than one might expect. They include nuts, bolts, or other parts and tools that a maintenance person might drop when working on an elevated platform or catwalk. Other falling objects might include an elevated fan that falls when the mounting bracket fails or a chunk of concrete that spalled off the concrete ceiling. When personnel are working on an elevated work surface, such as scaffolding, a scissors lift, or an aerial boom lift, they should barricade the area below them to keep others out of the area where a falling object could hit them.

Extent of Injury

The seriousness of an injury depends upon several factors. In the *Annals of the New York Academy of Science*, Vol. 107, William Haddon, Jr., proposed the concept of a three-phase sequence in any injury. This concept has been adopted by several others in research of injuries. The three phases are as follows:

1. The pre-energy release, or pre-injury phase, includes the factors leading to the accident.
2. The energy-release or injury phase involves the transfer of energy in a collision between a person and an object, and the resulting injury.
3. The post-injury phase involves the treatment of the injury.

A safety engineer should be concerned at least about the first two phases. The types of stresses placed on a worker and the types of errors a worker can make were discussed in Chapter 4. These stresses and errors are among the factors in the first phase.

The energy-release phase addresses the factors pertinent to this chapter. When an object and a person collide, the seriousness of the resulting injury depends on the following factors:

1. The sharpness of the object, or concentration of the energy
2. The momentum of the object in motion, whether it be the person or another object
3. The threshold of injury of the part of the body that contacted the object

The threshold of injury refers to the ability of the body to resist injury. Some areas of the body are much more sensitive than others and cannot tolerate transfer of very much energy.

Like flying objects, falling objects inflict injury by virtue of the momentum gained in moving at some speed. The falling object is usually thought of as a box, a tool, or some other object that vibrates off a shelf, a cabinet, or even a work bench. The falling object can also be a person falling onto the floor or against another object. In any case, the injury or damage inflicted depends on the same factors as those that are involved with flying objects.

In considering these mechanical hazards, it should be realized that all eight can cause injury to people but only the last two are likely to cause damage to equipment. The first six could cause damage if objects were placed at those points, but it is human injury that is the major concern.

Pressurized Containers

Another common source of injury and damage from mechanical hazards is the rupture of pressurized containers. This is usually not thought of as another classification of mechanical hazard, and perhaps could be considered under flying objects. However, the eruption of pressurized containers has certain unique characteristics.

Pressurized containers include aerosol cans, gas cylinders, compressed air tanks, as well as hydraulic and pneumatic devices, hoses, and fittings. All of these are common items in most industries. Some were mentioned in Chapter 11 when explosions were discussed. An explosion is usually associated with a detonation or deflagration involving a fuel, but the bursting of a container may also be considered an explosion. It is not clear just where the line of distinction lies between the rupture of a container and an explosion.

Rupture of a container may present several hazards, including flying objects, release of toxic material, and release of flammable materials. It is the flying objects that are of primary concern here. Among the flying objects may be fragments of the ruptured container, or the whole container, and in some cases the stream of liquid or gas released by the rupture.

The rupture of a hydraulic hose begins as a very small opening. The hydraulic fluid, which is under pressure, is forced out at a very high velocity. This jet of fluid has caused severe cuts and abrasions to people in its path. The force of such a jet is so great that it has been used experimentally to cut metal. The effects of a jet of fluid on human flesh are compounded by the fact that the fluid also spreads within the body, causing extreme pain and often serious damage to blood vessels, nerves, muscles, and other organs.

Very little research or experimentation has been done to determine the relationships among escape velocities and pressures and line velocities and pressures. Such calculations are complicated by the irregular shapes of ruptures. Generally the pressure coefficients for well-defined side openings, often called branches, range from about 1.0 to as much as 2.7. The velocity in the main line is usually very low or zero. It must suffice at this point just to realize that a jet of liquid or gas has a high velocity and is capable of inflicting a serious wound.

In any hydraulic or pneumatic system, there are many points at which bursting must be considered a possibility. There must be a pump or other device to provide pressure for the system. There must be some device that uses the fluid or utilizes the pressure, and there must be hoses or pipes, fittings, gauges, and valves. Each of these components must be designed to withstand the pressure to which it will be subjected. In addition, the joints are often the weakest part of the system, probably because they are not really designed but are merely selected as "suitable."

Another weak point is commonly found in hydraulic systems where the hose or pipe is stressed from continued flexing or from a bend made with too small a bend radius. These factors should be considered in the design or selection of hoses and pipes but are often overlooked. Although the hose may hold for a while, fatigue often weakens it to the point of failure. This is the kind of failure that often results in a very small opening and a high velocity jet.

Exposed hose or pipe, valves, and fittings are often damaged by forklift trucks and even by manual material handling. A flexible hose often has a woven metal shield or coiled steel wire around it to protect it from mechanical damage. This may also increase its bursting strength. Additional barriers may be required if crushing loads will be exerted against the hose or pipe.

In designing pressurized containers and pressure vessels, a safety factor is used to allow for flaws in materials and for other weaknesses that are not easily calculated. It is usually taken into account in determining the maximum allowable stress to which a given container may be subjected or in selecting the strength of a material to be used. The ASME Pressure Vessel Code stipulates that the maximum allowable stress to which a material may be subjected must be the lowest value of the following:

1. 25 percent of the specified minimum tensile strength at room temperature
2. 25 percent of the specified minimum tensile strength at the design temperature
3. 62 percent of the specified minimum yield strength at room temperature
4. 62 percent of the yield strength at design temperature

For example, if an air compressor tank is being designed to hold air at some given pressure and it will be kept approximately at room temperature, we must make calculations for its thickness and other characteristics based on some material strength. Let us assume a carbon steel will be used with a specified yield strength of 30,000 psi and a tensile strength of 55,000 psi. Since 25 percent of 55, 000 is 13,750 and 62 percent of 30,000 is 18,750, the 13,750 psi value must be used.

The use of tensile strength versus yield strength may be an important choice. Tensile strength is the maximum strength before failure, whereas yield strength is the point of stress at which the material begins to deform permanently. Ductile materials deform considerably before failure occurs, but brittle materials deform very little. In some cases, such as a cylindrical tank with no internal pipes or other devices, the tank may deform quite a lot without doing any harm. Another tank with pipes, fittings, and so on, may pull apart at joints as a result of very little deformation. The manner in which a pressurized container might fail must be considered in choosing appropriate strength or stress characteristics.

Most pressurized containers and/or systems used in industry should have some device to prevent excessive pressure from developing within the system. In addition to, or instead of, manually operated valves to release pressure, there should be an automatic device to release pressure when it rises above a given level. There are two types of automatic devices. One releases only the excess pressure; the other releases all the pressure. The first is a regulating device or a safety relief valve; the second is often referred to as a safety plug or rupture disc.

A relief valve generally has a spring or a weighted lever against which the pressure is exerted. When the pressure exceeds some predetermined amount, the valve opens, but it closes again when the pressure is reduced to the predetermined amount. Other pressure-sensitive devices are used when more accurate control of pressure is desired. It is important to ensure that relief valves are inspected, maintained, and tested as required by the manufacturer.

A safety plug or valve opens at a predetermined pressure and stays open. Usually a plug or disc is involved that must be replaced after it comes out. In some cases in which pressure results from increased temperature, the valve may be opened by a temperature-sensing device. It is important to ensure that these replaceable devices are inspected and maintained as required by the manufacturer.

It is also important to ensure that relief devices are not "valved out" so that there is a valve between the pressurized container or system and the relief device. If there must be a valve there, it must be locked in the open position.

Compressed air is used in practically all industries and is a common cause of accidents. Perhaps it is because its use is so widespread that it is so often misused. It is especially dangerous when used for cleaning machines and work areas. Debris, chips, and dirt are frequently blown into the air, into machines and instruments, and, more seriously, into people's faces and skin. OSHA Standard 1910.242(b) states: "Compressed air shall not be used for cleaning purposes except where reduced to less than 30 psi and then only with effective chip guarding and personal protective equipment."

Air pressure can be regulated at the compressor or closer to the point of use. The latter allows a main-line pressure of 80 to 100 pounds per square inch, which is common for the use of pneumatic devices and which is then reduced for cleaning purposes. In many plants, special air nozzles with vent holes are used instead of methods involving the regulation of line pressure. When the nozzle is held directly against a pressure gauge, or "dead-ended," 30 pounds per square inch or less is registered by the gauge and the excess is vented off. This is accepted by OSHA because there is no convenient way of measuring the pressure of a free-flowing jet of air. Nevertheless, it skirts the intent of the standard and still allows a hazardous condition to exist. Piping that is not designed for pneumatic pressure, such as PVC pipe rated for liquid pressure, must not be used for compressed air.

Hot surfaces and very cold surfaces are often included in the category of mechanical hazards. They present two hazards: burns and the element of surprise mentioned in connection with some other hazards.

As with any other hazard, the first attempt at correcting it should be to eliminate it. Soldering may be replaced in some cases by adhesives, which, in fact, have many advantages other than safety. Some adhesives, however, have hazards of their own, such as toxicity, flammability, and the potential for adhesion to skin tissue.

Some hot surfaces and very cold surfaces, such as steam pipes and liquid nitrogen pipes, can be covered, and such coverings can also be identified by word or color code. Other surfaces that cannot be covered during some process might be covered when the process is not in operation. Warning signs are of some help. Whenever there is a possibility that a worker may contact a hot or a very cold surface, the worker should wear gloves.

The unexpected release of energy in any form can cause serious injury or death. During the repair or maintenance of equipment that involves stored energy, that source of energy should be shut off by some valve or switch device. Furthermore, that valve or switch device should be controlled in such a way that other workers are prevented from inadvertently turning it on again while the repair or maintenance is being done. This is done by a lockout or tagout.

Lockout or tagout procedures have been common practice on electrical equipment for some time and are explained and illustrated in the chapter on electrical hazards. The OSHA Control of Hazardous Energy (Lockout/ Tagout) Standard (1910.147) requires such a lockout or tagout procedure for equipment that uses stored energy—electrical, mechanical, chemical, hydraulic, pneumatic, steam, gravity, springs, or any other form. Some of these forms of energy are overlooked, which increases the change of unexpected energization of the equipment.

Lockout is a procedure in which a padlock or other locking device is placed on the switch of the source of energy so that the circuit cannot be energized without removal of the lock. If two or more workers are doing maintenance on the same unit, each worker must place his or her own lock, and all the locks must be removed before the circuit can be started again. This prevents accidental energizing of the circuit. If there is more than one energy source present on a piece of equipment being locked out, there must be an equipment-specific lockout procedure that details the types of energy and how to lock them out.

Force of Flying Objects

In Chapter 7 it was mentioned that a railing around an opening in the floor must be able to withstand a force against it of 850 N, or 200 pounds. Given that value, one can design a railing to withstand that force. It is not so easy to determine the actual force against the railing. The force of an object in motion is expressed as $F = ma$, a fundamental familiar to any engineer. If it takes a force F to cause a mass m to accelerate at a rate a, then it must take a similar force to decelerate it at the same rate. If one could determine the length of time it takes a railing to reduce the velocity of a person walking into it to 0, one could then calculate the force.

Another, usually easier, approach is to determine the kinetic energy involved, using the familiar $E_k = mv^2/2$. If the energy that must be absorbed to stop a moving object is known, a railing can be designed that is capable of absorbing that energy with a given amount of deflection. In fact, in many cases the deflection may be a critical factor. Most engineers learn to recognize values of force, but a sense of value of energy is not as easy to acquire. This concept is generally a better one to use in calculations of moving or flying objects, however.

For example, if a lathe operator is struck in the face by a metal chip flying off the workpiece, what energy is absorbed by his face? Let us assume that the chip leaves the workpiece at the tangential velocity of the workpiece. If the work is rotating with a surface speed of 40 meters per second, and the mass of the chip is 0.04 kilograms, the kinetic energy is

$$E_k = \frac{0.04 \times (40)^2}{2} = \frac{0.04 \times 1600}{2} = 32 \text{ J.}$$

As a second example, suppose a worker is carrying a ladder 3 meters long on his shoulders, supported at the midpoint of the length of the ladder. As he turned a corner, the back end of the ladder hit another worker standing at the corner. If the ladder had a mass of 12 kilograms and the end of it was moving sideways at a rate of 1 meter per second, what energy did the second worker absorb? There are many other factors which might be taken into account but, ignoring them, the energy absorbed was

$$E_k = \frac{12 \times (1)^2}{2} = \frac{12 \times 1}{2} = 6\,\text{J}.$$

Calculations are extremely difficult in many situations involving flying objects, and, therefore, forces and/or energy involved are estimated. In some cases even estimating is difficult, and guards or barriers are designed from trial and error. For example, consider the energy that must be absorbed by a guard on a grinding machine when a grinding wheel breaks. If the rubber hose of a compressed air line is yanked off the nozzle at the end, the hose usually whips around erratically until the air pressure is dissipated. How much energy is imparted to a worker's leg by the whipping hose?

Falling objects dissipate energy when they strike a person or another object the same way that flying objects do. But falling objects accelerate as they fall under the influence of gravity. We can determine the velocity if we know the distance of the fall. The velocity is given by the relationship $V = \sqrt{2gs}$ where g is the acceleration of gravity, or 9.807 meters per second squared, and s is the distance of the fall in meters. A 5-kilogram wrench dropped from a scaffold hits the hard hat of a worker 4 meters below. The energy absorbed by the hard hat and/or the worker is

$$E_k = \frac{m \times (2gs)}{2} = \frac{5 \times (2 \times 9.807 \times 4)}{2} = 196.14\,\text{J}.$$

Threshold of injury was mentioned earlier in this chapter. Some parts of the body are able to absorb much more energy without suffering injury than other parts. How much energy can be absorbed without injury determines the threshold of injury. It has been the subject of quite a bit of discussion, but real data are not readily available.

Mechanical hazards can be a serious risk with the potential to result in catastrophic losses. There must be systems in place to identify and then eliminate or control these hazards. A regular inspection program performed by persons trained in hazard identification can identify many of the hazards. A program to identify process hazards, such as damaged or missing relief devices, can identify other hazards. A follow-up system for ensuring that identified hazards are corrected in a timely manner is also important. A system to ensure equipment reliability will also reduce exposure to mechanical hazards. If equipment is running as designed, there will be fewer times that employees might expose themselves to mechanical hazards to clear a jam, for example, without locking out the equipment. Effectively identifying and eliminating mechanical hazards requires more involvement than just that of the safety engineer. Employees, supervisors, engineers, and managers should be part of this process to ensure success.

QUESTIONS AND PROBLEMS

1. How is a pinch point distinguished from a squeeze point?
2. Try to locate a door on which the doorknob squeezes your fingers or hand between the knob and the doorjamb as you close it. How could it be redesigned to avoid the squeeze point?
3. How is a shear point distinguished from a sharp edge?
4. What are some examples of in-running or nip points?
5. How is a catch point distinguished from a sharp point?
6. When an object and a person collide, what are the factors that determine how serious the resulting injuries may be?
7. What hazards may be presented when a pressurized container ruptures?
8. Under what conditions would it be better to base calculations on the yield strength rather than the tensile strength of a material?
9. What is the distinction between a safety valve and a relief valve?
10. A worker was moving an overhead crane to a location where he wanted to lift a crate. He neglected to raise the hook while it was traveling and it hit another worker in the head. The hook and pulley just above it had a mass of 40 kg. If the crane was traveling at a rate of 4.8 km/h, what energy was transferred to the worker's head?
11. List some common in-running or nip points.
12. A 0.15 kilogram bolt was dropped by a worker onto the floor of a platform 10 meters above the floor below. It rolled off the platform because there were no toeboards and struck the floor next to a mechanic working on the floor below—a near miss. With how much energy did the bolt strike the floor?

BIBLIOGRAPHY

American Society of Mechanical Engineers, 2010. *Boiler and Pressure Vessel Code.* New York: ASME.

Chuse, R., 1993. *Pressure Vessels, The ASME Code Simplified.* 6th ed. New York: McGraw-Hill.

National Safety Council, 2009. *Accident Prevention Manual: Engineering & Technology.* 13th ed. Chicago: NSC.

_____. *Four Sides of Danger.* Chicago: NSC.

Robotic Industries Association. *Industrial Robots and Robot Systems—Safety Requirements.* Ann Arbor, Michigan: RIA.

15

Electrical Hazards

Steven W. Hays, CSP

Richard T. Beohm, PE, CSP, ARM

Terminology

Electrical circuits can be very complex, but the hazards associated with them derive from the most basic characteristics of electrical energy. It may therefore be appropriate to review quickly some of these basic relationships. The quantity of electricity, measured in amperes, is dependent on two basic characteristics. Electrical energy will not flow—that is, the energy is zero—unless there is a force, called the electromotive force (EMF), measured in volts, to move the energy through the resistance, measured in ohms, of the material.

Materials that offer little resistance to the flow of electricity are called conducting materials, and those with high resistance are called insulating materials, but there is no such thing as a perfect conductor or a perfect insulator. All materials will conduct some electricity and all offer some resistance. This includes not only copper wire and rubber, but the human body as well.

The amount of current that will flow through a given material is expressed as

$$I = \frac{E}{R}$$

where

> I is the intensity of electron flow in amperes
> E is electromotive force in volts
> R is the resistance in ohms.

The power produced by the flow of electricity is measured in watts.

$$P = E \times I = I^2 \times R$$

where

> P is power in watts.

If an electric current is flowing in a circuit in which there are several paths to follow, it will take the several paths and the amount of current in each will be inversely proportional to the resistance in each path. This is true in the human body also. The resistance of any path is dependent on three characteristics of the material in that path: the cross-sectional area of the path, the length of the path, and the conductivity (or resistivity) of the material. The total resistance of a given path, then, is expressed as

$$R = \rho \frac{l}{A}$$

where

R is the total resistance in ohms
ρ is the resistivity of the material in ohms/cm/cm^2 or ohm $\times$ cm
l is the length of the path in cm
A is the cross-sectional area in cm^2.

In many cases it is more logical to refer to the conductivity of a material rather than to its resistivity. Conductivity is the inverse of resistivity and is expressed in units of Siemens (S) and S = $1/\rho$. Conductivity, $\sigma = 1/\rho$. Table 15-1 shows the conductivity of several of the common conducting and insulating materials. The ability to conduct electric current depends on the existence of a continuous path. The conductivity of a material is greatly lowered by a reduced cross-sectional area, such as is created by point contacts.

When electric current passes through a material, it generates heat. The amount of heat generated is proportional to the amount of current and to the resistance of the material. Thus, forcing a large quantity of current through a small-diameter conductor will generate more heat than forcing the same quantity through a larger-diameter conductor. The National Electrical Code (NFPA No. 70), also available as ANSI C1, specifies the minimum wire sizes that may be used for various applications. For general wiring used to distribute electric power throughout a building, the wire sizes and maximum permissible current are listed in Table 15-2. These are for temperature ratings of 60°C.

Table 15-1. Conductivity of Common Conducting and Insulating Materials

Material	Conductivity in Siemens/cm
Silver	630×10^3
Copper	585×10^3
Aluminum	350×10^3
Ingot iron	107×10^3
Stainless Steel (301)	14×10^3
Graphite	1×10^3
Porcelain	3×10^{-15}
Acrylic plastic	1×10^{-14} to 1×10^{-16}
Mica	1×10^{-13} to 1×10^{-17}
Bakelite	2×10^{-18}
Rubber, vulcanized	1×10^{-20} to 1×10^{-22}

Table 15-2. Maximum Permissible Current for Given Wire Sizes of Copper and Aluminum Conductors to Maintain a Temperature Not to Exceed 60°C

Wire Size		Maximum Permissible Current (amps)	
Gauge no.	Dia.(in.)	Insulated Copper	Insulated Aluminum
14	0.064	15	—
12	0.081	20	15
10	0.102	30	25
8	0.128	40	30
6	0.162	55	40
4	0.204	70	55
3	0.229	80	65
2	0.258	95	75
1	0.289	110	85
1/0	0.325	125	100
2/0	0.365	145	115
3/0	0.410	165	130
4/0	0.460	195	155

If designed and installed correctly, electrical systems and their conductors are almost always safe from overheating and fire hazards. NFPA 70, the National Electrical Code (NEC), specifies the maximum safe current a conductor can carry without overheating. The safety of a system from overheating depends on the dissipation of the heat that results from electrical currents. The specified maximum current for a particular installation depends on the installation method (i.e., in air versus enclosed in a conduit or raceway), the type of insulation, and the ambient temperature of the environment in which the conductors are installed.

When specified currents are exceeded, the generation of heat becomes a hazard through the deterioration of the wiring insulation and the excess heat generated. Air circulation should be sufficient to prevent unsafe temperatures and the premature breakdown of insulation.

Many types of insulation are used on various conductors and involve different materials and combinations of materials. The electrical code specifies how these types of insulation are designated and the conditions under which they can be used. No material is a perfect insulator and all will allow electric current to pass through if the electromotive force is great enough. Furthermore, if the insulating material becomes thinner due to stretching around too sharp a bend, pressure against it, or wear, the resistance is obviously reduced. If the material deteriorates in any way, the resistance is also greatly reduced. This breakdown could cause arcing and short circuiting of the current.

Effects of Electricity on the Human Body

Like all other materials, human tissue has some conductivity. Unlike most engineering materials, human tissue is not uniform in its properties. Therefore, it is not meaningful or possible to refer to the conductivity of human tissue. Several

Table 15-3. Resistance of Human Body to Electric Current

Path through	Resistance in ohms
Hard, dry calloused skin on hand	500,000 to 600,000
Soft, dry skin on hand	100,000 to 200,000
Soft wet skin	1000
Internal body from hand to foot	400 to 600
Ear to ear	About 100

studies have revealed total resistance along common paths through the body, as indicated in Table 15-3. These values must not be taken as absolute; they vary a great deal.

Perhaps the most important fact revealed by these studies is that resistance within the body is very low, and electric current in the vicinity of several vital organs is extremely dangerous. It has been learned that many functions of the body are regulated by electric currents generated within the body. Probably the most important of these functions is the control of the heartbeat. If the muscles of the heart are disturbed by an extraneous electrical current, they go into a sustained contraction or, at the least, uncontrolled vibration, called ventricular fibrillation. If exposure is very short and the current is low, the heart will return to its normal rhythm after the current is removed. It has been found that in some cases a fibrillating heart can be restored to its normal rhythm with the use of a defibrillation unit.

Table 15-4 gives the current in milliamperes at which people experience different effects. Again, these are average values and can vary considerably. A current of 5 milliamperes is often considered the maximum "safe" current for

Table 15-4. Current at Which People Experience Various Effects from Electricity

Effect	Current in Milliamperes			
	DC		60 Hz AC	
	Men	Women	Men	Women
Slight sensation in hand	1	0.6	0.4	0.3
Perception threshold	5.2	3.5	1.1	0.7
Shock—not painful, no loss of muscular control	9	6	1.8	1.2
Shock—painful, no loss of muscular control	62	41	9	6
Shock—painful, let-go threshold	76	51	16	10.5
Shock—painful and severe, muscular contractions, breathing difficult	90	60	23	15
Shock—possible ventricular fibrillation effect from 3-second shock	500	500	100	100
Possible ventricular fibrillation from from shock of t seconds	$I = \dfrac{165}{\sqrt{t}}$			

skin contact under normal conditions. This obviously excludes any situation in which electric current passes in the vicinity of the heart. Any contact with electric current should be considered serious, but especially so if that current passes through the interior of the body.

A very important fact to remember is that the body, like any other material, will conduct electricity as determined by Ohm's law, $I = E/R$. If a worker's hand is placed across a live circuit of 120 volts, the current passing through that hand can vary from 0.2 to 12 milliamperes. If a worker becomes part of a live circuit through a defective tool or appliance, it makes no difference what current is drawn by that tool or appliance. The current flowing through the worker is determined by the resistance of the worker, not of the tool or appliance.

Another hazard of electricity that is often overlooked is the element of surprise. Even when the amount of current is too small to cause any physical harm, a worker is always startled when he or she contacts a live circuit. The reaction to this surprise is usually an impulsive flinging of the arm or moving of the whole body, often causing an "indirect" accident much more serious than contacting the electric current. Electrical hazards thus become even more serious in the vicinity of other hazards.

Special consideration must be given when using electrical power or equipment in hazardous locations. Hazardous locations are classified according to the properties of the flammable vapors, liquids, or gases and any combustible dusts or fibers that may be present. They are classified as Class I, Class II, or Class III and subdivided into Division 1 and Division 2, depending on the degree of likelihood that an ignitable atmosphere might be present. Combustible substances are arranged into seven groups—A through G—depending on the reaction of the substances upon contact with an ignition source, such as "highly explosive," "moderately incendiary," and so on. These are defined in the National Electrical Code (NFPA 70, articles 500–503), NFPA 497, and NFPA 30. A gasoline-dispensing island at a service station would be classified as Class I, Division 1 and 2, Group D. In an explosion-proof enclosure, the hot gases are cooled as they are forced past the threads of the cover.

Design of Electrical Circuits

Electrical Codes and Standards

All electrical installations in the United States should be made, used, and maintained in accordance with the NEC (The National Electrical Code), NFPA 70E (Electrical Safety Requirements for Employee Workplaces), and other standards, such as NFPA 70B, 79, 75, and 99.

The National Electrical Code (ANSI/NFPA 70) provides for the practical safeguarding of persons and property from hazards arising from the use of electricity by specifying appropriate installation requirements for various conductors and equipment in or on public and private buildings and selected vehicles.

The National Electrical Safety Code (ANSI C2) covers the practices of public utilities and others during the installation and maintenance of overhead and underground electrical supply and communications lines.

It is important to create a positive cultural awareness for an electrical safety program due to the inherent dangers of electricity. Compliance to

OSHA codes and current NFPA standards is paramount to safeguarding workers from the dangers of electricity. Training, supervision, good team morale, and corporate support are also keys to a good safety culture.

Today's emerging issues of alternative energy sources and green technologies present new hazards and safeguards for solar photovoltaic systems and fuel cells. As of November 2009, there were 1.5 million hybrids in the United States. Today's popular hybrids include batteries that surge with 500 volts of electricity, enough to cause serious injury or death. The new 2011 NEC will address this as well as requirements for wind systems and solar fuel cells.

All electrical equipment and conductors must be approved by the authority having jurisdiction and should be labeled and listed by a testing laboratory such as UL, FM, or CSA in Canada.

Building electrical systems, if properly designed, installed, and maintained, are both convenient and safe; otherwise, they could become a source of both fire and personal injury. Electrical arcing and overheating can occur if the system is unsafe, causing fires and injury or death through shocks and burns. An electrical arc could not only ignite nearby combustibles, wiring insulation, and coverings, but could also fuse the conductor metal. Hot sparks and metal could be thrown about, spreading the fire.

All standards governing electric equipment include requirements to prevent fires caused by arcing and overheating and to prevent accidental contact that could cause an electric shock.

After the functional aspects of a circuit are determined, the details must be specified. Determining the best conductor material and its size, as well as the type of insulation, are important details. For most applications, copper is preferred to aluminum as a conductor. Copper has a slightly higher conductivity, is stronger, and can be soldered more easily than aluminum. Copper has better resistance to corrosion and provides less creep under terminal screw heads and other pressure fasteners. Aluminum is lighter and less expensive and, therefore, more commonly used for outdoor transmission lines.

Aluminum and copper cannot be intermixed in a terminal or in splicing because they are dissimilar metals, unless the device is suitable for that purpose and condition. Aluminum is softer and less dense than copper. It has a higher thermal expansion rate than copper. Galvanic corrosion could occur at their junction. As the temperature of aluminum increases, it tends to "cold flow" or creep away from an area under high pressure. Repeated heating and cooling of the conductor could result in a high-resistance connection over time, which could lead to arcing and burning. An annual, nondestructive examination is recommended whenever aluminum conductors are used. Check for evidence of overheating at bolted joints. Bolts should be tightened where required, but not enough to overstress them.

Electrical circuits should be designed by electrical engineers and installed and tested by qualified electricians.

The size of the conductor should be determined by the amount of current that will have to pass through it to satisfy the functions of the circuit. The temperature rise is proportional to the square of the current

$$T = fI^2R$$

where the function f is a constant dependent on the temperature scale used.

This relationship is the basis for determining the current to be used in electrical resistance welding, such as spot welding.

While this heating effect can be used to good advantage, it can cause a fire if it is not used properly. In fact, it is a common cause of fires in buildings. A number 18 (18 gauge) copper wire will carry 10 amps without any appreciable heating, but if it is made to carry 30 amps, the temperature of the wire will be raised to well over 300°C (700°F). A number 14 (14 gauge) wire will carry 15 amps without overheating but if subjected to 100 amps, will reach a temperature of about 1650°C (3000°F).

A circuit that is not protected by a fuse or circuit breaker can become defective and draw large amounts of current without giving any noticeable clue. As long as no excess current flows through a circuit, a fuse or circuit breaker is not needed, but when a fault occurs, such protection is vital.

The choice of a single conductor versus a multiple-strand conductor depends primarily on the degree of flexibility needed. Generally, any available conductor that is number 8 (8 gauge) (0.128 inch) or larger will be a multiple-strand conductor. More important in many cases is the selection of insulation. Materials used for conductor insulation range from paper and cotton cloth to plastic and rubber or combinations thereof. Each has advantages. Factors to be considered in the choice include resistivity; strength; ability to resist abrasion, cutting, and penetration; and amount of deterioration in various environments. In some cases, resistance to solvents, gasoline, and other substances must be considered. Resistance to mechanical damage is often improved by adding a woven or solid metal shielding outside the insulation. The electrical code specifies applications for many types of insulated and shielded conductors.

The lack of proper grounding is one of the major causes of electrical shock to workers. There are two types of grounding: one involves grounding the whole electrical system, and the other involves grounding the equipment. System grounding protects the system from surges of overcurrent caused primarily by lightning. Equipment grounding protects personnel from contact with an energized circuit. The resistance of the grounding path in system grounding should be higher than that of the fuses or circuit breaker for that system, so that a short to ground will open the fuse or circuit breaker.

It is equipment grounding that is of primary interest in safety engineering. Equipment should be bonded electrically to a conductor that is continuous to a suitable ground. For best protection, this grounding conductor should not be bonded to any part of the normal current-carrying circuit. It is a common practice, but a poor one, to connect the neutral conductor terminal in the equipment to the grounding conductor terminal, as shown in Figure 15-1(a). When this design is used, if there is a break in the neutral conductor, the grounding conductor and the equipment itself become energized to full-line potential.

Article 250 of the NEC contains general requirements for grounding electrical installations. All cable metal-armor, raceways, boxes, cabinets, and fittings must be properly grounded. Certain cord/plug-connected appliances, such as refrigerators, freezers, air conditioners, and portable tools, must also be grounded.

The connections shown in Figure 15-1(b) are the proper ones. In this arrangement, the equipment—that is, the housing or other noncurrent-carrying members—will be energized only when a current-carrying conductor

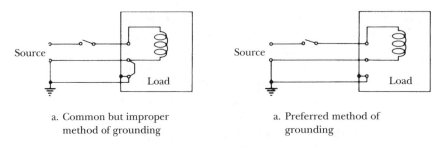

a. Common but improper   a. Preferred method of
   method of grounding   grounding

Figure 15-1. Wrong and right ways of connecting the equipment grounding
terminal.

shorts to the equipment. When such a short occurs, the grounding conductor
carries the current to a grounding electrode, which carries the current into the
earth. The grounding conductor for the system and the grounding conductor
for the equipment may use a common grounding electrode.

The method of grounding described above is best accomplished by providing a separate wire from the equipment through conduits, junction boxes,
switches, and so on, to the source. The source is considered to be the service
panel from which power is distributed through the building.

There are other methods of achieving a grounded equipment circuit. One
involves the use of the conduit or channel carrying the wires as the grounding
conductor. This practice saves an extra wire, but it is often not a satisfactory
method. Any grounding conductor must be continuous and free from high-
resistance points. Conduits and channels require frequent connections and
joints, which are prone to corrosion, mechanical stress, and breakage. It is
difficult, therefore, to maintain good continuity. Conduits or channels used
as the grounded conductor should be inspected regularly.

Another method involves running a wire directly from the housing or frame
of the equipment to a grounding electrode. This is a satisfactory method if that
electrode is nearby and the grounding conductor is protected from mechanical
damage.

Any grounding system is rendered useless if the continuity of the grounding conductor and the grounding electrode is interrupted. The u-shaped or
round pin on the plug of a three-wire cord should never be removed to allow
the plug to be used in a receptacle designed for two-wire cords.

At the other end, grounding electrodes are susceptible to corrosion from
the moisture of the earth. Corrosion-resistant materials should therefore be used
for grounding electrodes. Periodic inspections should be made and corroded
electrodes replaced.

There are some installations in which the system or equipment is not
usually grounded. These include industrial electric furnaces and systems
separated or isolated from the primary system, such as secondary systems used
in healthcare facilities or for control circuits only, where access is limited
to qualified service personnel. In the latter situation, ground detectors are
required; they will be discussed later on.

Separated or isolated systems are so named because no continuous conductor exists from the source to this isolated circuit. Instead, current is induced
into the isolated circuit by means of a coupler or by the use of a battery, as shown
in Figure 15-2.

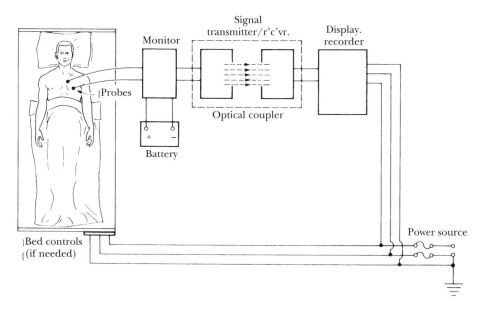

Figure 15-2. Battery and optical coupler that completely isolate electrical probes from 120v system. Nonmetallic bed frame should be used if electrical bed controls are needed.

The need for grounding should not be confused with the need for fuses or circuit breakers. Grounding by itself will not provide any protection to equipment or to the system if an overcurrent surges through the circuit. That is the function of fuses and circuit breakers. A fuse is a thermal device containing a conductor with a low melting temperature. It is designed so that if a current exceeding a given amount passes through the conductor, it will heat to its melting point, melt, and break the continuity.

A circuit breaker may operate on the thermal principle or, as is more often the case, on a magnetic principle. If too much current passes through the device, a magnetic relay opens the circuit without destroying any component.

The most common form of electrical shock and electrocution is the line-to-ground fault. With a ground fault circuit interrupter or GFCI, the ground-fault current must pass through its electronic sensing circuitry, causing it to trip and preventing injury and even death. Each year, an estimated 300 electrical deaths occur in general industry throughout the United States. Many more workers suffer serious injuries from electrical shock. These accidents usually occur from unsafe conditions and unsafe work performance.

The National Safety Council reports that in 2006 there were about 392 deaths due to electric current. NFPA reports that shock, electrocution, arc flash, and arc blast are responsible for one fatality every workday in the United States, and some 8,000 workers are treated in emergency rooms for electrical contact injuries each year. In August, 1997, two girls, ages 7 and 9, were electrocuted when a hairdryer fell into the bathtub in which they were trying to escape 100-degree temperatures.

The Occupational Safety and Health Administration (OSHA), in its Part 1926, Subpart K (Electrical Standards for Construction), requires the use of GFCIs and proper equipment grounding for construction sites to help reduce the number of injuries and accidents from electrical hazards. Hazards are created when cords, cord connections, receptacles, and cord and plug-connected equipment are improperly used and maintained. Cords are easily abused and damaged by activities on the job site. GFCIs are required for any temporary wiring receptacles.

Fuses and circuit breakers are designed for the protection of the electric circuit or the equipment on that circuit. They are activated by current in excess of a given amount, often 15 amperes up to 100 amperes or more. This amount is far more current than the human body can tolerate. If a worker's hand touches a 120-volt line, the current can vary from .2 to 12 milliamperes, which can result in a slight sensation on the hand to a painful shock, possibly holding the hand to (not letting go of) the shorted device or machine. As few as 60 milliamperes will kill a person if the current flows into certain parts of the body. So fuses and circuit breakers must not be considered devices to protect people. Grounding is a great help in protecting people, but it is not as dependable as might be desired. It becomes defective too easily and the defects are not readily seen. It takes about a full minute for the heart to recover after an electrical shock.

A more reliable method of protecting people is the use of a device called a ground fault circuit interrupter or GFCI. A GFCI detects very small unbalances in the line current, implying a leak to ground, and opens the circuit. It is designed to trip when an unbalance of 5 milliamperes ±1 milliampere occurs.

GFCIs are readily available for any given range of amperes and volts normally used in industry, homes, or other installations where such protection is needed. The National Electrical Code, enforced by OSHA, requires their use on construction sites where installation of electrical equipment is not permanent. GFCIs are required by the NEC in new homes at all outdoor and bathroom receptacles. They are also required at swimming pool electrical installations.

Ground Fault Receptacles (GFRs) can replace the standard receptacle in your bathroom or shop. A GFCI can protect a number of receptacles on a circuit downstream of the protected receptacle. Some are installed on the main electrical panel and some are an integral part of the extension cord. Always use them when working in and around a wet environment.

GFCIs are to be used in addition to, not instead of, conventional grounding. Grounding protects the electrical system, equipment, and personnel, but conventional means of grounding are not sensitive enough, nor do they act fast enough, to provide adequate protection for personnel. Several contractors complained when they were required to use GFCIs, saying that these devices were tripping unnecessarily and causing delay in their work. They called them "nuisance tripouts." When an investigation was made, it was found that the GFCIs were doing just what they were supposed to do. The contractors were working with faulty equipment, dragging power cords through mud and puddles, and ignoring grounding connections.

Some GFCIs are made as a permanent part of a circuit, while others are made to plug into any existing receptacle. One model is referred to as a ground fault receptacle in a 15–20 amp, 120-volt circuit. Figure 15-3 shows a simplified diagram for a typical GFCI. Practically all GFCIs made in this country use a solid-state circuit and include electronic components to control the operation of the unit. GFCIs should be tested regularly.

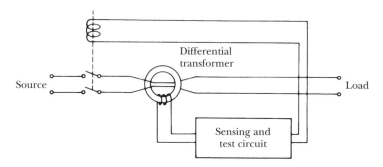

Figure 15-3. Schematic of a typical differential-type ground fault circuit interrupter (GFCI).

A GFCI obviously does not protect a person who contacts the two sides of a circuit, nor will any grounding system. Both systems protect people only from line-to-ground current. Nor will a GFCI or a grounding circuit provide protection against ground faults in other parts of a circuit not included in the circuit in which the GFCI or grounding system operates.

Small household appliances and hand-held power tools have become very popular, and many of them are used in conditions that are far from ideal. These appliances and power tools are subjected to a considerable amount of vibration and other abuse. It is not surprising that conductors inside the housing are pulled loose or broken. When this happens, the loose or broken wire very often comes in contact with the inside surface of the housing, energizing it if it is metal.

This energizing of the housing can be avoided in two ways. A layer of insulating material, such as fiberboard, can be placed between the wires and the housing, or the housing itself can be made of an insulating material. Since the wires themselves are insulated, except for ends and places where insulation becomes broken, this additional layer is called "double insulation." Most tools and appliances so protected are labeled "double insulated" or carry this symbol ▣.

Tools and appliances that have double insulation do not have a grounding conductor. But this does not mean that any tool or appliance that has no grounding conductor therefore has double insulation. Tools and appliances made in recent years should carry labels of approval from Underwriters Laboratories(UL) or Factory Mutual (FM). If a model is so approved, that is evidence that its design and construction have been inspected and tested and found to be acceptably free of hazards.

Frequently, a tool or appliance will have a power cord that bears a UL or FM label, but the tool or appliance itself does not. This means that only the cord has been approved. If the tool has been approved, the label will be on the tool itself.

Many electrical circuits contain elements that, if shorted or opened, create hazards at other elements of the circuit. An open or short will cause more current to flow through other components that were not designed for that much current. This may not only be harmful to other components in the circuit, but may also create hazards for personnel. Maintenance personnel are often plagued by charged capacitors in a circuit that failed to discharge the capacitors. The design of any electrical circuit should include a check to see what would happen if a short or an open were to develop at each component and at each terminal. If a hazardous condition is found by this check, the circuit should be redesigned to eliminate the hazard.

Arcing is a problem in some electrical circuits. It occurs whenever two current-carrying members come together or separate. Common examples include switches, relays, and commutators on motors and generators. Each of these, by nature of its purpose and function, must involve making and breaking contact. There is no way to eliminate the arcing without eliminating these components.

Arcing creates the potential for three undesirable events. The most serious is the ignition of flammable gases and dusts; another is burns to personnel; and the third is the pitting and distortion of parts which, in turn, increase the problem of arcing. Arc Flash protection clothing is required to meet NFPA 70E, Annex M, ASTM F1506, NFPA 2112, and NESC standards.

Electrical Failures and Preventive Maintenance

A fault arc can release enormous amounts of energy, with heat so intense that it vaporizes copper or aluminum conductors and destroys surrounding steel enclosures. Electrical faults can cause serious fires. Preventive maintenance is the key to preventing or reducing such failures. Portable infrared detection equipment has proved to be invaluable in detecting potential electrical and mechanical failures which can lead to fault arcs. These units can pinpoint "hot spots," which are characterized by increased thermal activity—the first sign of a potential problem. Typically, an infrared scanning unit consists of a hand-held scanner, a monitor screen, and an attached camera. This produces a picture called a "thermograph."

Most electrical failures begin at electrical connections, which usually do not fail without warning but deteriorate over a period of time. They can fail as a result of wear, fatigue, improper or loose connections, overloads, or corrosion. In the early stages of an electrical failure at a connection, heat is generated due to a localized increase in impedance. This heat accelerates the deterioration of the connection and the surrounding insulation. If not detected and corrected, a major failure can occur.

Several years ago the author had a rather disquieting experience. On two occasions the lights in the house went out, then after a while came back on again. This did not seem to be an unusual event, but when it happened again one evening, he went to the basement to examine the fuses in the main fuse box. On opening the cover, he saw one of the main cylindrical fuse holders glowing bright red. The copper holders had become heated and expanded, allowing arcs between the holder and the fuse. The arcing further heated and distorted the holder each time it occurred. A temporary repair restored proper contact, but the fuse box was soon taken out of service and replaced with a breaker panel.

Arcing can often be reduced by decreasing the amount of current or by shortening the duration of the arc. A slow make-or-break will result in an arc of longer duration. The effects of arcing can be reduced by selecting proper materials for the members involved in the make-break event. Gold, silver, and platinum are among the most resistant materials to arcing. It is interesting to note that the principle of arcing is put to good use in the metal-removal process known as electrical discharge machining (EDM).

When switches, relays, and motors must be used in hazardous areas, such as in the presence of flammable gases and dusts, these components must be enclosed so that the gases and dusts cannot reach the area of arcing. Relays are often encapsulated in rubber, plastic, or an elastomer.

Encapsulation is usually accomplished by dipping the component, with conductors or leads attached, into a container of molten material. Another method is to place the component in a mold and pour molten material around it. This method is usually called "potting" and may be used for enclosing not only one component but even a whole circuit. Potting and encapsulation may also be done to protect the electrical components from moisture or other environmental elements harmful to them.

Motors are not encapsulated or potted, but some models are available that are enclosed by virtue of the design of the housing and seals. When motors are to be located in hazardous atmospheres, enclosed motors designed for the hazardous location must be specified. It is usually more desirable to locate switches outside the hazardous atmosphere. Mercury switches contain an enclosed switching unit that can be used in such an atmosphere, but it must be oriented in a given manner. Other types of switches cannot be enclosed satisfactorily.

In hospitals and other installations where very small amounts of current are used in instruments, the control of that current is often very critical. This is especially true when electric probes are used on or in the body of a patient in a hospital. The current is in the order of microamperes, ranging up to a few tenths of a milliampere. At these levels of current, leakage current becomes a problem.

Leakage current is current that is transferred by natural means, not by fault, to a conductor outside the circuit in which the current originated. There are two sources of leakage current: resistive leakage current and capacitative leakage current. The word "leakage" is not a good choice because it implies some phenomenon that should not happen, which is not the case.

It was stated earlier that all materials are capable of conducting some electric current. It is impossible to find a material for use as an insulator that will not allow a small amount of current to pass through it. If another conductor is in contact with the outer surface of an insulated conductor, some current will pass through it. This is resistive leakage current. Current derived in this manner is usually so small that it is not a serious problem.

When two conductors lie parallel to each other, they constitute a capacitor. If they are close enough, alternating current will pass from one to the other. Thus, a conductor outside a live, alternating-current circuit will pick up current from the live circuit. This is capacitative leakage current. This is a more serious problem than resistive leakage current and should be avoided when circuits for very small amounts of current are designed and installed.

Arc Flash

Arc flash is a short circuit where a high level of current passes through the air between conductors. Arc flashes can be caused in a number of ways, including accidental contact by dropping a tool or touching a test probe to the wrong energized part, insulation failure, dust buildup between conductors, and so forth. The energy generated during an arc flash creates a bright flash of light, extreme heat up to 35,000 degrees F (19,425 degrees C) that can cause severe burns on unprotected skin and set clothing on fire, very high noise levels, flying metal parts equivalent to shrapnel, flying molten metal, and a very strong pressure wave that can knock people down or propel them across a room. This can be a catastrophic event that can cause very serious injury and death to anyone nearby as well as cause destruction of equipment.

An arc flash hazard analysis should be conducted on the electrical system at a facility by a qualified person. This analysis will determine the Arc Flash Protection Boundary and the required personal protective equipment for people inside the boundary. The arc flash hazard analysis should be updated whenever a major modification takes place but no less than every five years. An arc flash hazard analysis is not required on circuits when all of the following conditions exist: (1) the circuit is rated 240 volts or less, (2) the circuit is supplied by one transformer, and (3) the transformer supplying the circuit is rated less than 125,000 volts (125 kVA).

The arc flash hazard analysis will determine the level of hazard that exists at each electrical disconnect, panelboard, switchgear, etc. It will establish the arc flash protection boundary in inches, the available incident energy in calories per centimeter squared (cal/cm^2), and PPE requirements (Category Level 0 – 4). The analysis will also generate labels with at least the above information plus the identity of the enclosure that should be posted on the enclosure. Personal protective equipment for arc flash protection should be worn when there is a possible exposure to an arc flash above the threshold energy level for a second degree burn (1.2 cal/cm^2). The PPE is designed to stop burning when the ignition source is removed and to prevent the wearer from receiving any worse than a second degree burn. Refer to NFPA 70E Standard for Electrical Safety in the Workplace for more information on arc flash.

Installation of Electrical Circuits

In general, when electrical circuits, especially those of over 300 volts, are installed in industrial and commercial buildings, the conductors should be placed in a rigid pipe, or conduit, or in a cable tray. A conduit or cable tray is needed to provide mechanical protection for the conductors. Conductors without a conduit or cable tray are permitted in some situations. There are many variable conditions, and the National Electrical Code should be consulted. When conductors are carried in conduit or closed cable trays, no other materials or substances are allowed to be carried in that same conduit or tray. All conduits and trays must be joined rigidly for mechanical strength and also to form a continuous electrical path for use as a grounding conductor.

In industrial and commercial buildings, when these conductors are used without a conduit tray, they are referred to as open conductors or flexible conductors. These open conductors must not pass through holes in a wall, ceiling, or floor, or through doors or windows. Neither may they be concealed behind walls, ceilings, or floors. Only conductors with approved types of insulation may be used.

In any terminal, panel, or other place where there is exposed wiring, the conducting components must be covered in such a manner that a person's finger or a metal probe cannot reach the bare element. Bare conductors are not permitted in any exposed location. An exposed location means any spot within 2.5 meters (8 feet) vertically above the floor or platform where people are expected to walk. Bare conductors are permitted in rooms or enclosures restricted to authorized personnel and in work stations where bare conductors are necessary and access is restricted.

In any installation of bare conductors, whether in an enclosed area or in a work station, there must be space in front of the installation to allow a worker free movement. Table 15-5 shows the required clear distance for several conditions. In all cases, the free space must be at least 0.9 meter (36 inches) wide and 2.0 meters (6.5 feet) high, both dimensions allowing for a normal standing position in relation to the bare conductors.

All live parts operating at 50 volts or more must be guarded against accidental contact if they are in an exposed location. Approved electrical boxes or panels, including receptacles and switch boxes, provide sufficient guarding as long as they are covered. Receptacles, plugs, and other insulated covers do not provide sufficient guarding if they are broken.

In the installation of any electrical circuit that includes a grounding conductor, the grounding conductor must be distinguished from any other wire. It may be bare or insulated, but if it is insulated it must be colored green. Most equipment has terminals and connections that are colored for ease in identification of circuits and conductors. Grounding terminals should always be green.

The layout of wires within a junction box or panel of any type is an important but often overlooked matter. For at least two reasons wires should be arranged neatly and free from any pressure that could crush a conductor or squeeze two conductors together. Even when resistive and capacitive leakage current are of no significance, excessive pressure against the conductors will damage the insulation. Furthermore, identification becomes very difficult when wires are not in reasonably neat order, as illustrated in Figure 15-4. Figure 15-5 illustrates a good layout.

In most industrial plants there are at least three levels of disconnects, or switches. The first level is the one that the worker uses at the work station or on the machine. The second level is a switch controlling the power to a given

Table 15-5. Required Clearance in Front of Exposed Live Electrical Components

Nominal Voltage to Ground	Working Clearance in Meters & (Feet)		
	Condition 1	Condition 2	Condition 3
0 - 150	.914 (3)	.914 (3)	.914 (3)
151 - 600	.914 (3)	1.07 (3.5)	1.22 (4)
601 - 2,500	0.9 (3)	1.2 (4)	1.5 (5)
2,501 - 9,000	1.2 (4)	1.5 (5)	1.8 (6)
9,001 - 25,000	1.5 (5)	1.8 (6)	2.8 (9)
25,001 - 75,000	1.8 (6)	2.5 (8)	3.0 (10)
above 75k	2.5 (8)	3.0 (10)	3.7 (12)

Condition 1—Exposed live parts on one side and no live or grounded parts on other side of working space, or exposed live parts on both sides effectively guarded.

Condition 2—Exposed live parts on one side and grounded parts on the other side.

Condition 3—Exposed and unguarded live parts on both sides of working space.

Figure 15-4. An example of a haphazard and potentially dangerous layout of conductors.

Figure 15-5. An example of an orderly layout of conductors.

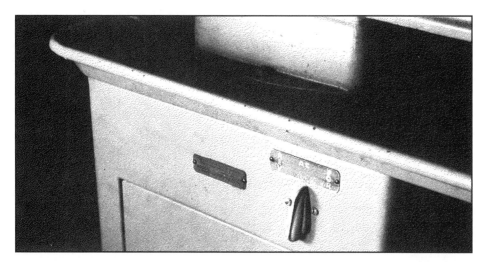

Figure 15-6. An example of a switch that is poorly designed and poorly located for the application.

area. This is often thought of as an emergency shutoff, but in many plants it is used for shutting off the power during nonworking periods. In large plants there may be a disconnect controlling several area switches. The last level, or in the minds of some the first level, is the main disconnect for the entire plant.

Most disconnects should be located in prominent places and should always be accessible. Access is frequently blocked by stored materials, especially in the case of emergency switches, which are seldom used and therefore forgotten. Continual supervision is required to maintain this access; however, the choice of a good location in the first place may preclude the need for such supervision. A clearance of about three feet should be maintained in front of electrical panel boxes, transformers, and switchgear to provide safe access for working and to prevent the ignition of nearby combustibles.

The design, location, and orientation of the switch is important both for emergency switches and for switches at the work station or machine. In Chapter 4, human responses to signals or stimuli were discussed. In order for a worker to respond quickly and accurately when it is necessary to use a switch to open or close a circuit, that switch should be designed and installed properly. Figure 15-6 illustrates a switch that is poorly designed and poorly located for this application. It is underneath the pan of the machine and the operator cannot see it without stepping back. It has "forward," "reverse," and "off" positions on a rotary switch. If the operator is not careful, the switch could be turned from the "forward" position right through the "off" to the "reverse" position. A push-button switch located at waist level would be much safer.

Switches on high-voltage circuits are often to be used by authorized persons only. They are usually located at a height well above normal reach, and authorized workers use a pole with a hook at the end. Such switches are usually knife-type switches and arcing results when they are either opened or closed. The operator of such a switch should be protected by wearing rubber boots, rubber gloves, a head covering, and a face shield, as shown in Figure 15-7. The pole must be insulated also.

High-voltage transmission lines must always be treated with care. Outdoor power lines must be located a safe distance above the ground or buried

Figure 15-7. Worker properly attired for switching on high voltage lines.

a safe distance beneath the surface of the ground. In spite of careful installation, many power lines are contacted every year by mobile construction cranes. Distance is always the best method to protect the workers on this type of equipment. However, the OSHA standards require crane booms and all equipment to be kept a minimum of 3.05 meters (10 feet) away from energized power lines. Protective devices for this type of hazard include the following:

1. Insulating devices
2. Proximity detectors
3. Motion-limiting devices

There are two types of insulating devices. One type is inserted in the lifting cable just above the work-holding hook. This insulator electrically isolates the work from the rest of the crane, but people on or in contact with the crane are still exposed to the hazards of contacting power lines. The insulator is quite large and is sometimes allowed to fall in mud; thus covered with semiconducting material, it loses much of its insulating ability.

The second type of insulating device consists of a frame or cage made of insulating material attached to the top of the crane's boom. It must be restricted in size and location to permit versatile use of the boom. This device does protect people in the crane as well as the work. Figure 15-8 illustrates a crane boom with both types of insulators.

Proximity detectors are attached to the top of the boom and can detect an electrostatic field that is produced around the power line conductors. They have limited value because of weak fields and many distortions of the field caused by other conducting objects, including the crane's boom.

Figure 15-8. Two types of insulators used on a mobile crane. Courtesy Saf-T-Boom Corp.

Motion-limiting devices can be attached to the boom to restrict motion physically in each of several directions. Such a restriction often results in sudden stops, which can cause undue side forces from a heavy, swaying load, twisting and even buckling the boom.

The subject of electrical installations should not be left without the mention of the use of extension cords. These cords are very convenient and serve a good purpose, but they are often misused and abused. The same rules and practices that pertain to open or flexible conductors mentioned earlier should be applied to extension cords, except that extension cords should not be used as a permanent installation.

Extension cords should be selected with the same criteria in mind as those used for other conductors. Copper is always used because aluminum will not withstand the abuse given to these cords. Several wire sizes are available, however. Number 18 wire is a common size, but it should not be used when more than 10 amps is to be drawn through it. Extension cords are commonly used with a two- or three-place receptacle with several appliances plugged into it. The sum of all of these should be less than 10 amps. Only heavy-duty extension cords should be used in industrial locations.

Several types of insulation are also available on extension cords, each type meant for certain types of use, taken from the National Electrical Code Table 400-5.

Extension cords should not pass through doors, windows, or openings in a wall, ceiling, or floor. They should be used only where they are protected from crushing, cutting, and moisture. Like any other conductor, extension cords should be maintained in good condition. Any force against any part of the cord except for the plug and the receptacle should be avoided. These are "commonsense" rules for use with any electrical installation, but extension cords seem to be abused more often than other installations. Defective cords could lead to accidental fires, electrical shock, and hazardous arcing.

Maintenance

Electrical preventive maintenance is the practice of conducting routine inspections, tests, and servicing of the electrical system to detect and reduce or eliminate problems. When designing new facilities, conscious effort is required to ensure optimum maintainability. A good program is helpful in discovering deterioration and other causes of electrical failure. A well-administered electrical preventive-maintenance program will reduce accidents and minimize costly shutdowns. A defective extension cord in a booth caused a $160 million loss to an exhibition hall and the economy of the area. The failure of electrical equipment could endanger or threaten the safety of personnel as well as result in costly repairs and loss of production.

NFPA 70B, The Electrical Equipment Maintenance Code, provides the recommended practice to reduce the hazards to life and property that can result from the failure or malfunction of industrial-type electrical systems and equipment. It applies to institutional and commercial buildings, large residential complexes, and similar structures. NFPA 70E 2009 Edition, Electrical Safety in the Workplace, has additional requirements about employee training—the most important part of an electrical safety program.

The code stresses electrical preventive maintenance (EPM), which is the practice of conducting routine inspections, tests, and servicing so that impending troubles such as normal deterioration can be detected and corrected.

OSHA's workplace standard, 1910 subpart S, covering electrical hazards was taken from the NEC and simplified. It includes Electric Utilization Systems (1910.302) for Buildings and Structures; Wiring Design and Protection (1910.304); and Electrical Safety-Related Work Practices (1910.331), which includes the installation of electric conductors and equipment within or on buildings or other structures.

One of the most common causes of electrocutions in industrial plants is maintenance work on circuits that are mistakenly believed to have been turned off. Some maintenance workers, furthermore, will attempt to work on circuits that they know are energized. In a few situations it is necessary to work on circuits that are energized in order to make electrical measurements, but this practice should be limited.

There are well-established procedures for de-energizing an electric circuit and keeping it de-energized while anyone is working on it. If one worker is alone and no one else is likely to turn the circuit on again, a simple tagging procedure can be used. Tagging involves the placing of a tag on the switch that has been turned off, indicating the reason it was turned off and stating that it must not be turned on again without the knowledge and consent of the maintenance worker who is also identified on the tag.

When more than one person is, or may be, working on the de-energized circuit, a lockout should be used. Each worker must have his or her own padlock in place so that no one can throw the switch until all workers have removed their padlocks. Figure 15-9 shows a typical lockout/tagout notice.

When a circuit is to be de-energized for repair work, all persons affected must be notified ahead of time so that the appropriate plans can be made. When people are not notified, attempts are often made to re-energize the circuit. Removal of fuses is never an acceptable method of de-energizing a circuit. Even when lockouts are used, the circuit must be checked to be sure it

Figure 15-9. An affixed lockout/tagout notice. (Photo courtesy of MasterLock.)

is de-energized completely. OSHA 29 CFR 1910.147 requires a lockout/tagout procedure for all equipment that utilizes stored energy—electrical, mechanical, hydraulic, pneumatic, steam, gravity, or any other form. Such a lockout or tagout was explained under the section titled "Pressurized Containers" in Chapter 14. It involves placing a padlock or other locking device on the source switch so that the circuit cannot be accidentally activated. (See Appendix D for Electrical Safety checklist.)

Static Electricity

If a direct current circuit includes a capacitor, the plates of the capacitor become charged: one, called negative, with an excess of electrons, and the other, called positive, with a deficiency of electrons. Now, if those plates could be removed from the circuit without being grounded, each plate would be charged with static electricity.

Static electricity can build up on nonconducting materials as well as on conducting materials. It is caused by the motion of two or more objects of dissimilar materials that are in contact with each other. This action causes a transfer of electrons, which results in the states of excess and deficiency. Objects that are charged with an excess or deficiency of electrons have this charge on the surface only; static electricity does not flow through the object.

Reference is therefore made to surface resistance rather than to volume resistance. If a low-resistance surface path is provided from either a negative or positive static charge to ground, the charge is quickly discharged. In fact, a charge can become so great that it will discharge through space in the air to reach a path to ground. Potentials as high as 15,000 volts have been measured in a static charge.

The greatest hazard of static electricity is this great potential for discharge. The surprise and involuntary response of a person who receives the static charge could result in injury to the person. If a static charge can be drained to ground without any part of that path being through the air, there is generally little harm done. The danger lies in the spark created when the charge passes through an air gap. Such a charge is capable of igniting many types of fuel and has been the cause of many disastrous accidents.

There are several things that can be done to reduce the buildup of a static charge. The object on which a static charge accumulates is called an accumulator; a material is a good accumulator if it has a high surface resistance and thereby prevents the rapid flow of electrons. To avoid the accumulation of a static charge, the material must be able to conduct the charge away faster than it can accumulate.

Several materials are known to be excellent accumulators and should therefore be avoided when static electricity could be a problem. Most plastic materials, especially nylon and polyethylene, are good accumulators. Wool is also a good accumulator. Many nonconducting liquids are good accumulators. They include gasoline and other petroleum products, as well as water if it is free of minerals and other conducting particles.

Other objects become good accumulators by virtue of their shape or the type of contact they make with other materials. Screens in a moving steam of liquid usually accumulate charges. The inside of pipes carrying moving liquid, dust, or small solid objects frequently become charged.

When static charges cannot be avoided by the use of proper materials, it is often necessary to provide grounding conductors. A bare copper wire or braided band is clamped to the surface of the accumulator and to a suitable grounding electrode, as shown in Figure 15-10.

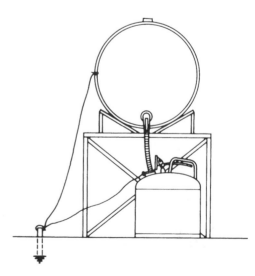

Figure 15-10. Drum and safety can are both grounded during the filling of the safety can.

Other actions can be taken to reduce the buildup of a static charge. The humidity and/or temperature can be increased to reduce ionization. Conductive coatings can be sprayed onto the surface of accumulators to reduce the charge that can accumulate; however, a continuous path to ground is still needed to eliminate a static charge. The presence of a radioactive atmosphere or a high-voltage field will reduce or eliminate the buildup of a static charge but may present other hazards.

The following list gives examples of situations in which a static charge is easily generated; they should be avoided by one or more of the methods described above:

1. Transfer of liquid fuels from one tank to another
2. High-velocity flow of any liquid in a pipe
3. Flow of liquids through a screen
4. Flow of dust or small, solid objects in a pipe
5. Conveyor belts made of rubber, leather, or plastic
6. Movement of material in a pile of dust or of fine, solid particles
7. Rubbing of explosives
8. Plastic covers and containers of flammable materials
9. Metal or plastic chutes and funnels used with flammable dusts or liquids
10. Steam-cleaning of tanks in which there are flammable vapors
11. Nylon or wool carpet in the vicinity of flammable materials
12. Personnel-borne charges
13. Conductors in the vicinity of magnetic fields
14. Atmospheric-induced charges (thunderstorms)

Two serious accidents involving identical rocket motors occurred within six months of each other in 1963 and 1964; the latter resulted in three fatalities. In both cases the rocket motor, the X-248, had been covered with a polyethylene bag. In both cases someone had brushed or rubbed this cover and generated enough static electricity to ignite the very sensitive fuel.

Lightning is static electricity of enormous magnitude. Charges accumulate on the surface of clouds and discharge either from one cloud to another, from earth to cloud, or from cloud to earth. People are cautioned to seek refuge from lightning in a place where the resistance of the path the electrical discharge may follow is high. The charge may be so great that the discharge may follow a number of paths simultaneously.

The NSC reported that lightning killed 25 people in 2008.

Lightning is a phenomenon of nature and follows the rules of physics, chemistry, electricity, and meteorological principles and theories. It imparts benefits to the earth by helping to balance nature's complex ecological systems. The harmful effects to man and his created environment can be reduced and even prevented by the application of engineering principles to loss-control measures.

The most important lightning protection is a good, low-impedance grounding that diverts the induced or direct currents away from sensitive and costly equipment and shunts, or short-circuits them into the earth to be neutralized.

Some of the ways that lightning engineers and consultants accomplish this are by using the following:

1. a UL "Master Label" system with roof-mounted air terminals or rods
2. overhead masts to provide protected zones

3. ionized roof terminals
4. dissipation arrays on high points to "bleed" charges back to the cloud, preventing a discharge to earth
5. a good (10 ohm), low-impedance earth-ground, using two or more 8- to 10-feet ground rods
6. "chemical" ground rods for poor, high-impedance soil or rock
7. ground mats
8. equipment bonding
9. the building's grounded steel framing as a shield
10. lightning and surge arresters

They also use the following codes and standards for lightning protection: the National Electrical Code (Sections 250 and 280), NFPA 780, and IEEE and ANSI standards for lightning protection. Other methods to further reduce damage by lightning include eliminating overhead electrical and cable exposures, eliminating electrical dead ends, and housing equipment in well-protected buildings.

Lightning risk susceptibility depends on a number of factors, such as the type of building (its construction, use, contents, and height), the area around the building, the topography of the land, and the frequency of storms in the area. NFPA 780 is the standard for lightning protection.

Although static electricity is a phenomenon generally to be avoided, it has been put to good use. Electrostatic painting, for example, involves charging the object to be painted as well as the droplets of paint as they leave a spray nozzle. The paint is attracted to the oppositely charged object and sticks to it. The result is a uniform paint thickness and practically no waste. Electrostatic painting saves paint, eliminates flammable paint vapors outside the immediate painting area, and gives a better quality paint job as well.

QUESTIONS AND PROBLEMS

1. What is the importance of designing and installing electrical systems correctly to the NEC?

2. What is the power (watts) for a 110 volt, 15 amp electrical circuit?

3. What are the three classes of hazardous locations? What is the class, division, and group for a gasoline-dispensing island?

4. What is the importance of using ground fault receptacles in a bathroom or shop?

5. What are some safeguards for electrical tools against electrical shock?

6. What is the key to preventing electrical faults and resulting fires?

7. Name several standards and codes for safe workplace maintenance practices.

8. What are some of the hazards of static electricity?

9. Name several ways to minimize the effects of lightning damage.

10. Name several new technologies and their electrical hazards.

11. What is one of the most common causes of electrocutions in industrial plants?

12. What is the allowable current in amps for an AWG 18 wire with thermoplastic C insulation at 30 degrees C?

13. Name three protective devices to protect cranes from contact with electrical power lines.

14. How does a circuit breaker operate to protect its circuit?

15. What is one of the major causes of electrical shock to workers?

BIBLIOGRAPHY

American Society of Safety Engineers, Engineering Division Newsletter, Fall 1997, "Building Electrical Safety Engineering," Richard T. Beohm; Winter 1998, "Ground-Fault Circuit Interrupters (GFCIs) Can Save Lives," Richard T. Beohm; Summer 1996, "Protecting Your Buildings from Lightning," Richard T. Beohm, Des Plaines, IL: ASSE.

Browne, Jr., 1984. *Circuit Interruption: Theory and Techniques.* New York: Marcel Dekker.

Fordham, Cooper W., 1986. *Electrical Safety Engineering.* Stoneham, MA: Butterworth.

Levenson, Harold and John A. Allocca, 1982. *Electrical and Electronic Safety.* New York: Prentice-Hall.

Lüttgen, Günter and Norman Wilson, 1997. *Electrostatic Hazards.* Oxford, England: Butterworth-Heinemann.

Magison, E. C. and W. Calder, 1983. *Electrical Safety in Hazardous Locations.* Research Triangle Park, NC: Instrument Society of America.

National Fire Protection Association, 2009. *Electrical Safety, Requirements for Employee Work Places.* Quincy, MA: NFPA.

_____. *The National Electrical Code,* 2008. Quincy, MA: NFPA.

_____. *The National Electrical Code, 2008 Handbook,* 2008. Quincy, MA: NFPA.

_____. *NFPA 780, Standard for the Installation of Lightning Protection Systems,* 2008. Quincy, MA: NFPA.

National Safety Council, 1988. *Electrical Inspection Illustrated.* 2nd ed. Chicago: NSC.

Winburn, D. C., 1988. *Practical Electrical Safety: Occupational Safety and Health.* New York: Marcel Dekker.

16

Tools and Machine Controls

Dennis Cloutier, CSP

Robert N. Andres, CSP, CPE, CMfgE

Use of Anthropometry

The root cause of many accidents is often found to be behavior (human error), when in fact, a deeper assessment would likely uncover a faulty design of a tool or a control. A tool should be thought of as an extension of a human capability, enabling a person to work more easily, more quickly, or more accurately. A control is an accessory on a machine that provides the point of interface between the machine and the human operator (HMI). Any tool or control exists for the convenience of and extension of the capability of the human worker.

It should follow, logically, that a tool or control should be designed to best suit its user, but that is not always easy and has not always been the case: A screwdriver slips out of the slot in the screw and gouges the worker's hand. A hammer handle or axe handle slips out of the user's hand and goes flying through the air. Some control systems are difficult to operate, and their identification is sometimes confusing. The list of examples could go on and on. A great many tools and controls have been designed for universal application, with consideration for only the most average worker and/or working conditions in which they are intended to be used. Recently, more attention is being given to training design engineers in human factors, and the results are tools that are more comfortable for individuals to use—a greater variety of shapes and special designs for left- or right-handed persons. This trend should continue and—if it's not already in place—become required study for designers of tools and controls.

Many studies of human capabilities have been made. Data on human dimensions are called *anthropometric data* and are available from the following sources, among others:

Human Engineering Design Criteria for Military Systems, Equipment and Facilities, Military Standard MIL-STD-1472D (01 Mar 1989), available from:

DODSSP–Customer Service
Standardization Document Order Desk

700 Robbin Avenue, Bldg. 4D
Philadelphia, PA 19111-5094
https://assist.daps.dla.mil/quicksearch

The Measure of Man and Woman

John Wiley & Sons
111 River Street
Hoboken, NJ 07030-5774
www.wiley.com

Table 16-1 shows some of the major body dimensions.

Anthropometric data is almost always presented in three sets of values. The 50th percentile represents the average dimensions; the 2.5th or the 5th percentile represents practically the smallest dimensions; and the 95th or the 97.5th percentile represents practically the largest dimensions.

The choice of which set of figures to use depends on how the design item is to be used. One should choose whichever will be the safest or easiest for the greatest number of people. A clearance, such as the space between two control buttons, should always be designed for the 95th or 97.5th percentile. However, if the intended work environment will require personnel to wear hand protection (gloves), such accommodation should be included in the design. A reach, such as the height of an emergency switch, should always be designed for the 2.5th or the 5th percentile. The size of a handle is normally designed for the 50th percentile, unless a specific application will require a larger or smaller size.

The greatest amount of work in gathering and applying anthropometric data, and human factors in general, has been done by the military. In addition, the insurance industry has had a significant impact on the private sector. Adherence to MIL-STD-1472 is required in the design of military equipment. The insurance industry has worked through its own loss-control departments and the voluntary consensus standard activities in the United States. Some of the military data that has been gathered over time may be too restrictive for application to the entire population. In general, the body of knowledge has expanded greatly over the past two decades and, with Internet search capabilities, this information is almost limitless.

Anthropometric data is very useful in designing hand tools and operator controls. There are several sources of dimensions available for typical hand-manipulated tools and controls. These dimensions were derived either from anthropometric data or from extensive studies of these tools and controls actually in use. At this point, it seems more useful to look at these specific dimensions, recognize where they can be applied, and resort to anthropometric data when necessary for special applications.

The human hand is the most versatile tool available, but it does have limitations. Although we can rotate the hand through quite a large arc around several axes, the most efficient activity is limited to rather small angles. Recommended dimensions have taken this into account. The greater the surface area of contact between the hand and a tool, the greater the transfer of torque or force. The use of the first and second fingers (see Figure 16-1) will permit the greatest accuracy and speed. If continued control over a tool is desired, the thumb must be in contact with the tool at all times and must not be

Table 16-1. Some Pertinent Anthropometric Data, Taken From a Variety of Sources

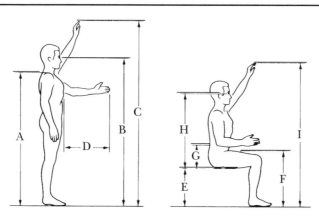

Dimensions in centimeters and in inches					
	Men			Women		
Dimension	5th	50th	95th	5th	50th	95th
A (cm)	133.3	143.8	154.2	123	131.3	142.2
(in.)	52.5	56.6	60.7	48.4	51.7	56
B	152	165	173	143	150	160
	60	65	69	56	59	63
C	195	211	225	175	193	208
	76.8	83.1	88.5	69	76	82
D	57.1	61	65.3	51.8	57.6	61.7
	22.5	24	25.7	20.4	22.7	24.3
E	40.6	45	50	38	40.5	44.1
	16	17.7	19	15	15.9	17.4
F	52.3	59.2	65.3	49.5	54.6	61
	20.6	23.3	25.7	19.5	21.5	24
G	20.3	22.9	24.9	18.8	21.1	21.8
	8.0	9.0	9.8	7.4	8.3	8.6
H	72.6	78.7	84.6	68.8	73.7	78.7
	28.6	31	33.3	27.1	29	31
I	156.7	171.4	185.7	141.5	156.5	168.9
	61.7	67.5	73.1	55.7	61.6	66.5

removed to operate a switch or other control. When the hand is raised above shoulder level, it becomes fatigued much more easily than at positions below that level. The hand and arm can exert much more force when pulling toward the body or pushing away from the body than they can when moving sideways.

The foot is not nearly as versatile as the hand. We depend on our feet to keep our balance in a standing position, and, therefore, the foot is not a good body member to use for operating tools or controls in most cases.

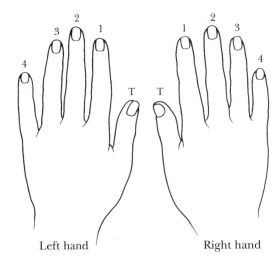

Figure 16-1. Conventional numbering of the thumb and fingers.

When a worker is sitting and a control can be operated by a simple up-and-down motion, the foot can be used to relieve the hands. Only when a tool requires that the mass of the body be applied to it, such as a shovel, should the foot be considered preferable to the hand.

It is not uncommon to find tools and controls that require too much force to be easily operated. The strength of the hand and arm muscles varies a great deal from one person to another—and even in one person from one time to another. Furthermore, the force needed to operate a tool or control should never be the maximum force a person can muster. Table 16-2 gives the maximum force that can be expected of the 5th percentile of the male population for various activities. Designing for the 5th percentile is better than designing for the 95th percentile when strength is concerned.

Tools and controls should always be thought of as extensions of a person's own capabilities. It was mentioned in Chapter 4 that many of our actions are based on what we anticipate. When an object does not respond the way we thought it was going to, an accident often ensues. This phenomenon is a very important factor in the design of tools. Accordingly, a tool or control that is used to move large objects, exert large forces, or cause large changes should be large and may or may not require a relatively large amount of force to operate. Power-operated machines greatly enhance the human capability and, many times, require little effort on the part of operating personnel. On the other hand, a tool or control used for delicate work should itself be relatively delicate.

In human-factors engineering, the term *control/display ratio* (C/D ratio) is used to describe the relationship between the tool or control and what it accomplishes. Some tools and controls are moved a small distance and with a small force to cause large motions and forces—for example, the controls in a power shovel. This is a low C/D ratio. Others involve a relatively large movement of the control to effect a small movement—for example, the controls on a high-powered microscope. This is a high C/D ratio. However, the C/D ratio has limitations, imposed by the space that is available, by the limits of hand and arm motions, and by human-sensory capabilities.

The C/D ratio is related to the sense of compatibility. Controls should be located in an expected place and operate in an expected manner.

Table 16-2. Maximum Hand/Arm Strength of 5th Percentile Males in Various Positions

Dimension	Maximum Force	
	in N	in lbs.
Arm pull (arm straight forward)	231	52
Arm pull (elbow at 90°)	165	37
Arm push (arm forward)	222	50
Arm push (elbow at 90°)	160	36
Arm lift (arm straight forward)	62	14
Arm right to left (right-handed person)	89	20
Arm left to right (right-handed person)	62	14
Hand grip, momentary	260	59
Hand grip, hold	155	30
Thumb-finger grip, sustained	35	8

The speed of motion, especially repetitive motion, may be important in some applications. A worker's hands and fingers should not be required to perform at a speed beyond that which is comfortable. At the same time, the mechanism in a tool or control should operate at a speed faster than its human operator. An automatic door, for example, must open faster than the normal walking speed of an individual once the sensor signals the presence of an approaching individual.

In designing work activities, planners often mistakenly assume that a worker can and will use two controls simultaneously. As mentioned in Chapter 4, a person can act on only one stimulus at a time. If the operation of a control requires mental activity, it cannot be done simultaneously with another operation requiring mental activity. If it is very important that two controls be operated simultaneously, they must be identical in operation and located side by side. Figure 4-4 is a guide to the types of hand motions that can be performed simultaneously.

Design of Hand Tools

The subject of hand-tool design could fill several books; this section will be limited to a few common tools that are most often involved in accidents.[1] Screwdrivers have caused a great many hand punctures, and several alternatives are now available to reduce or eliminate the risk associated with this hazard. The accident almost always occurs when the screwdriver slips out of the slot in the screw head or when it is used for something other than turning a screw.

These accidents are largely due to inattention or behavior leading to unsafe acts. A straight-blade screwdriver should be kept blunt with square edges and used only in slots in which it fits snugly. A worn, rounded-edge tip may more easily slip out of the slot, which could result in an injury. A shallow socket, such as that of the Phillips-head screw, provides an improved fit between the tool and the fastener, but even the Phillips screwdriver must be maintained to fit properly.

Figure 16-2. Shown at left are slow head, square head, hex head, and tourx head driver tools.

Figure 16-3. Hammer with steel handle covered with shock-absorbing coating.

The socket type of screw head and screwdriver provides superior fit and hold, resulting in low risk of injury from the tool slipping out of the socket. Figure 16-2 shows a variety of socket heads commonly available today.

All these and more are commonly available today. The slot and hex heads have been around for awhile. The tourx was first found in the automobile industry, and the square head has become very popular in the construction industry. There is little chance that the screwdriver will slip out of the socket-type fasteners. Because of the common availability of these fasteners today, it has become a personal preference as to which design to use.

Open-end wrenches are often misused, and bruised knuckles are usually the price paid for this misuse. Although they are convenient, adjustable wrenches are the most often misused. The jaws of open-end wrenches should be inspected periodically to make sure that the surfaces are flat and parallel. When they become bell-mouthed, they are dangerous and should be discarded.

All cutting tools, such as knives, wood chisels, and cold chisels, should be kept sharp so they will cut with minimum effort. It is often said that the most dangerous tool in the home kitchen is a dull knife. A dull knife requires more effort to cut through the product. But with more force being used, slipping is more likely, which could result in an injury. Some fixed-blade knives have a safety shield separating the handle and the blade. The head of a cold chisel should be maintained to prevent it from becoming mushroomed. The edges of such a mushroomed head present two hazards. When they are hit by a hammer, fragments break off and become flying objects. The edges also create sparks, which can start a fire if flammable gases or vapors are present.

Wood handles are commonly used on hammers, mallets, axes, and other tools because they are inexpensive, provide good balance of mass for the various heads, and will not conduct electricity. However, they do get slippery when wet and present slivers when abused. Modern-day handles are made of composite materials or steel coated with shock-absorbing materials. Handles are now shaped to better fit the human hand (see Figure 16-3).

In locations where sparks might start a fire, it is helpful to use tools made of non-sparking materials. Beryllium copper and aluminum bronze are used for this purpose. These materials are also nonmagnetic and quite resistant to corrosion. Almost all common tools are available in these materials.

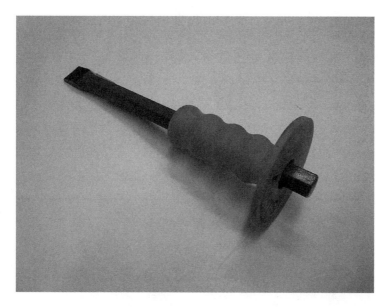

Figure 16-4. Shield built onto a chisel to reduce the possibility of hitting the hand holding the chisel.

Sufficient clearance for the fingers or, in some cases, for the whole hand or arm, should be provided to prevent pinch points. The chisel shown in Figure 16-4 has a built-on shield that prevents a hand injury should the operator miss striking the head of the tool.

Tools with two handles that close toward each other, such as pliers, should have the handles far enough apart at their closest point to allow room for whatever fingers are likely to be between them. This is especially true not only for tools normally operated by two hands but also for pliers, tweezers, and cutters in which one or more fingers may be inserted between the handles to gain more accurate control. These tools should always have some mechanical stop to prevent the two handles from coming completely together at a point where a finger could be pinched.

In many cases, a clearance must exist between the tool and another object. This design requires anticipating how and where the tool will be used. Many tools are designed for very general use, so this condition cannot be anticipated, but tools with more specific uses should be designed to provide sufficient clearance. Table 16-3 shows some typical tool handles and recommended minimum clearances. The size of the handle itself should be in proportion to the forces to be transmitted through it and normally varies from a minimum of 6 millimeters up to 25 millimeters.

Any edge on a tool with which the user's hand may come into contact should be rounded. If pressure is exerted on an edge, it should have as large a radius as can be provided.

Tools that are designed for a specific use should allow the hand to be maintained in a relaxed and normal position during use, as illustrated in Figure 16-5. Tools that require or encourage poor posture should be avoided, especially if they are to be used for an extended period of time.

Table 16-3. Minimum Clearances on Handles and Knobs

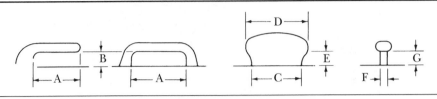

	Dimensions in millimeters						
	A	B	C	D	E	F	G
Bare Hand	115	50	50	70	25	8	20
Gloved Hand	135	90	45	65	30	8	25

Figure 16-5. Insulated pliers with offset handles or gripping jaws are now available. These allow the hand to be held in a natural position and provide flat or contoured jaws. Some are constructed with a rachet arrangement to available jaw force.

Since hand tools are used so frequently, any feature that will prevent or reduce accidents or any physical harm should be incorporated into their design.

In many industries, special hand tools are used to grasp and carry objects that cannot or must not be picked up with the fingers. Manufacturers of electronic components have the task of handling integrated circuits (chips) and other small components as small as 2 millimeters across and 0.01 millimeters thick. Furthermore, the chips would be contaminated by contact with human hands. Small "vacuum pencils" are used to pick up these chips. They consist of small tubes with a larger, cylindrical handle and an on-off valve. They are attached to a vacuum line with a regulator to control the vacuum. The tube is bent to allow the hand to maintain a normal position. The end of the tube is flat to provide good contact with the chip.

Figure 16-6 (courtesy of Rockford Systems) illustrates a hand tool commonly used in the metal-forming industry. The large vacuum cups fitted with handles are used for picking up sheets of steel and plates of glass. The thumb button on the handle vents the vacuum, allowing the operator to release the material.

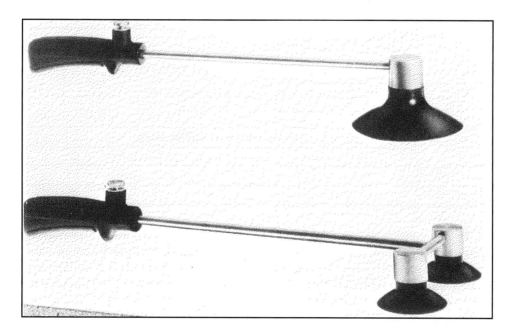

Figure 16-6. Vacuum-lifting tools for objects with a flat surface.

Specialty tools are available from numerous sources, including popular retail hardware stores and online manufacturers or outlets. One such tool, shown in Figure 16-7 is a long reach, flexible gripping device with a built-in LED light—an indispensible tool for retrieving dropped hardware from tight, dark places.

Figure 16-7. This is a long-reach, flexible lighted retrieval tool. (Courtesy of HomeDepot.com.)

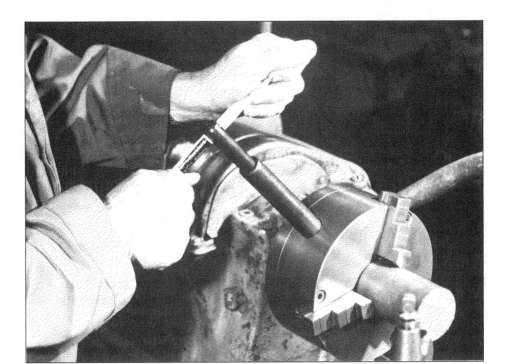

Figure 16-8. Chuck wrench with a spring to prevent its being accidentally left in the chuck. (Courtesy of Rockford Safety Equipment Co.)

Many accidents have been caused by chuck wrenches left in lathe or drill chucks. When the switch is turned on to start the spindle rotating, the wrench flies through the air. Figure 16-8 shows a chuck wrench that has a spring in the shaft. The wrench shaft must be pressed into the chuck to use it, and the spring makes it impossible to leave the wrench in place.

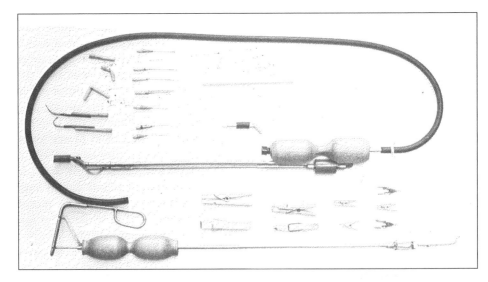

Figure 16-9. Specially designed tools for manipulating parts and other tools inside a bottle, for use by a model maker. This picture first appeared in the *Journal of the Nautical Research Guild* and is shown here by permission of R. W. Preston.

Many special tools have been designed to enable a worker to reach places that he or she could not or should not otherwise reach. Figure 16-9 illustrates tools that were designed to enable a worker to reach into a small-neck bottle.

Portable Power Tools

There are several features on small, handheld power tools that can make them hazardous to use. The type of switch and its location are among these important features. A person using a power tool should not have to lose control of the tool by removing a thumb or finger in order to operate the switch. Depending on the normal position of the hand on the tool, the switch may be operated by either the thumb or the first finger without relinquishing control. Most hand drills are held in such a manner that the first finger is by far the easiest to remove from the grip without losing any control over the handle. This finger can, therefore, easily be used to operate a switch if it is a trigger-type switch and conveniently located. Any other type of switch or location might require awkward or dangerous movements. Figure 16-10 shows a power jigsaw with a switch that is actuated by the index finger. This allows the hand to remain in a natural position while operating the tool.

If the tool has a motor capable of developing a high torque, two handles should be provided. The two handles should be easy to grasp and as far apart on the tool as feasible.

Figure 16-10. Illustration of a power scroller saw with a "trigger" type operating switch. The black knob on the red switch is a speed-adjust that, when adjusted in, allows the operator to limit the speed yet still be able to grip the tool firmly. Without this solid stop adjustment, the operator would have to hold the switch in a mid position while cutting, which is sometimes difficult to do with a vibrating tool.

Figure 16-11. Heavy-duty double-insulated belt sander with two convenient handles. (CourtesyofSears/Craftsman)

Most power tools provide a handle to control the tool, but handles frequently are not located in the best place. They should be located in positions where they will provide a balance for the mass of the tool and permit the application of force in the direction needed. Figure 16-11 illustrates well-designed handles.

When any tool with a power unit is being designed, whether that power is generated by an electric motor or by a hydraulic or pneumatic device, suitable guards should be provided. In some tools, such as power wrenches, only the rotating portion that is not engaged with the work needs to be guarded. A saw, grinder, or any other tool that presents a hazard should have a guard over that hazard. A guard designed to be an integral part of the tool is always more effective than one attached to the tool as an accessory.

Some power tools have been designed with a lock-on device that enables the user to keep the power on without having to hold the switch on. Hand drills often have this feature. This feature is not recommended, because it bypasses a fail-safe feature. If a tool is awkward or tiring to use without such a lock-on, either the handle or the switch, or both, have not been well designed in the first place. It is especially important that heavy-duty tools have a fail-safe switch, so that if the user loses control for any reason, the power will be shut off.

Design of Controls

The design of controls and the operator-interface control panels for machinery is a different design discipline than that for hand power tools. Engineers called upon to design controls and control panels must have a thorough understanding not only of human capabilities and limitations but also of the machine's capabilities and intended use. Without this understanding, inappropriate design could lead to operational errors in use that could result in serious consequences. Chapter 4 provides an introduction to the subject, and the reader is encouraged to study further in this field.

The overall objective in the design is to make the controls easy to locate and easy to operate. This serves to minimize errors that might arise from their use or misuse. In Chapter 4, there is a discussion of intentional errors and unintentional errors. An intentional error is the result of a deliberate action that was believed to be the proper action but was not. An unintentional error is the result of an inadvertent action. Every effort should be made to design controls to avoid both types of error. Controls should be located and oriented in a way that will cause minimal confusion. They should fit whatever body member and body motion the operator is expected to use, and they should be as intuitive as possible. They should require the proper resistance to be operated, and they should produce the results that the user expects.

Push Buttons

Contact button switches are a common control device, and there are numerous configurations. They are advantageous when a simple motion is desired for an on-off or yes-no selection and when space for the control is limited. In this discussion, we will distinguish between the traditional push button and soft-touch buttons.

Push buttons, as distinguished from touch buttons, require a deliberate application of force to actuate the switch (see Figure 16-12 for example). The amount of force needed should be in proportion to the force that the switch controls, but it should be within the minimum and maximum designated in Table 16-4. Special applications may require dimensions or resistance outside the recommended range, but such deviations should be given careful consideration.

Figure 16-12. Typical push-button on-off controls.

Table 16-4. Recommended Dimensions for Various Types of Controls Illustrated

	Size (A × B) in mm		Displacement (D) in mm		Spacing (S) in mm		Resistance (R) in N	
	Minimum	Maximum	Minimum	Maximum	Minimum	Optimum	Minimum	Maximum
Touch button	20 × 20	50 × 50	0.5	2	1.1 × A		0.3	10
Push button	10 dia.	20 dia.	3	40	15	50	1.5	11
Emergency-stop button	30 dia.	80 dia.	2	5	2 × A		1.5	8
Keys on keyboard	10	20	1	5		1.1 × A	1	4
Toggle switch	2 × 10	10 × 50	30°	120°	20	50	3	10
Rotary switch	16 × 25	75 × 100	30°	90°	25	50	0.1 N•m	1.0 N•m
Rotary knob	12 × 10	25 × 100			B + 25	B + 50	0.01 N•m	0.05 N•m
Handwheel	8 × 50	50 × 500			B + 25	B + 80	20	200
Crank	40 × 50	100 × 400			2B + 75		10	200
Lever	12	75		360/960	50	100	10	130/90
Foot pedal	25 × 75		12	65	B + 100	B + 150	40	90

SOURCE: Adapted from MIL-STD 1472D.

Touch buttons are used when a casual touch by any finger, by a hand, or even by an elbow is considered a satisfactory way to use the control. They usually have a large, flat surface and require very little force and motion to operate. Touch buttons are used to good advantage for selecting one of several conditions on a control panel; however, they should not be used in any situation in which accidental contact could cause a serious error. The large size of touch buttons offers the advantage of permitting an identification label to be placed on the buttons themselves. This avoids confusion, provided the identification word, number, or symbol is well chosen.

Push-Button Guarding Requirements

The necessity for guards has been mentioned but needs further discussion. If something is needed to prevent the finger from slipping off a given button onto an adjacent button, a raised partition between the buttons just high enough to prevent this slipping can provide the needed divider or barrier. The fingers will still have enough freedom of movement to find a series of locations. If the need is to prevent accidental actuation of a button or switch, a larger barrier is required. In this situation, each button will often have a circular barrier around it; that is, the button is contained in a cup. This is well illustrated in Figure 16-13 in two areas.

The control console in Figure 16-13 has five controls along the front edge. The buttons at the extreme left and right are used simultaneously to start the press cycle. The high cups around these buttons have two purposes: they prevent the operator from inadvertently contacting the buttons, and they prevent the operator from using the elbow and a finger of the same arm to press the two buttons, leaving the other hand free to reach the hazard area. There are also two buttons mounted on the machine on each side above the die. These operating buttons are also enclosed in cups. The controls on the machine are used when the press is utilized for secondary forming operations

Table 16-4. Cont'd.

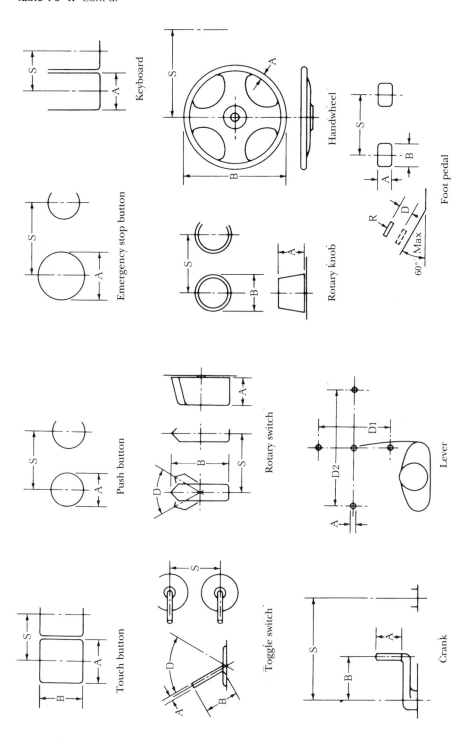

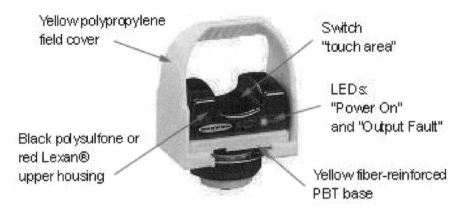

Figure 16-13. Guarded capacitive finger-sensing actuator. (Courtesy of Banner Controls)

that require manual loading and removal of the workpiece. Depending on the forming system, these two control pushbuttons could be utilized as a point-of-operation safeguarding device. When the console is utilized for machine-operation control, alternative safeguarding for the press point of operation (the die area) is required.

Two-Hand Trips and Controls

Two-hand machine actuators can be used on otherwise unguarded machines to effect a machine cycle (a two-hand trip) or sustain motion that is hazardous to hands and fingers (a two-hand control). The placement and configuration of the two palm buttons is as important as their physical guarding to prevent unintended and fail-safe operation (see Figure 16-14 for an example).

The important feature is that the buttons and circuitry must be designed so that:

- Concurrent depression of both buttons is required.
- There is no way that one button can be held down, enabling the machine to be operated with only one button (anti-tie-down capability).
- Guarded to minimize the possibility of unintended activation
- Spaced sufficiently distant from each other to assure that both buttons cannot be simultaneously depressed by one hand or arm. This generally means spacing the buttons at least 24 inches apart if they are on the same plane.

Additionally, when applying a two-hand trip to meet the requirements for a point-of-operation safeguarding device, such as on a full-revolution power press, the designer must make sure that the buttons are located so they are at least the minimum safety distance required by the safety-distance formula in Figure 16-15 below. This means that the palm buttons must be located far enough away so that, after the control is tripped and the operator releases both palm buttons, the operator cannot "beat the ram" or reach into the point-of-operation or other pinch point while the hazard still exists.

Figure 16-14. Optical two-hand actuators on control bar.
(Photo courtesy of Break-A-Beam, Inc.)

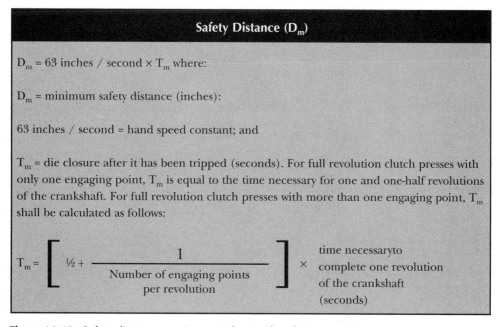

Safety Distance (D_m)

D_m = 63 inches / second × T_m where:

D_m = minimum safety distance (inches):

63 inches / second = hand speed constant; and

T_m = die closure after it has been tripped (seconds). For full revolution clutch presses with only one engaging point, T_m is equal to the time necessary for one and one-half revolutions of the crankshaft. For full revolution clutch presses with more than one engaging point, T_m shall be calculated as follows:

$$T_m = \left[\frac{1}{2} + \frac{1}{\substack{\text{Number of engaging points} \\ \text{per revolution}}} \right] \times \substack{\text{time necessary to} \\ \text{complete one revolution} \\ \text{of the crankshaft} \\ \text{(seconds)}}$$

Figure 16-15. Safety distance requirements for two-hand trip actuation.

Emergency Buttons

The emergency-stop buttons on the machine and on the console that meet current design requirements are large, mushroom-shaped buttons well separated from the other controls and readily accessible in the event of an emergency. Emergency-stop buttons or switches are not allowed to be guarded with ring guards, but consideration must be given to their location to prevent incidental contact leading to a false shutdown.

An emergency button should protrude out from the surface of the panel or mounting surface at least as far away as any other control. It should be large enough and located in such a position that the worker can reach and operate it from any location if the need arises. Emergency buttons should be located wherever necessitated through the performance of a risk assessment. In many situations, the emergency button is rounded on the top (mushroomed shaped), so that it can be distinguished from other controls, easily located by personnel and operated easily from several directions and angles. It should be a red button on a yellow background—and should require a reset in order to restart the machine (see Figure 16-16).

A question sometimes arises about what the emergency control switch should do: should it remove power from the entire machine, or just the workstation or area/zone, or the hazardous-motion-control power? The designer has to consider what emergencies could arise, and why and to what elements power should be removed at all. In some cases, removing power to exhaust fans, cooling fans, or some control devices could create undesirable consequences. An emergency situation might best be managed if power is removed from only those elements that need to be stopped.

Sometimes, the process cannot tolerate an instantaneous disconnection of power or similar abrupt response. In such cases, the E-Stop switch may initiate a "controlled stop."

Figure 16-16. Typical locking emergency stop button.

Part No. CTL-507

Part No. CTL-502

Part No. CTL-525

Figure 16-17. Various machine actuation palm button configurations. (Courtesy of Rockford Systems)

(1) Category 0 is an uncontrolled stop by immediately removing power to the machine actuators.
(2) Category 1 is a controlled stop, with power to the machine actuators available to achieve the stop, then removal of the power when the stop is achieved.
(3) Category 2 is a controlled stop, with power left available to the machine actuators.

One of the first steps is determining where the E-Stop fits within your machine-control system and whether your particular application requires Category 0, 1, or 2 emergency shutdown. The intended application often determines the placement, size, electrical specifications, mechanical characteristics, ergonomics, color/legends, and number of E-Stops required. So a thorough understanding of the machinery and associated control systems is key to making the right E-Stop choice.

The photos in Figure 16-17 show a typical two-hand actuation system using dual ring-guarded palm buttons (black), an emergency stop (red button on a yellow background) and top-stop or cycle-stop (yellow on black background).

Foot Switches

Foot pedals should be mounted rigidly, when practical, except for the one plane of motion. Some foot pedals are removable in order to eliminate the

Figure 16-18. Typical guarded foot switch. (Courtesy of Triad Controls)

risk of inadvertent actuation or a tripping hazard when not in use. The axis of motion should be horizontal or nearly so. Pivoting the pedal at the heel provides the best control and the least fatigue. The pedal should be located so that force can be applied by the ball of the foot in a downward direction. The foot becomes fatigued rather quickly if repeated motions are required. A pedal should have a large enough foot pad to allow even distribution of the actuating pressure as applied by the worker's foot. The pad should have a flat, nonslip contact area.

In any safety application, the foot pedal must be guarded to minimize unintended actuation, but care must be taken to select a foot switch that will allow easy entry of a shoe type being worn in the facility. Depending upon the risk, the foot pedal may also be designed with an electrically interlocked flap that must be raised to allow the pedal to function. This assures that a foot must be in place before the pedal can be activated (see Figure 16-18).

Mat Actuators

Mat switches are typically used to activate an alarm condition, enable operation of a machine requiring operator attention, or disable a function should someone come too close to a hazard area. The most common type of mat switch uses two steel plates separated by compressible spacers (which may

Figure 16-19. Two-wire mat with ramped edge. The safety version uses four wires for redundancy. (Photo courtesy of Bircher America)

vary in density according to the desired actuation force/weight), each with a separate wire or, for greater safety reliability, dual wires, that either make or interrupt an electrical circuit. The entire assembly is molded with cut-resistant rubber or plastic.

The operation of the safety mat is easy to understand. The safety mat is a simple, normally open switch. When a specified minimum weight is applied to the safety mat, the switch closes. This sends a signal to the safety mat controller, which in turn sends a stop signal to the guarded machine. A mat specifically designed for safety applications has four wires to the safety mat controller. This provides the redundancy required to monitor the wires for open circuits due to incorrect wiring or physical damage to the safety mat wires (see Figure 16-19).

In a safety application, depression of the mat closes the contacts, but this actuates a self-checking relay that interrupts power to the machine control. This is not a fail-safe system, but it does provide a reasonably high degree of reliability. Safety mats should not be used in high-risk applications.

Safety Interlock Switches

When hazardous exposure is controlled by the use of a physical barrier, the risk level will determine whether the barrier needs to be electrically interlocked to allow access only when the hazard has been neutralized. One common method of accomplishing this is to interlock the barrier to assure that it cannot be opened until the hazard condition has been removed. This is typically accomplished by using a solenoid-latching safety interlock, as shown in Figure 16-20.

Toggle Switches

Toggle switches are available in a wide range of styles and sizes. A toggle switch is basically an on-off switch. Many remain in the open or closed position after being actuated. Others are spring loaded to return to an initial or neutral

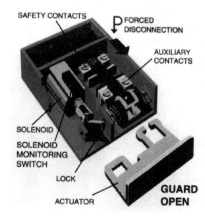

Figure 16-20. Forced disconnect switch cutaway illustrations showing unique key tripping of internal switch components. (Courtesy of STI)

position. Some toggles have three positions that enable the user to select any one of three conditions. This arrangement is convenient and usually quite satisfactory but requires special consideration when designing which position of the switch performs a function. For example, if the "off" command is the middle position of the switch, a hazard may exist if it is possible to push the switch past the "off" position into the opposite position.

Toggle switches may be prone to inadvertent operation and unintentionally starting something into motion. A person's hand or sleeve can easily flip a toggle switch to the opposite side. The probability of an accident happening can be reduced by mounting the switch on a recessed surface or by placing barriers on each side of it. Toggle switches that perform critical functions often have a protective cover over them. This spring-loaded cover requires the operator to first lift the cover and then operate the switch.

Safety Pull Switches

Where an electrical switching function must be easily accessed, particularly for an emergency stop serving a wide area, as a processing line, a cable-operated safety

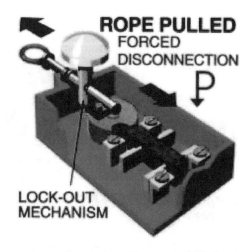

Figure 16-21. Modern pull switches. (Courtesy of STI, Inc.)

pull switch provides an inexpensive and effective actuator. The only acceptable type of switch for a safety application is designed to be set on a detent—so that the switch will trip the system to a safe condition if the tension on the cable increases or decreases (if the cable is pushed, pulled, or cut). See Figure 16-21.

Rotary Switches

Rotary switches perform the same function as push buttons and toggle switches. They have the added advantage of having multiple positions as the switch is rotated around the circle. These devices are routinely used for mode selection and can have as many as 12 positions—and sometimes more. Rotary switches snap into a detent at each position and are never able to move past the last selection to the first selection when rotated in the same direction. Some are locking and require the operator to push down or lift up on the actuator to rotate it to another position. For example, the cycle selector on a standard home-laundry machine must be pushed in to rotate it and then pulled out to start the selected cycle.

A design mistake often found on rotary switches is the marking of a pointer and other information on a flat skirt around the bottom of the switch. As soon as the worker's hand is on the switch knob, the pointer and other information are covered up, and the worker must use the switch without being able to see them. A rotary switch actuator should always be the pointer, and some even have an arrow molded into the plastic.

A rotary knob is distinguished from a rotary switch by the fact that the rotary knob is capable of a full 360° rotation. Rotary knobs may also be used as multipoint switches, but most often are utilized as infinite controllers, such as speed controls, dimmer controls, sound-level controls, and so on. Some may have specific points to which they could be set, or they may have smooth movement providing unlimited settings.

A rotary knob may have a skirt, but, as with switches, the skirt should not contain any needed information or pointer unless it is visible outside the area covered by the fingers using the knob. Generally, a calibrated scale or other information is more useful on the fixed panel behind the knob, with the pointer on the knob itself. The pointer must extend beyond the fingers holding the knob, so that it can be seen easily.

Other Types of Devices

Handwheels, cranks, and levers are used to accomplish one of two things: to amplify motion and thereby provide a more sensitive control, or to provide leverage to reduce the force needed to move something. The proper design must recognize the purpose and provide the needed amplification or leverage.

The major characteristic of these devices that affects safety is the design of the knob or handle. It should be made to fit the hand or fingers expected to use it so that contact can be made over the greatest surface. The greatest amount of torque can be obtained with a cylindrical handle diameter, which enables the user to just close his or her fingers around that handle. Shallow flutes, light knurling, or a rubber surface also helps. In many cases, it is not torque that is desired, but a firm grip to allow pushing or pulling. A cylindrical handle is usually the best design for a crank, but a sphere is a better design on a lever because the orientation of the handle changes as it is pushed or pulled. A spherical handle about 50 millimeters in diameter is an optimum size.

Location and Arrangement of Controls

Controls on a machine should be located where the operator can reach them easily, and they should be arranged in some logical order. This is not always the case, particularly on older machines but even on some new ones. Figure 16-22 shows an older machine with a less-than-ideal array of controls. This, in large part, is due to all controls being mechanical in operation and located at their applicable axis. Some controls are conveniently located on the front of the machine, but some require that the operator walk around the end of the table to reach them. The start/stop lever is located at the top of the machine. Such placement was common on machines made prior to the 1960s. Progress has been made as electrical/electronic devices and controls have evolved. Today, machine controls are almost all electronic, with servomotors and other types of drives replacing almost all the manual motions once performed by operating personnel. But still, we occasionally find the need for further improvement.

Figures 16–22 and 16–23 shows two milling machines; one manufactured in the 1950s and the other in the 1990s. This comparison shows the evolution of similar machines over 50 years. For the 1954 lathe shown in Figure 16-23, each black knob visible in this photo is a control used by the operator; there are many controls, yet they are all located where the individual normally stands when operating the machine. The newer machine in Figure 16-24 illustrates how all these manual controls have been replaced as technology has evolved. Many advances in safety, as well as production, are seen in this comparison.

Figure 16-22. Old-style milling machine.

Figure 16-23. An old machine with many manual controls located on the axis that they control.

Figure 16-24. New milling machine. Note control placement and enhanced (electrically interlocked) physical guarding.

When a large number of controls must be located on a given machine or control panel, they should be arranged in some logical order. In their book Human Factors in Engineering and Design, E. J. McCormick and M. S. Saunders list four guiding principles for arranging controls and displays. They are referred to as:

1. The Importance Principle;
2. The Frequency-of-Use Principle;

3. The Functional Principle;
4. The Sequence-of-Use Principle.

If one control is significantly more important than others, it should be located in a prominent, easy-to-reach location and oriented in such a way as to make it easy to use. Emergency buttons fall into this category. The factors that are considered important should be given careful consideration. The safety of the worker should certainly be given top priority, but other factors may also be important.

Frequently used controls should be given preference to a convenient location over those less frequently used. This principle does not generally affect the safety of the operator, except as it relates to minimizing fatigue. Very often, other principles will take priority over this one.

Controls should be located in a place, and in a manner, that will help the operator to associate them with what they do. Frequently, a gauge or dial will indicate some particular characteristic, such as pressure or temperature, while another device controls that characteristic. The gauge and the control should be placed together, so that there is no confusion about which control regulates the characteristic indicated on the gauge. Generally, the control should be located first and the gauge then placed with respect to the control.

Any visual display of labels or other markings should be located above a control, never below it. The worker's hands should never have to reach above eye level for routine operations and should never cover up any display that needs to be seen. If a control can be made to resemble the device it controls, or if any association between them can be suggested, confusion and errors will be reduced. A diversified workforce increases demands in this respect that might not otherwise be foreseen.

Physical location may be part of that association, although no location or position that is unsafe for the worker should ever be created. In some cases, a series of controls is used in a sequence, and controls, as well as tools, may be arranged to aid this sequence. Such arrangements will help to reduce errors, but, again, they must never create an unsafe condition for the worker.

Each of the four principles should be considered by itself, and then priorities should be assigned. The order of priorities will differ from one situation to another. The major factor in setting priorities should be the influence each principle will have on the safety of the worker and others in the area, as well as on the machine and other property. The "importance principle" will certainly be a contender for top priority.

Figure 16-25 shows a complex control panel in which these principles have been applied. Note that the screen is placed at the upper left-hand corner where it will not be covered by the hand or arm of a right-handed worker. Touch buttons on the left side have identification marks on the buttons. Most of the other controls have identification labels and other markings above the control. A few controls in the center of the panel have labels on the right-hand side that may be momentarily covered by the right hand. The emergency-stop button is very prominent and protrudes beyond the other nearby controls.

Figure 16-25. A well-designed array of controls and displays on a control panel.

The controls and displays in the cockpit of modern aircraft are a good example of design and arrangement that enable the user to identify and use the controls quickly and without error. Designers of aircraft, particularly military aircraft, have been the leaders in the study and application of human factors. Most of the data that has been developed for use in this field has come from military studies and applications.

Many industrial companies are learning that it is very costly and often impossible to design tools and controls that will sufficiently reduce a worker's exposure to hazards. They are finding that mechanical manipulators and robots can do the work in hazardous locations more satisfactorily than humans. The distinction between a robot and a manipulator is somewhat unclear, but, essentially, a manipulator has a very limited motion—usually three or fewer axes—whereas a robot can move objects in almost any direction and can follow a more complicated sequence of motions.

Figure 16-26 illustrates the use of a robot to automatically weld a workpiece assembly. This application moves the workpiece into various positions so the robots can perform the welding operation on all surfaces, load, and unload a machine. A manipulator or robot is particularly useful for working in hazardous atmospheric environments, for moving hazardous materials, and for performing highly repetitive work with very high precision.

Figure 16-26. A common application for robots is welding.

QUESTIONS

1. When is a graphic representation of a control function considered appropriate?
2. Why is it safer to use a Phillips-head screw and screwdriver than a conventional slotted screw and screwdriver?
3. Which of the following statements are false?
   a. An emergency-stop button should be prominent and easily accessible.
   b. An emergency-stop button should be protected by a guard to prevent unintended actuation.
   c. An emergency-stop button should be red on a yellow background.
   d. An emergency-stop button should require reset before the machine can be restarted.
   e. An emergency-stop button should be recessed.
4. Design handles for a bolt cutter that will not allow the user's fingers to be pinched or hit when the cutter is closed.
5. Steel strapping on crates is usually cut while it is under tension, as is wire on a bale of hay. Design a tool to cut strapping or wire that will prevent the ends from flying when cut.
6. Design a handle that can be attached to a 12-in. flat file to enable the use of two hands on the file.
7. A stack of aluminum discs 0.5 mm thick and 20 mm in diameter are brought to a worker who must pick them up one at a time and feed them into a die for another power-press operation. Design a tool that will enable the worker to pick up one disc at a time.
8. Design a physical control layout for a machine that includes two-finger capacitive actuation and an e-stop.

9. Which method of machine actuation provides the highest degree of safety reliability and why?

10. In a large package-distribution facility, there is a long, fast-moving conveyor.

    Occasionally, packages will jam up, requiring that the line be temporarily shut down to clear the jam. Assuming that the jam cannot be automatically detected, what is the best way to manually shut down the line?

    a. Telephone to the conveyor control station operator.

    b. A series of e-stop stations along the line.

    c. A power switch on the wall midway along the line.

    d. A cable push/pull switch with the cable running the full length of the line.

11. Which type of actuator would be appropriate for use as an e-stop?

    a. A push button that instantaneously removes power from a standard relay.

    b. A cable pull switch.

    c. A positive-acting, locking, mushroom-shaped, red push button with a yellow background.

    d. A toggle switch.

NOTES

1. Readers are also referred to Vern Putz-Anderson's *Cumulative Trauma Disorders.* New York: Taylor & Francis, 1988.

BIBLIOGRAPHY

Adams, Jack A., 1989. *Human Factors Engineering.* New York: Macmillan.

Burgess, John H., 1989. *Human Factors in Industrial Design: The Designer's Companion.* Blue Ridge Summit, PA: TAB Books.

Clarke, T. S. and E. N. Corlett, 1984. *The Ergonomics of Workspaces and Machines: A Design Manual.* New York: Taylor & Francis.

Galer, Ian A., ed., 1987. *Applied Ergonomics Handbook.* 2nd ed. Stoneham, MA: Butterworth.

National Safety Council. *Hand-Held Hazards.* Chicago, IL: NSC.

Woodson, Wesley E., 1981. *Human Factors Design Handbook.* New York: McGraw-Hill.

Principles of Risk Assessment and Machine Safeguarding

Robert N. Andres, CSP, CPE, CMgfE, BCFE

Every machine or process creates hazards during its life cycle, which, in turn, pose risks of harm either directly to people or indirectly through damage to the environment. The degree of risk is determined by the severity of injury that might result and the probability of that injury occurring.

There appear to be two distinct philosophies concerning the safeguarding of machinery to reduce risk: *prescriptive hazard control* and a concept brought to the forefront in the mid-nineties called *task-based risk assessment and reduction*. In the first, the goal is to identify all hazards associated with the machine and eliminate the hazards or guard them to the highest practicable level to effect a condition of minimal risk. The assumption is generally made that the hazard exposure is the "worst-case scenario"—and that all hazards pose an unacceptable risk.

In the second, there is a recognition that not all hazards pose a risk—and that no operation is risk free. In the iterative process of risk assessment and reduction, the first task is to identify all those hazards that may pose risk to any person involved with the machine. This is done by determining the limits of the machine and identifying all of the tasks to be performed in and about it. Then, the level of each hazard (severity of injury that could result) and the probability of the occurrence of a hazardous event are evaluated to arrive at a level of risk. Protective measure(s) (following the "safety hierarchy") are then taken to reduce that risk to an acceptable level. More specific information may be found in ANSI B11.0–2020, "Safety of Machinery—General Requirements and Risk Assessment."

We have found that the problem with the *hazard control* approach is that, by not assessing tasks, many hazards that may present themselves during the lifetime of the machine may be overlooked. On the other end, determining risk-reduction options, the *best possible safeguard* approach may squander valuable resources on minimal risks and create additional risks, as excessive or inappropriate safeguards are removed or purposefully defeated because they impede production or maintenance functions.

Design Responsibility Overview

Safety begins during the product-design phase. The designer who ignores or minimizes the application of sound safety principles may be, in whole or part, responsible for an equipment-related deficiency that could cause serious injury and also result in significant economic loss. Even if the designer does not actually commit errors in design, he or she may still inadvertently omit factors needed to protect users of the equipment.

We cannot overemphasize that, as machines and processes become more sophisticated, it is imperative that the designer exercise due diligence to *identify and control all the hazards that affect people, either directly or indirectly*. Safe design may well end up being a complex, multidisciplinary effort, involving many people and functions on whose input and cooperation the designer must rely. Proper design that prevents defects involves a host of considerations, including:

- Material stress factors
- Human behavior
- Construction materials
- Safeguarding measures
- User needs
- Foreseeable emergencies and unplanned maintenance
- Accident or crash worthiness
- Proper operating instructions
- Codes, standards, and accepted practices
- Manufacturing specifications
- Product-service experience and facilities
- Properly trained and motivated users
- Design envelopes
- Financial constraints

The design of equipment and products is essentially evolutionary—always changing as new materials, processes, and applications are developed. Designers must constantly acquire wider experience and knowledge to keep abreast of these changes. The designer's position is critical, since he or she is key to eliminating problems and avoiding forestalling unsafe situations. In addition to doing design work, the designer must assess safety and health hazards, be knowledgeable about safety and health regulations, know the environmental demands of the workplace, and be aware of the various engineering tools, controls, and equipment to solve problems.

If it is to be approached properly, equipment-and-process design must include the concept of *risk assessment and analysis* from the onset. Risk assessment, which dictates the design and inclusion of appropriate safeguards, can be significantly improved by knowing:

- loss histories (from loss-control personnel or legal counsel)
- maintenance requirements (from both internal- and user-maintenance personnel)
- anticipated use and foreseeable misuse of the machine (from users, field-installation-and-maintenance personnel, and sales personnel).

Fundamental Steps in the Risk Assessment and Reduction Process

As defined in ANSI B11.0–2000, the risk-assessment process includes the following series of logical steps to systematically examine the hazards associated with machinery:

- Prepare for and set the scope (parameters) of the assessment and establish the level(s) of acceptable risk.
- Identify tasks with that scope and the hazards associated with those basks
- Assess the initial risk
- Take steps to reduce that risk
- Assess the residual (remaining) risk
- Repeat the iterative process until acceptable risk has been achieved
- Validate the solution(s)

Establishing the Limits of the Machine

The limits of the machine or process will determine which tasks can logically be performed. For example, a power press is defined by its physical size and weight, tonnage capability, the side of the bed, its stopping time, and so on. Tasks exceeding these limits create a much greater risk of a hazardous event. The limits establish the framework of operation for determining which tasks can be reasonably performed—and what types of use and misuse are foreseeable.

Determining the Tasks to be Performed

Task identification focuses on how people interact with the machine or process in order to identify how they may be harmed. Tasks involved with a machine or process are not limited to setup and operation. The designer must envision all of the tasks that will be done with respect to the machine or process, including:

- Packing and transportation
- Unloading and unpacking
- Preparation of the area
- Provision of services—such as air, electrical, water, or fuel
- Commissioning—including handling and filling of fluids and connection of services
- Setup—including installation of tooling, fixtures, and accessory items
- Operation—including foreseeable use and *misuse*
- Maintenance—including planned and *unplanned* maintenance
- Tooling changes
- Housekeeping
- Decommissioning—including safe removal and disposal of fluids
- Removal—including packaging
- Disposal (including handling and transportation).

People who might be affected include, but may not be limited to:

- The cleaning crew
- Contract/service personnel
- Observers and passers-by

- Installation-and-removal personnel
- Setup-and-maintenance personnel
- Temporary or stand-in operators or trainees

Identifying the Hazards Associated with Those Tasks

The most critical step in safe design involves identifying all hazards posed by the equipment in its environment. Hazards are a product of some form of energy or toxicity. Things to look for are:

- Mechanical hazards due to machine parts or workpiece shape; relative location; stability; mass and velocity; limits in mechanical strength; accumulation of energy from springs, liquids, or gases; and the effects of a vacuum—such hazards as may result in crushing, shearing, cutting or severing, entanglement, entrapment, puncture, abrasion, fluid injection, striking, impact, or slips and falls.
- Electrical hazards due to contact of persons with live wires or components, either directly or as a result of a faulty condition; approach to parts with high voltage; electrostatic phenomena.
- Thermal hazards, which may result in burns, scalds, and damage to health from exposure to a hot or cold environment.
- Noise and vibration hazards, which may result in interference with communication, hearing loss, or other physiological disorders, neurological and vascular disorders, and physical trauma.
- Radiation hazards from low-frequency, radio-frequency microwaves; infrared, visible, and ultraviolet light; alpha, beta, gamma, neutron, and X-ray production; electron or ion beams; lasers.
- Ergonomic hazards from unhealthy and unnatural postures and movement or excessive effort; inadequate placement of controls and displays; poor lighting; and so on.
- Material and substance hazards caused by harmful fluids, gases, mists, fumes, and dusts; fire and explosion hazards; chemical, biological, or microbiological (viral or bacterial) hazards.

Keep in mind that such hazards may be presented by or associated with:

Unexpected startup caused by failure of the control system
Restoration of energy supply after interruption
Software error
Environment and surroundings
Human error
External influences

Reasonably foreseeable hazards that are not related to tasks must also be identified. Some examples are: an explosive or oxygen-deficient atmosphere, noise, instability, or equipment failures.

Establishing Hazard Levels

All hazards are not equal—they vary by degree—and although it is essential that all hazards be *identified*, it is not always necessary that they be *addressed*.

The designer must keep in mind that hazards posing a threat of hazardous exposure to people or the environment—in any reasonable manner—must be evaluated.

Hazards are generally rated, either qualitatively or quantitatively, according to the *severity of consequence*, measured in terms of the severity of injury and the economic damages caused by exposure to a hazardous event (see Table 17-1). All machine accidents have associated costs, both direct and indirect, which must be taken into account in determining the severity of consequence.

One example of this rating, shown in ANSI B11.0, is derived from MilStd 882C. Assuming, for our purposes, that we are concerned only with the effects upon people, this example establishes four qualitative hazard categories relating to the effects that *could* result from a hazardous event.

Table 17-1. Hazard Rating/Severity of Harm

	The Effects of a Hazardous Event
SEVERE	Death or seriously debilitating long-term injury, such as multiple amputation, coma, or permanent confinement, where a person is unable to return to work.
SERIOUS	Permanent and nonreversible injury significantly impacting the enjoyment of life, and which may require continued medical treatment, but person is able to return to work at some point
MODERATE	Permanent and nonreversible minor injury that does not significantly impact upon the enjoyment of life, or a reversible injury, either of which requires medical treatment. The person is able to return to the same job.
SLIGHT	Reversible injury requiring only simple medical treatment with no confinement. No lost work time.

Determining the Elements of Probability of a Hazardous Event

Once the hazard level has been determined, the other factor of risk is the probability of occurrence of a hazardous event. There are several significant elements to consider in this analysis:

- The frequency, duration, and extent of exposure. This is probably the most important single element of probability (see Table 17-2). As an example, a rabbit crossing a busy highway (the hazard) has a much better chance of survival if he does so rather infrequently, rather than several times per hour. Also, the rabbit running quickly across the busy highway has a much lower risk of being struck than the much slower possum or the even slower turtle.
- The motivation to be exposed to the hazard. This often-underestimated element is strongly determined by the tasks to be performed and is the basis for determining the foreseeable use and misuse of the machine during its life cycle. For every hazard, the question must be: "Does anyone have a reason to be exposed to it, and to what extent?"

Table 17-2. Probability of a Hazardous Event

VERY LIKELY	Very likely to occur. The frequency of exposure is high and/or the protective measures are nearly worthless.
LIKELY	Occurrence is likely. The frequency of exposure is significant or control measures are inadequate.
UNLIKELY	Occurrence is possible, but not likely.

With a remote probability, the occurrence is so unlikely as to be considered nearly zero.

Establishing a Level of Risk

Using the *severity of consequence* and the *probability of occurrence*, the *level of risk* can be determined with some degree of accuracy using a model. Although proven very useful in trial testing under NIOSH auspices, this is *only* a model. It takes into account subjective inputs and gives a subjective output, which must be interpreted and modified by the subjective application of "non-safety-related" items such as:

- Public perception
- Productivity
- Economic costs—direct and indirect
- Politics

One simple model, shown in Table 17-3 and derived from Mil-Std 882C, allows the risk level to be determined using the two basic inputs and may logically establish a "tolerable risk" as a condition where either the *severity of consequence* is slight or the *probability of occurrence* is so remote as to be considered nearly zero.

Table 17-3. Risk Levels Relating to Severity of Consequence and Probability of Occurrence of a Hazardous Event

Severity				
Probability	**Severe**	**Serious**	**Low**	**Slight**
VERY LIKELY	High	High	High	Medium
LIKELY	High	High	Medium	Low
UNLIKELY	Medium	Medium	Low	Negligible
REMOTE	Low	Low	Negligible	Negligible

In assessing risk for a given task/hazard pair, you may find that the answer to a single question—"Is the person in the hazard zone, and how often?"—is the primary factor in determining both the level of risk and the appropriate protective measures.

Once protective measures have been taken, it must be determined whether:

- Any protective measure introduces new hazards
- The protective measure is difficult to use or prevents the task from being performed effectively
- All affected persons have been considered
- Safeguards are appropriate for all affected personnel

Documenting the Findings and Determinations

Nowhere is the old phrase "it ain't over until the paperwork's done" more appropriate than in risk-assessment activity. Somewhere in the follow-up documentation, the following should be noted, where possible:

- The machinery for which the risk assessment has been made
- The information available at the time
- Any relevant assumptions
- The identified tasks and hazards
- The initial risks associated with the machine or process
- The applied protective measures
- Residual risks
- The validation of the protective measure(s)—including applicability and effectiveness

Safeguarding the Hazard—Application of the Safety Hierarchy

If the application of the risk model says that the risk is acceptable and the results can be validated, the job is complete. If the risk is *not acceptable*, it is necessary to apply safeguarding (protective) measures in accordance with the *safety hierarchy*, in this order of preference:

- Eliminate the hazard
- Modify the hazard to reduce the energy, toxicity, or its presence
- Safeguard the hazard using a guard (barrier) or safety device
- Warn of the hazard

The purpose of safeguarding is, simply, to reduce the risk to an acceptable level. The application of guards, safety devices, or warnings will do nothing to modify the *severity of consequence*—**only the** *probability of occurrence.*

It is important to realize that the reduction of risk to an acceptable level may be accomplished by the application of a single safeguard or, if required, several. For example, a particularly high-risk operation may call for a combination of modifications of the hazard, one or more safeguards, and warnings. It is the cumulative effect of all of the incremental protective measures that ultimately achieves the goal.

Eliminating the Hazard

Often a hazard can simply be eliminated. A good example is adjusting an opening to 1/4 inch or less. This approach is often taken with spot welders, presses with an adjustable stroke, or press brakes using unitized tooling (where the ram can be closed to the "plunger" and the tooling itself only affords a 1/4-inch opening).

Automatic feeding and ejection of parts may make it unnecessary for the operator to reach into the point of operation at all. In addition to increasing safety, such feeding methods are usually much faster than manual loading and unloading and cause less damage to parts.

Hazard elimination should be fully explored before considering other protective measures.

Remember: "A hazard without exposure does not constitute a risk!"

Isolating the Hazard

Once a hazard has been recognized, the first priority must always be to attempt to eliminate the hazard or reduce the severity of consequence of a hazardous event to an acceptable level of risk. If this attempt fails or for some reason is not feasible, the second priority should be to isolate the hazard.

Isolation means the prevention of contact between the hazard and a person or, in some cases, between the hazard and another object. There are two general ways to achieve effective isolation. One is to separate the hazard from the person or other object by distance. The second is to place a barrier or guard between the hazard and the person or other object. Again, it must be stressed that neither of these methods eliminates the hazard nor reduces the severity of the hazard.

When isolation by distance is chosen, there is always a question about how great a distance is necessary to provide satisfactory isolation. The distance may depend on the probability that the person will come into contact with the hazard by extending a part of his or her body. It is also conceivable that a hazard might be capable of moving so as to come into contact with a person located in a position too far away to reach the hazard. For example, an overhead coupling might break or a chip or broken part might fly out of a machine.

This problem raises another question. In a given situation, will it be more effective to prevent the person from reaching the hazard—or prevent the hazard from reaching the person? This is somewhat analogous to the policies taken at public zoos. In some cases, dangerous animals are caged to prevent them from getting out. In others, animals are allowed to wander freely within a large area, and fences and moats prevent people from reaching the animals.

Isolation by Distance

If it is deemed more desirable to prevent people from reaching a hazard, we turn to anthropometric data to determine how far is enough. It seems very unlikely that anyone would bump his or her head on a point or surface 2.4 meters (8 feet) above the floor. Because the risk is considered negligible, an overhead shaft coupling 8 feet or more above the floor or normal walking surface does not need to be guarded, according to the ANSI B15.1 Standard. OSHA 1910.219 currently establishes this "safe height" at 7 feet.

Similarly, it is unlikely that anyone would extend any body part more than 1.5 meters (5 feet) horizontally beyond the position of his or her waist. Thus, if a pipe barrier is 42 inches high and 5 feet horizontally from a point of hazard, it is considered sufficient distance to provide isolation from the hazard.

These distances are obviously not sufficient to provide effective isolation in all cases. Some adults can easily reach a point 8 feet above the floor if they stretch or if they inadvertently extend an arm upward.

The safety engineer must consider what body positions are possible or likely in a given situation, as well as the seriousness of an accident if a body member and a point of hazard should meet. There has been extensive discussion within the ANSI B11.3 "Safety Considerations for Press Brakes" committee regarding the appropriateness of distance as an acceptable safeguard.

When distance is chosen as the means of safeguarding, a great deal of judgment must be used to determine the possibilities and probabilities involved. The following should be considered:

- Is the hazard relatively fixed in location, so that it will not move to contact a person or other object?
- What are the people like who are to be protected? Are they all of reasonably uniform body size? Will they be continually aware of the existing hazard and refrain from extending any body member to a point at which it might come into contact with the hazard, either intentionally or unintentionally?
- What body positions will people assume in their activities near the hazard? In any and all of these positions, what distances in all directions will be needed to prevent them from coming into contact with the hazard?
- Is distance meaningful in this instance? Hazards posed by lasers and other types of radiation, or biological or chemical agents, may not be effectively addressed by distance!

Distance can be effectively used to isolate the hazard only when there is no need or motivation to access the area near the hazard or when such access can be controlled in some way. Moving parts on continuously running machinery, such as compressors, turbines, pumps, and so on, are often guarded in this manner.

Distance is generally the least expensive method of hazard isolation. It permits visual observation of the work area, but it cannot contain flying particles or splashing liquids, and its maintenance may interfere with needed access. Also, distance in itself cannot restrain a person from intentionally or perhaps inadvertently accessing the hazard area or being affected by it.

Ergonomic Considerations

Ergonomics is the mating of the machine with the person. ANSI B11.TR1, "Ergonomic Considerations for Machine Tools," provides a useful guideline for addressing ergonomics in machine design to reduce risks.

Characteristics of Safeguards

When distance is not a feasible means of protecting people from a hazard, some form of safeguarding should be explored. The word "safeguard" will be used here to refer to devices that are attached to or surround a piece of equipment to prevent or restrict a person from coming into contact with a hazard. "Guards," as defined in the B11 Definitions Manual and in OSHA 1910.217, are the highest form of safeguard, designed to *prevent* access to a hazard area. Other devices that restrict a worker's movement or control the machine to prevent an accident will be discussed here, but these should not be called guards.

The term *point of operation* is used to describe the point or area where a tool comes into contact with the workpiece. This may be a very small area, as in a drilling operation, or a relatively large area, as in the operation of a large power press. The point of operation should be guarded when feasible. The type of guarding usually depends upon the operation being performed; the performance level of the safeguard depends upon the desired degree of risk reduction.

Criteria for evaluation, restrictions, and objectives should always be established before an engineer designs or selects a safeguard. It is helpful to ask such questions as

1. What is the hazard to be safeguarded?
    a. How serious an injury could result if a person came into contact with the hazard?
    b. Is the hazard fixed in location, or does it shift from one moment to another?
    c. Is the hazard at a specific point, or does it exist as an area or a volume of space?
2. Why is a safeguard needed?
    a. To prevent anyone under any circumstances from coming into contact with the hazard?
    b. To prevent anyone who is untrained or unauthorized from coming into contact with the hazard?
    c. To prevent accidental contact with the hazard?
    d. To prevent objects, particles, or liquids from coming out of the area?
3. Will it be necessary to pass workpieces or materials into and out of the hazard area?
    a. What sizes and shapes will be involved?
    b. Would it be feasible to provide a means of passing materials into the area without requiring a worker's hands to enter the hazard area?
4. Will it be necessary to reach into the hazard area to manipulate controls or adjustments?
    a. Could these controls and adjustments be moved outside the hazard area or eliminated?
5. Will it be necessary or desirable for the worker to see inside the hazard area?
6. What forces may be imparted to the safeguard?
    a. Will people be leaning against or hitting the guard?
    b. Will materials be pushing against or hitting the guard?

7. Will the safeguard be attached to an existing piece of equipment?
   a. How can it be attached without interfering with the operation of that equipment?
   b. What method of attachment will provide sufficient strength?

Two other general requirements of a safeguard are very important:

1. The guard or device must not interfere with the normal activities of the worker or equipment.
2. The guard or device must not in itself, or in concert with other elements, create a hazard.

Guards

Sometimes referred to as an *enclosure guard*, a guard, as defined in the ANSI B11 Definitions Manual, prevents a person from reaching over, under, around, or through it to reach a hazard. It offers the most complete isolation of all protective measures. A guard is mounted permanently to the extent that tools must be used to remove or displace any part of it in order to reach the hazard. In higher-risk situations, special tools should be required to remove the guard.

This raises a question that must be considered in the choice or design of any guard: why is it necessary to make a guard difficult to remove? There are numerous reasons why a worker will try to remove a guard, but very often such action is based upon a legitimate need to have access to the hazard area. Some guards are overly restrictive and thus encourage the worker to remove or alter them to reduce the restriction. The need for access and the provisions for it are among the most important considerations in the design of a guard.

Fixed guards are used in applications that will remain in the same condition for a length of time and in which little or no access is needed for normal operation or maintenance. Typical of such installations are flywheels, belts and pulleys, gearing, or fully automated and reliable processes.

The material to be used in making a guard should be selected on the basis of the properties needed. The major characteristics that must be provided by the material in various applications include:

- Sufficient strength to resist forces exerted against it.
- Sufficient rigidity to minimize vibration and/or deflection.
- Containment of flying particles and splashing liquids.
- Visibility within the work area.
- Insulation from electrical current.
- Flexibility for use on moving machine parts.
- Resistance to corrosion.
- Ability to be formed or fabricated.
- Low cost.

When the guard needs to permit visibility inside the work area, several additional factors must be considered. A person's ability to see is greatly affected by illumination, contrast, glare (or the lack of it), color, and visual distractions.

The guard should certainly not reduce visibility any more than is necessary. If a guard must be placed where it will block needed light, another source of light should be provided, but not where it will produce glare. Exposed surfaces of the guard, as well as other nearby surfaces, should not be reflective. They should have a dull finish and a neutral color.

Positive Features:
- Positive protection against catastrophic control failure or tool shattering.
- Highly reliable in keeping the body, hands, and fingers out of the hazard zone.
- Easily fabricated—need not be purchased in many cases.
- Moderate cost.

Negative Features:
- Inflexible—a guard is generally designed and fitted for a specific purpose.
- Must be secured so that it cannot be removed. This may require tamper-proof fasteners.
- There is a possibility that the guard might be removed for setup or maintenance and not replaced or reinstalled properly.

Adjustable guards are similar to fixed guards in that they prevent a worker from reaching over, under, through, or around the guard to the hazard area (see Figure 17-1). However, adjustable guards are used in applications that are temporary or in which variations are necessary. Thus they are made of components that can be changed in position—*but not by the operator!* Setup or maintenance personnel normally make adjustments. The most common application for adjustable guards is point-of-operation protection on a machine in which the workpiece is fed through an opening in the guard. If a part of a different size is to be used, the opening in the guard must be changed, but once the change is made, the configuration becomes fixed for the duration of that job (see Figures 17-2, 17-3A, 17-3B, and 17-4).

Figure 17-1. Typical fixed guard used to control access to a flywheel. (Courtesy of Rockford Systems)

Figure 17-2. Adjustable guard with "piano wire," fixed-front
vision panel. (Courtesy of Panelbar Barriers, Inc.)

Positive Features:
- Positive protection against catastrophic control failure or tool shattering.
- More flexible—can be adjusted for a variety of applications.
- Highly reliable when adjusted properly.
- Component parts can be separately purchased.

Negative Features:
- May be adjusted improperly, resulting in a compromise of reliability.
- May be removed for setup or maintenance and not replaced or reinstalled properly.
- Must be secured so that it cannot be removed by the operator.
- Is generally purchased and is more costly than a fixed guard.

Another type of guard is one with hinged covers or removable sections. Such a guard is fixed, but it has sections that can be moved or removed to provide access for adjustment or maintenance. This access should be no more than is necessary. The designer must assume that any guard that can be removed or left open will be. In all but the lowest-risk situations, guards should be electrically or mechanically interlocked to prevent operation of the machine with the guard open or missing. In lower-risk situations, a sign on the machine or *under* the guard (not *on* the guard) should warn that it has been removed and that the machine should not be operated.

Positive Features:
- Minimizes the chance that the guard will be removed.
- Allows for access where needed.
- May be fabricated.

Negative Features:
- May allow access to a hazard zone if not interlocked.
- Is more costly than a fixed guard.

Figure 17-3A. Box barrier guard with adjustable end panels and inter- locked front panel. (Courtesy of Panelbar Barriers) Inc

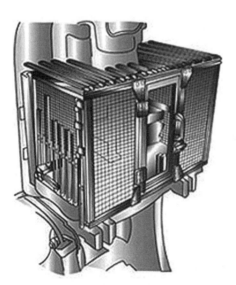

Combination Fixed and
Adjustable Guard on Press

Guard Mounted to Conveyor System and Machine

Barrier Guard—Adjustable Panels with Hairpins
for Material Feeding, Chutes for Ejection, etc.

Figure 17-3B. A variety variety of guarding solutions are available to suit specific needs. (Courtesy of Rockford Systems)

Figure 17-4. Complex acrylic barrier provides protection against access and liquid discharge while permitting full visibility. (Courtesy of Bosch)

Interlocked Guards

Frequently the cover or sliding gate will be made to contact a switch that stops the flow of electric current to the machine. In some cases, the interlock can be made to operate a clutch or brake device, thus controlling the operation of only certain parts of the machine. In some cases, the interlock is tied into controls that will only allow the machine to operate in a safe manner. A guard with an interlocked cover or gate is very desirable in applications in which the machine operator needs access to the point of operation to load and unload the workpiece but must not have access while the machine is in operation. Power presses and special assembly machines often have such a guard (see Figure 17-5). Even simple machines, like grinders, may be equipped with an interlock to assure continued of the guard.

Figure 17-5. Interlocked grinder shields. (Courtesy of Rockford Systems)

An interlocking device adds some complexity to the design of a guard. A cover, a sliding gate, or a removable section requires loose pieces that must, nevertheless, be attached securely. They must be as foolproof as the rest of the enclosure (Figures 17-6A and 17-6B). The switch or control device must be mounted so that it will be actuated by the cover or gate when it is closed but, at the same time, be resistant to tampering by the worker.

Figure 17-6A. Interlock switch on sliding gate.
(Courtesy of Schmersal USA)

Figure 17-6B. Interlock switch on swinging gate.
(Courtesy of Banner Engineering)

Positive Features:
- Allows opening or removal of the guard while providing greater assurance of replacement before machine can be operated.
- Costs little more than a good standard switch.

Negative Feature:
- Adds to the cost of the guard.

There are two factors to consider when choosing an electrical safety interlock switch: reliability and tamper resistance. Thanks to European Standards, the de facto state of the art for safety-reliable interlock switches dictates that the switch shall:

- have dual (redundant) contacts that can be monitored, continuously or at each cycle, or
- have those contacts positively driven to assure a change of state.

Commonly used are simple push-button or lever-action switches operating in a *normally open* (NO) mode. Typically these are wired in series with the operator's start/stop switch. These switches often rely upon spring action to trip, use only a single contact, and are usually not tamper resistant. They are unsuitable in all but the lowest-risk applications.

So what is suitable? There are now several brands of "safety-rated" interlock switches on the market. Where tamper resistance is a consideration, they provide special, keyed actuation (see Figure 17-7). Where stored energy is a problem, they incorporate solenoid pin locking of the enclosure and motion (or other) sensing. Acceptable units are tamper-resistant and incorporate redundant, positively driven, forced contacts that can be monitored. With proper control interfacing, they are suitable for high-risk applications.

We have all seen instances where a worker wires or tapes an interlock switch closed to keep the machine operating, whether the cover or gate is in place or not. There are many ingenious methods to prevent this, but just as many to circumvent the safeguard. Unfortunately, we often spend time and effort trying to prevent the circumvention, rather than analyzing why it is occurring. Think of this before you merely replace an interlock switch with a tamper-resistant unit.

Figure 17-7. Coded, keyed, positive-forced-
contact interlock switch.
(Courtesy Schmersal-USA.)

Automatic Guards (Gates)

An automatic guard is a device that either prevents the hazard from occurring or moves the worker's body out of the way of the hazard. It is actuated by the component of the machine that creates the hazard. Most automatic guards are used on power presses. They operate before, or as, the ram of the press starts its descent.

The most reliable, because it does not require human intervention, is the gate device (see Figure 17-8). This is generally a framed, clear shield which, when activated, drops by gravity. If the gate completely closes, hidden limit switches activate the machine. If the gate is held open for any reason (such as the presence of a hand or arm), the machine is not activated.

Positive Features:
- Provides the same degree of protection as a barrier guard.
- Allows full access to the point of operation when open.
- Protects in the same manner as a full barrier guard when closed, while affording good visibility.

Negative Features:
- Relies upon many moving parts; requires regular maintenance.
- Rather costly, particularly for large gates ($1800–$3000).
- Cannot protect against a catastrophic mechanical or control failure.

Figure 17-8. Movable gate device shown with side panels swung away for die change. (Courtesy of Gate Devices, Inc.)

Figure 17-9. A trial-ring guard in operation on a riveting machine.

Trial-Ring Devices

Used almost exclusively on riveters, staking machines, and spot welders, a trial-ring device relies upon a properly adjusted ring dropping to sense the presence of a finger in the point of operation (see Figure 17-9). If the thickness of a finger is sensed, the ring is held up high enough to prevent actuation of the machine cycle. The unit is to be adjusted for each part and is, therefore, unacceptable for use on power presses.

Positive Features:
- Provides the same protection against inadvertent point-of-operation access.
- Inexpensive and easy to adjust and use.
- Provides an adjustable spark shield on welders.

Negative Features:
- Requires adjustment and relies upon proper adjustment and use!
- May interfere with some operations.
- Requires specific rings to be fabricated for various operations.
- Not to be used in high-risk applications.

Light scanning devices are also generally appropriate for this type of application.

Pullbacks and Holdouts (Restraints)

Generally used only on older, full revolution presses and drop hammers, a *pullback device* (Figure 17-10) involves a chain-and-sprocket mechanism attached to the top of the ram at one end and to a set of harnesses at the other. The harnesses, in turn, are attached to the worker's wrists and hands. When the device is adjusted properly, the cables can pull the harness and the worker's hands away from the point of operation as the ram descends. The worker is free to move about within the normal area of activity until the ram of the press starts its descent. Then, if the worker has not moved his or her hands out of the way, the cables and harnesses will pull them away. These are typically used on hand-fed power presses or on press brakes used for small-part or punching applications. (see Figure 17-11).

Passive *holdout* devices are similar to pullouts, but they do not move. They are designed to simply prevent hands from entering the danger zone—and do so very well when properly adjusted and used.

Positive Features:
- A pullout, when properly adjusted and used, will withdraw the hands even in the event of catastrophic control or mechanical failure.
- A holdout will keep hands out of the point of operation and force the use of hand tools.
- Provide visible evidence of proper use and function.

Negative Features:
- Require proper adjustment to fit the tooling and the operator!
- Must not be used where the wristlet could snag on tooling.
- Do not provide the same level of protection (against tool breakage or similar occurrence) as a guard or gate.
- Some operators may feel "tied to the machine." To some extent, this is a real problem, not just a psychological one, since the harness must be unfastened before the worker can leave the work area. Although it takes only a few seconds to unfasten the harness, such a delay can be serious in case of an emergency.

Figure 17-10. Overhead pullout for a large press. (Courtesy of Woodbury Devices, Inc.)

Figure 17-11. Sit down pullout for smaller press.
(Courtesy of Woodbury Devices, Inc.)

Since these devices rely so heavily upon proper setup and use by the operator, they are recommended only where the workforce is solidly "sold" on their unique benefits. The complete access to the point of operation and nearly foolproof protection they provide make pullout devices desirable in many situations despite their disadvantages.

Sweep and Pushout Devices

Now considered obsolete, some special mechanical safety devices still exist in industry. The most common is the *sweep device*, found on a power press, which uses a mechanical linkage from the ram. As the ram descends, the device "sweeps" across the face of the point of operation, pushing hands or arms out of the way and blocking reentry. Because the sweep must traverse the entire face of the machine before the die's pinch point is reached, it must move very fast on a press with a short or fast stroke—too fast to be safe and not to pose a hazard in itself.

Although quite uncommon, a *pushout device* usually consists of a paddle that is mechanically linked to the ram of a press, pushing hands away from the point of operation as the ram descends.

Both of these devices may create pinch and impact hazards and have limited application. There may be instances, however, where the user can use such a device as a secondary safeguard for positive protection against an unintended descent of the ram.

Positive Features:
- Provide positive protection against catastrophic failure of the control or braking system.
- Do not require user action.
- Are relatively low cost.

Negative Features:
- May create significant hazards alone or in concert with the machine!
- Should not be used on fast- or short-stroke presses.
- Not accepted by OSHA as a primary safeguard in most cases.
- Require proper installation, adjustment, and continued maintenance.
- No longer manufactured commercially.

Shields and Awareness Barriers

Shields and so-called awareness barriers are designed to deter, rather than prevent, access or entry into a hazard zone. A swing-away shield on a vertical mill serves the purpose of creating awareness of the hazard and, to some extent, prevents inadvertent contact—but its primary purpose is to deflect chips and lubricant (see Figure 17-12). A lathe chuck "guard" protects against inadvertent contact with extended chuck jaws but cannot be considered a guard in the same sense as those discussed previously, since one can easily move it or reach around it (see Figure 17-13).

A good example of an awareness barrier is a series of steel rollers stretched across the face of a shear. The rollers will accept passage of sheet stock, but the weight of the rollers on fingers makes the operator aware of the impending hazard. A *post and rope* or a *railing* is a type of awareness barrier.

Figure 17-12. Swing-away, clear plastic shield being used on vertical mill. (Courtesy of Bristol Engineering, Inc.)

Figure 17-13. Chuck shield being used on engine lathe with chuck. (Courtesy of Flexbar, Inc.)

While shields and barriers have important functions, they should not be used in high-risk situations where a guard, as discussed earlier, is feasible. For this reason, ANSI and OSHA do not recognize this type of safeguarding for power presses, but find it suitable for accessories such as roll or hitch feeds, lathe chucks, and many production or assembly machines. Again, the choice of safeguarding must be based upon the posed risk and feasibility of use.

Positive Features:
   • Provide protection against inadvertent injury.
   • Carry a low cost.
   • Are easily installed.

Negative Features:
   • Unless interlocked, rely upon the operator to adjust and use!
   • May impede operations if not properly selected.
   • Should not to be used in high-risk applications.

Presence-Sensing Devices

Would a worker intentionally place a hand in the point of operation when the machine is operating or starting? Yes, for several reasons. Any one of the following may be sufficient:

   • To relocate the workpiece in the die if the worker notices that it is not correctly located.
   • To save a fraction of a second on the cycle time, thus enabling the worker to increase productivity and earn a higher wage on an incentive system.
   • To impress a fellow worker or supervisor by showing how fast the worker can operate the machine.

There are several types of devices that will detect the presence of hands or a human body and can be used to render a hazard harmless, such as stopping a press ram.

A presence-sensing device is only as good as the control system with which it interacts. If the purpose of the device is to stop motion, it is only as good as the electrical, mechanical, pneumatic, or hydraulic control elements and the braking capability (see Control Reliability). It must always be placed a sufficient distance away from the hazard to assure that the worker should not be able to reach the point of operation before the hazard ceases to exist. In the United States, the accepted speed of hand or body motion—the speed *constant*—is 63 inches per second. It is on this basis—and the cumulative switching and braking time of all the elements of the system—that the *safe distance* is calculated.

Two-Hand or Captive Controls

The most common presence-sensing device is a two-hand control, which requires that both hands be placed at a safe distance when activating hazardous motion. The requirements are:

- That the buttons, paddles, or levers be protected against unintended activation.
- That they are designed and mounted in such a way that only the hands or fingers can be used to activate them.
- That concurrent action of both hands is required for activation (generally within 0.5 or 1 second).
- That the machine cannot operate with one button continuously depressed (anti-tie-down provision).

While most two-hand controls are nested electrical palm buttons, many units that use air logic are in use, and, most recently, low-pressure, capacitance, optical, and hybrid palm buttons are being used to address ergonomic concerns. To provide maximum reliability, each of the two palm buttons should typically have a normally open and normally closed contact, which, when used with an anti-repeat, anti-tie-down control system, will be self-checking (see Figure 16-14).

In addition to or instead of two-hand controls, it is sometimes desirable to locate the controls of a machine far enough away from the machine itself so that the operator has no direct contact with it. This method is appropriate whenever the operation of the machine presents hazards that cannot feasibly be isolated by guards. The major concern in this situation is to make the machine operable only when the operator is in a remote position or location. In some cases, the operator may be in an enclosure or in a separate room. In may cases, the control can be placed in the same general area but far enough away so that the operator cannot reach the hazard.

Positive Features:
- Are relatively low cost for regular palm buttons and control circuitry—slightly higher for ergonomic palm buttons (optical, capacitance, or hybrid units).
- When properly installed, assure that the body or hands are a safe distance from the point of operation during the cycle.

Negative Features:
- Require repetitive motion in some instances that may cause cumulative trauma disorders (CTD).

- The location of mounting must be carefully selected to minimize CTD, prevent unintended actuation, and maintain productivity.

Photo-Optical Sensors

Most photo-optic light curtains use an infrared light source generated by high-intensity LEDs (light-emitting diodes). They are programmed so that the beam travels directly from a transmitter to its respective receiver. If the receiving sensor at a given position does not "see" its respective light source, the unit sends a dual solid-state output to a set of control relays to stop hazardous motion. One manufacturer assures "hand-shaking" of the machine control with the light control on each cycle, for maximum fail-safe reliability (see Figures 17-14 and 17-15).

Because of changes in both European and domestic standards, retroreflective light curtains (those using a transmitter/receiver on one end and a mirror on the other) are no longer acceptable in many situations and are no longer manufactured.

Positive Features:
- Require no adjustments and little maintenance.
- Are highly reliable.
- Provide unencumbered, clear field.
- Allow for feeding of thin stock between light beams without compromising safety.
- May be set up for self-actuation of the pressure-sensing device (PSD) cycle or foot actuation.
- Most are suitable for high-risk applications with a proper control system.
- May be programmed to provide selective and automatic blanking of beam.

Negative Features:
- Cost about twice as much as guards, in most instances.
- Must use mirrors to go around corners; cannot be used to guard complex areas with protrusions into the field without compromising safety.
- Where beams can be blanked, lack of strong supervision and control invites misuse and compromise of the safety reliability.

A relatively new product, the safety-rated, infrared-scanning device has earned a special niche in guarding wide areas from a point location. These units rely upon reflection of a scanning beam but are quite sensitive and have a range of 60 feet or more (see Figure 17-16).

Positive Features:
- Have single-point installation.
- Are programmable to recognize the static surroundings and create alarm and stop points.
- Create a visible "picture" on a computer screen of what the sensor "sees."
- Are highly flexible and reliable.
- Provide cost-effective safeguarding for large or complex areas.
- Are suitable for high-risk applications.

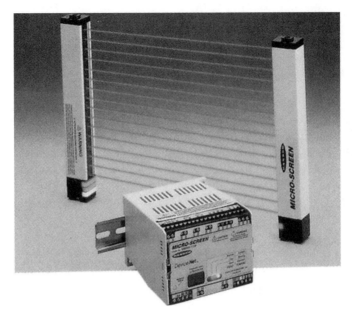

Figure 17-14. Typical light-curtain system with controller.
(Courtesy of Banner Engineering, Inc.)

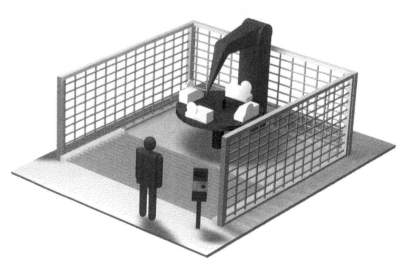

Figure 17-15. Area guarding with light-curtain scanning entrance. (Courtesy of
Sick-Optic Electronic)

Negative Features:
   • Require programming.
   • Are relatively expensive compared to other alternatives, especially for
     small areas.

Capacitance Sensors

No longer recommended for safety applications since the development of the
more reliable laser scanner, this type of presence-sensing device senses body

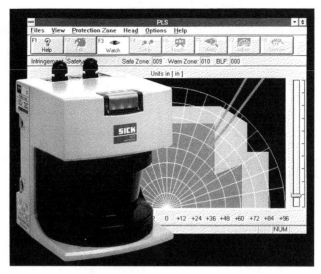

Figure 17-16. Programmable area sensor showing 60-foot radius, dual-detection scan. (Courtesy of SICK, Inc.)

capacitance by means of a field generated around a metallic- frame antenna. A change in the field is created when any mass of material, such as a hand, enters it. This change in the field sends a signal to a control unit, which stops the machine.

Positive Features:
- Are very cost effective in safeguarding complex machine configurations.

Negative Features:
- Rely upon proper adjustment of a *null* point.
- Are affected by individual capacitance, grounding, and humidity!
- Must use an antenna system, which may be difficult to design and fabricate.
- Use an antenna that may encumber the point of operation and interfere with maintenance.
- Are not recommended for high-risk applications.

Safety Mats

A safety mat consists of two metal plates separated by bumpers, compressible washers, or foam encased in an impervious, molded flat plate and has a redundant set of output wires that go to a control unit (four-wire configuration). A force of as little as a few kilograms will trip the control, and a "safety-reliable" output stops the machine (see Figure 16-19). The best units use non-corroding flat plates, as opposed to a wire mesh.

Positive Features:
- Do not encumber access to equipment or operator movement.
- Can be used to protect a wide and complex area by mating up to ten mats into one control unit.
- Are relatively low cost, particularly with smaller systems.

Negative Features:

- Are subject to deterioration from dropped material, floor debris, or solvents.
- May be damaged if exposed to extremely high weight or the twisting of forklift tires.
- Have a relatively long switching time, precluding their use to protect against imminent danger.
- Are not recommended in high-risk applications.

When used in place of a physical guard, a presence-sensing device is installed in such a manner that the worker cannot reach over, under, or around this device. Although it does not physically prevent entry into the hazard zone, the presence-sensing device must render the hazard harmless. (This type of guard must not be used on a power press or other similar machine that cannot stop immediately. This generally precludes its use on a full-revolution press—one in which the ram continues through a full cycle once it is tripped.)

In calculating a safe distance for any presence-sensing device, the cumulative effects of the device's switching time, the control's reaction time, and the machine's stopping time must all be taken into account. With light curtains, the spacing between beams affects the amount of penetration before the device will trip.

Control and System Reliability

Depending on the degree of risk reduction that is required of a selected safeguard, both it and the control system that facilitates its performance must meet a standard of performance appropriate to that goal.

The term *control reliability* has been around for years and has generally been accepted to mean a system which is *fault tolerant*—one that will "fail safe" in the event of a failure of a single component or several components. But there are degrees of control reliability, which, again, must be considered, depending upon the degree of reduction of risk that is required. For more information, consult ANSI B11.TR6.

Table 17-4 looks at the safeguard and control-performance guidelines for various levels of desired risk reduction.

Table 17-4. Safeguard and Control Performance Guidelines

HIGH	Highest possible level of reliability or highest possible fail-safe structure under one or more faults
MODERATE	High level of reliability or high fail-safe structure under single-fault condition
LOW	Use of well-tried components, well-tried safety principles—may not be able to detect a single fault
NEGLIGIBLE	Little need for consideration as part of safety measures

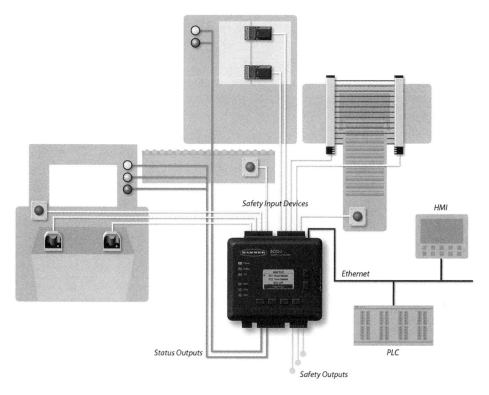

Safety Input Devices

HMI

Ethernet

Status Outputs

PLC

Safety Outputs

Figure 17-17. Schematic of safety-reliable control system. (Courtesy of Banner Engineering)

It is essential that all safety functions and equipment have an equal or better reliability than is required for the risk reduction desired. A safety system consists of a number of elements, each one a link in the chain. As the adage goes, the weakest link will likely fail.

For control systems and e-stops, it often is necessary to provide a self-checked, independent output. In such applications, the input from devices such as two-hand controls, light curtains, interlocks, or e-stops is fed to a safety-control module (see Figure 17-17), which performs the self-checking functions to assure continued redundancy and a safety-reliable output to one or more elements of the system. A safety-control module can be used to center hydraulic spool cylinders, dump hydraulic pressure to the tank, set off an alarm, apply a brake, turn off a motor, shut off an air supply, or other things.

Unless specifically designed for safety applications, programmable logic controllers (PLCs) should *not* be used for safety functions. A safety-rated PLC will have two or three independently functioning processors with cross-checking to assure sustained proper functioning. The best will have the processors built by different manufacturers and can be programmed independently.

In most cases, multiple inputs (safety interlocks, e-stops, light curtains, etc.) and/or multiple outputs (to apply a brake, to close an air valve, to disconnect power to a hydraulic pump) can be easily processed by a safety-control module (see Figure 17-18). The module can also advise the PLC of a "stop" condition and cycle interrupt. Such modules are highly reliable and suitable for use in high-risk applications.

Figure 17-18. Typical safety control module (din-rail mountable safety relay) with three safety-reliable outputs. (Courtesy of Pilz-USA)

QUESTIONS

1. An electric heating element on the side of a large machine has to be left exposed for a certain operation but should be protected from accidental contact by people passing by it. It is located 1.6 meters (58 in.) above the floor. How far away should a 42-inch-high pipe railing be located to be effective as a barrier?

2. A drive belt runs at a 45° angle from a motor on the floor to a pulley 3.8 meters (138 in.) above the floor beside an aisle. How far up should the belt be protected by a guard, and what type of guard is recommended?

3. Would a presence-sensing device be feasible as a guard for a portable chipper used to cut up tree branches and limbs?

4. A machine has a stopping distance of 300 m/sec and a control response time of 15 m/sec. Safeguarding is by a vertical light curtain with a depth penetration factor (DPF) of 850 mm. What is the minimum safe distance of the operator to the light curtain? $[D_s = 63(T_s + T_c + T_r) + DPF]$

5. Describe a simple guard to cover as much of the exposed blade on a band saw as feasible. Consider how it will be attached to the machine and the fact that the operator should be able to see the point at which the blade enters the workpiece.

6. What type of guard would be most suitable for the chuck and chuck jaws on an engine lathe?

7. Would a guard be required for a large coil of strip stock being fed automatically into a press?

8. A worker whose mass is 70 kg (155 lbs.) was reading a book while walking and failed to see a barrier around some construction work. If he was walking at a speed of 1.0 m/sec (2.24 mph), what energy must the barrier absorb to remain in position? $(E = mv^2/2)$

9. Do awareness barriers prevent entry into a hazard zone?

10. Under what circumstances would a fixed guard be appropriate?

BIBLIOGRAPHY

American National Standards Institute, 1988. *ANSI B11.1-1988*, "Safety Requirements for the Construction, Care and Use of Mechanical Power Presses." New York: ANSI.

————. *ANSI B11.3*, "Safety Consideration for Press Brakes." New York: ANSI.

————. *ANSI B11.0-2020*, "Safety of Machines: General Requirements and Risk Assessment New York: ANSI.

National Safety Council, 1992. *Accident Prevention Manual for Business and Industry.* 10th ed. Itasca, Illinois: NSC.

Noro, K., ed., 1987. *Occupational Health and Safety in Automation and Robotics.* New York: Taylor & Francis.

U.S. Department of Labor. *Essentials of Machine Guarding,* OSHA 2227, Washington DC: Government Printing Office.

————. *Concepts and Techniques of Machine Safeguarding,* OSHA 3067, Washington DC: Government Printing Office.

————. *General Industry Standards Part 1910.* OSHA. Washington DC: Government Printing Office.

18

Material Handling and Storage

Kirk E. Mahan, PE, CSP

Manual Handling

Many on-the-job industrial injuries are sustained while the worker is handling materials. Material handling includes not only the handling of materials with mechanized equipment, but also the manual handling of materials. Refer to Chapter 4: Work Systems and Ergonomics for additional information on body mechanics. Even when mechanized equipment is used, it is seldom completely automatic; manual labor is often needed for loading and unloading and for controlling this equipment. Injuries may occur during both manual and mechanized handling.

The manufacturing process may leave sharp edges or burrs on parts. These parts may cut the hands of workers who must handle them in subsequent operations. Another source of cuts is the sharp edges and corners of totes in which parts are carried. Ideally, these should not have sharp edges and corners, but the normal wear and tear of manufacturing makes it very difficult to prevent sharp edges from developing. Wooden boxes or pallets do not burr, but they do have splinters or nails, which are also hazardous. In many cases, workers may wear gloves when handling totes or parts with rough or sharp edges. Refer to Chapter 5: Personal Protective Equipment for additional safety considerations when selecting gloves.

Dropping objects on one's toes is sometimes something to laugh about, but it is a common incident. The use of safety shoes is required by many employers and has prevented many injuries. Still, such injuries do occur. Dropped parts, tools, boxes, and so on, may also injure knees, legs, and hands.

The dropping of objects may be caused by carelessness, and correction of the problem is then a matter of training and supervision. However, poor design and maintenance of containers is also a contributing factor. Many totes and boxes that will obviously be lifted and carried manually are designed without handles or hand-holes. Convenient handles should be provided on containers whenever possible. If a large mass is to be carried, the handles should enable the worker to wrap all fingers firmly about the handle, not just the ends of the fingers. When protruding handles interfere with stacking, holes may be provided to serve the same function; however, such holes should have rounded edges.

A worker carrying a tote or other large object may be unable to see the floor directly in front of him or her. It then becomes especially important to keep aisles free of tripping and slipping hazards. A person who falls can be seriously injured in any case, but when carrying heavy or sharp-edged objects, he or she can incur additional, more severe injuries.

Incidents involving material handling often affect other workers in addition to, or instead of, the person doing the handling. It is quite natural that a worker's attention will be focused on materials that he or she is carrying, and other people and objects are often overlooked. Typically the worker carrying objects concentrates on guiding the material at intersections of aisles and at corners. As a result many people have been hit or run into by carried objects or mechanically transported loads. Long objects, such as ladders, long lengths of pipe, and lumber, are often involved in such incidents. Whenever long objects are manually handled, more than one person should move the object.

In a similar manner, doors frequently open into walkways. Mechanized material handling equipment is often moved into intersections by workers who did not think about the possibility that someone might be approaching along the cross-aisle. These situations might appear at first to be a matter of training—that is, a management problem—but if the engineers who designed these doors and intersections had given some thought to normal human behavior, and material handling patterns, and had provided better visibility, for example, many of these incidents could have been prevented.

Handling with Powered Equipment

Many people have been injured while working on or around conveyors. Conveyor systems are used in almost all industries. It is important to design out as many hazards as possible and to provide effective guarding and adequate warnings. Belt conveyors and roller conveyors account for most of the injuries. Both of these conveyors have in-running nip points where either the belt or the workpiece approaches a roller. It is often difficult to guard the nip points on a roller conveyor, since the rolls must be exposed in order to function. A powered roller conveyor, called a "live roll conveyor," must have some form of chain or belt to transmit power to each roller. The chain and sprocket power transmission creates another in-running nip point. ANSI/ASME 20.9-2009, Safety Standards for Conveyors and Related Equipment, describes the guarding requirements of conveyors.

Lubrication of the rollers and other mechanisms on conveyors may be done while the conveyor is running if oiler tubes are extended outside of any hazard area. Some adjustments, such as alignment of belts, must be done while the conveyor is in motion. These adjustments should be made possible by extending adjusting screws outside any hazard area. The ASME standard states that no maintenance except for lubrication and adjustment shall be done while the conveyor is running. Nevertheless, serious injuries have resulted when workers attempted to do some maintenance on moving conveyors.

Occasionally, heavy objects are set on a conveyor with the edge of the object protruding over the edge of the conveyor. The object then hits something beside the conveyor, which snags the object and pulls it off, hitting a worker, damaging the object, or damaging property. Care should be taken when objects are placed on a conveyor. In many cases, a railing could be installed to keep objects from falling off.

Workers should never attempt to walk over a conveyor or stand on it. Cross-over steps should be used where there is a need to cross a conveyor.

Any equipment with exposed moving parts should have emergency-stop controls located within reach of any place where a worker is likely to be working on or around that equipment. Emergency cables may be installed along the length of conveyors to allow stoppage at any point along the line. Start buttons and speed controls should have guards to prevent accidental operation. As with any installation, the emergency-stop control should be conspicuous and easy to operate. Other conveyor hazards include idler rollers and the counterweights. Employees may become entrapped and strangled on idlers. Often workers sit on the framework beneath conveyor counterweights. Idlers and counterweights should be enclosed to prevent workers from being crushed by the weight if the belt should break.

Cranes and Hoists

Cranes and hoists pose several potential hazards. These devices should conform to the ANSI/ASME B30 series and OSHA 29 CFR 1910.179 and 29 CFR 1926.550 standards, Cranes and Derricks. One of the most serious hazards is the load falling from the crane or hoist. This can result from one of several failures, among which is a broken or bent hook. Crane hooks are made from relatively ductile material with a high yield point to allow for work hardening. A crane hook is subjected to a great deal of abuse and should be inspected frequently and thoroughly. When there is evidence of cracks, a hook should be discarded. Attempts to straighten or recover the original throat openings are not recommended. Hooks with more than 10 degrees of twist or throats opened more than 15 percent of the original must be removed from service. Property damage and minor injuries frequently result from hook failures. Hooks should conform to ANSI/ASME B30.10-2009.

There are several types of safety latches used to prevent a cable or sling from slipping out of a hook. One type is shown in Figure 18-1. Some of these latches will slip by the throat section if the hook is deformed; they provide evidence of deformation but do not prevent the deformation.

Cables or chains on a hoist should be inspected frequently for evidence of weakness and, of course, replaced when that evidence is seen. Overload indicators are often used to sound an alarm when a worker attempts to lift a load that is over a predetermined limit. A better device is one that actually prevents

Figure 18-1. Safety finger or latch mounted on a crane hook.

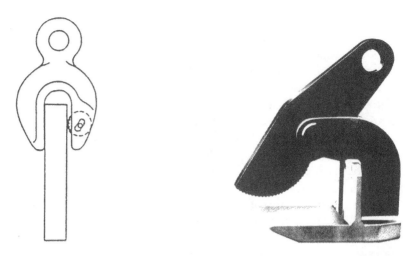

Figure 18-2. Lifting devices used on a crane. Illustration (right) courtesy Inter Product bv.

the lifting of too great a load. A strain gauge or other sensing device trips a switch when a given limit is exceeded, and the switch shuts off the power. The load capacity of an overhead crane or hoist must be clearly marked on both sides of the crane or hoist.

The choice of a work-holding device for a given type of load is important. Figure 18-2 shows two types of clamps designed for workpieces in a variety of shapes. Slings made of steel cable, chain, rope, flat fabric straps, and nylon straps are commonly used. Special racks and other work-holding devices can also be used with cranes and hoists. Ensure that the work is well-supported and secure.

Some types of work-holding devices enclose the work or secure it in a fixed position, but others require judgment and care on the part of the operator. For example, slings are used to lift bars of steel, and it is logical that the sling be placed at midpoint of the length of the bar. If the bar is not centered on the sling, or if something knocks it off center, one end of the bar will drop. If the bar has already been raised, the dropping of one end can cause serious damage. A nonskid sling can be used or, better still, two slings placed as far apart as feasible, to keep the bar stabilized.

A common error in using a cable or chain for lifting a heavy load is illustrated in Figure 18-3. Two cables lifting vertically will each carry half of the downward force created by mass M assuming that the force F is centered between the two cables. However, when the cables are at an angle a, each side of the cable will carry load $L = (F/2) \times \operatorname{cosec} a$. It is easy to see that as angle a gets smaller, the tension in the cable becomes much greater than the force exerted by the mass being lifted.

Controls for an overhead crane are often mounted in a box suspended by a cable from the top of the crane. This device is called a pendent switch, and it allows the operator to walk around and control the crane from any point within quite a large circle. The controls normally include at least six switches—up, down, right, left, forward, and reverse—and often include a "jog" switch as well. The "jog" allows the operator to raise the hoist in very small increments to apply the load gradually.

The switches on the pendent control box should be momentary-contact or spring-release type buttons. These buttons must be clearly identified. If the

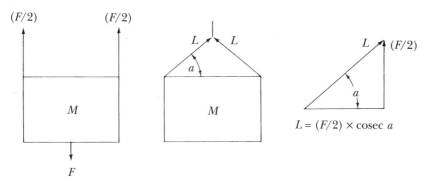

Figure 18-3. Illustration of loading on hoist cables.

pendent control-box has a cable, it must also have strain relief. Rotary switches and toggle switches may not be used. In hazardous locations, a crane may be operated by remote control. It is important, however, that the operator have a clear view of the entire area in which the load is moving.

Powered Industrial Trucks

There are many types of powered trucks used in industry, including platform trucks, forklifts, tractor trailers, front-end loaders, straddle carriers, and many other special types. There are many variations within each of these categories. ANSI/ITSDF B56.1-2010 and 29 CFR 1910.178 contain requirements for powered industrial trucks.

One problem common to all powered trucks is interference with pedestrian traffic. Aisles must be wide enough for their usage, along with pedestrians. Flashing lights and/or horns help caution pedestrians of the presence of moving vehicles. Powered trucks should be easily visible, and must be equipped with a horn. Lifts must also be equipped with seat belts to prevent drivers from falling off while turning or overturning the lift.

Another problem common to all powered trucks is the possibility of improper loading of materials. Loose parts or boxes should be stacked so that they will rest on one another and help hold the stacks together. Otherwise, they are likely to fall off when the truck goes around corners. It is recommended that, wherever possible, all objects being moved together should be in one container, strapped together or shrink wrapped, so that they can be handled as one object. This is called a unit load and is illustrated in Figure 18-4.

Figure 18-5 illustrates unit loads with a poor choice of container. Wirebound boxes, as shown, are very good for lighter objects but do not withstand abuse as well as heavier wooden boxes, or steel or plastic tubs. It is unsafe to move boxes in the condition shown here, but such a practice is not uncommon.

The platforms that constitute the base of the wirebound boxes in Figure 18-5 are one form of skid. A pallet has boards on the bottom as well as on the top. This provides a better bearing-surface than the two or three upright supports of a skid. However, the pallet cannot be picked up by a platform truck. A fork truck is required.

A platform truck has a solid, flat surface on which to carry a load, as distinguished from a fork truck. Forks are much more versatile, but do not provide as stable a surface and cannot accommodate loose objects. Many platform trucks are fixed in height, however; some have a low-lift (4 to 8 inches) capability and some have a high-lift (10 to 13 feet) capability. Fork trucks are always lift trucks, since they would have little use as fixed-height trucks. In spite of the fact that they provide less stable support than platform trucks, they are used for higher storage. This is because platform trucks cannot handle pallets, but only skids and loose objects. Pallets are used almost exclusively in high storage. Skids, which are open on the bottom require more careful placement on racks, and this becomes difficult at high elevations where visibility is poor from the driver's position. Industrial powered trucks can pose a fire hazard due to the type of fuel they use, such as LP, CNG, diesel, and gasoline. NFPA 505 is the Fire Safety Standard for Powered Industrial Trucks. Fire safety requirements are also located in 29 CFR 1910.178 and ANSI/ITSDF B56.1-2010.

In horizontal travel, the load carried on a truck severely reduces visibility. If the load obscures the driver's vision, the lift must be driven in reverse. Even then, it is important that the operator be able to see around corners as far ahead as possible. Aisles and corridors for vehicular traffic should have as few blind corners as possible. Figure 8-4 in Chapter 8 illustrates one way of increasing visibility in pedestrian corridors, but this type of construction is not appropriate when vehicular traffic is involved. Figure 8-5 illustrates open corners reflected in an overhead hemispherical mirror. Open corners provide good visibility, as does the mirror.

Floors for all kinds of vehicular traffic should be smooth and level and maintained in good condition. Rough floors cause vibration of the load, often to the extent that a load falls off the truck. Floors must also be strong enough to bear the load of the trucks and their cargo.

Special precautions must be taken at docks when trucks are carrying loads into and out of railroad cars and highway trucks. There is always some gap between the dock and the railcar or truck. The wheels of trucks, trailers, and railcars must be secured to prevent the vehicles from rolling away from the dock location. There must then be some sort of bridge arranged between the dock and the vehicle. Figure 18-6 illustrates such a bridge, referred to as

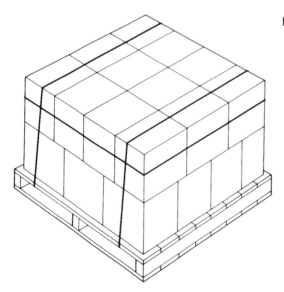

Figure 18-4. Proper stacking and strapping of loose objects to make a unit load. Objects of irregular shape should be carried in boxes.

Figure 18-5. Dangerously damaged pallet boxes still being used.

a dockboard. The use of a simple flat steel plate for this purpose has caused many serious injuries. The driving wheels of the forklift put a horizontal thrust on the plate and kick it out of position, leaving no supporting surface for the truck. Another common type of dock bridge is called a dock leveller, which consists of a heavy plate attached to the dock with a hinge set back from the edge of the dock. The plate is raised by either a hydraulic or a mechanical jack to rest on the floor of the railcar or highway vehicle.

There are four types of motive power units used on industrial trucks: diesel engine, gasoline engine, liquefied petroleum engine, and battery-powered electric motor. Diesel and gasoline engines are seldom used inside industrial buildings because of the toxic exhaust and the noise. LP gas engines produce some exhaust but not as much as diesel or gasoline engines, and they are somewhat quieter. However, LP gas engine vehicles should be used only in large buildings with good ventilation systems. Electric trucks have no harmful exhaust or noise and are the most commonly used trucks in small plants or in confined areas of large plants.

Electric vehicles have limitations and hazards of their own, however. The major problem is the necessity to charge the batteries frequently. Batteries may be charged while they are in the truck or they may be taken out of the truck. Lead acid batteries, generate hydrogen gas, especially while they are charging. Therefore, the charging area must be located away from any source of ignition. No smoking and no electrostatic discharges may be allowed. Figure 18-7 shows a typical battery-charging area. Eye-washing facilities must also be present in the charging area.

Another hazard of most electric trucks is the potential for ignition of flammable gases and dusts by sparks and/or hot surfaces on the motor. Some types of electric trucks are designed to prevent exposure to these sources of ignition.

Figure 18-6. A magnesium truck dockboard in use. Courtesy Magnesium Company of America.

Gasoline, diesel, and LP gas engines also involve hot surfaces and electrical sparks. Some of these trucks also have enclosed engines to prevent exposure to these sources of ignition. Table 18-1 gives the designations and brief descriptions of the several types of power units used in industrial trucks.

While speeds in excess of about 5 miles per hour (8 kilometers per hour) are considered too fast for industrial trucks in aisles used by other vehicles and pedestrians, slow travel is a hazard on public highways. A special placard, as shown in Figure 18-8, must be displayed on all vehicles traveling on public roads at speeds of 25 miles per hour (40 kilometers per hour) or less. The use of this emblem is not limited to agricultural equipment, as is often thought. It should be used on slow-moving industrial vehicles on public roads as well.

Many rules of the road are applied to the use of industrial vehicles in industrial plants the same way that they are applied to vehicles on public roads. The training and supervision of the operators of powered trucks is a very important factor in the safety of all personnel and equipment in the area where the trucks are used. Training and supervision are management functions and, therefore, are not covered in this text. Manufacturers of industrial trucks have provided pamphlets and even training courses for truck operators. Training requirements are described in 29 CFR 1910.178. The emphasis in this text is on factors that are of primary concern to engineers who work, or are training to work, in industrial plants.

Figure 18-7. Batteries are removed from trucks in this charging station. Chargers are stacked to save space and the area is ventilated. Courtesy of Hobart Brothers Company.

Storage Racks

It becomes even more important to move and store materials in unit loads when they are being stored at higher elevations. It is more difficult to see and to maneuver pallets at these heights, and loose objects are more likely to get bumped and fall off a pallet. Since they have farther to fall at these heights than at lower levels, they can cause more damage when they do fall.

The racks themselves can also become hazards. They must be sufficiently rigid and strong enough to withstand the loads put upon them. In addition, they must also be able to withstand being hit by trucks. Of course, cross-bracing must be used to maintain rigidity. Racks should be secured to the floor or a wall, when possible.

Racks and other equipment located at aisle corners are frequently subjected to impact from vehicles maneuvering around the corner. Barriers have proven helpful in reducing damage to these racks and equipment. One of the best ways to reduce this type of incident, however, is to design the aisles wide enough so that there is less chance that vehicles will hit things at the corner. Extra aisle width is often considered wasted space, but aisles should not be made so narrow that they increase the probability of incidents.

There are two types of forklift truck design, and the choice of one or the other affects the design of the storage racks they will serve. One type is called a balanced truck because the load being carried on the forks must be balanced by the mass of the truck itself. The other type is usually referred to as a narrow-aisle truck. It has wheels on outriggers extending forward to the front of the forks. This provides support without the need of a massive truck, so the truck can be shortened; it is thus able to maneuver in narrower aisles

Table 18-1. Type Designations of Powered Industrial Trucks

D	Diesel
DS	Diesel with safeguards to exhaust, fuel, and electrical systems
DY	Same as DS but have no electrical equipment and have temperature-limitation features
E	Electric
ES	Electric with safeguards to prevent hazardous sparks and to limit surface temperatures
EE	Same as ES but with enclosed motor and electrical equipment
EX	Electric, designed for use in areas containing certain flammable gases and dusts
G	Gasoline
GS	Gasoline with safeguards to exhaust, fuel, and electrical systems
LP	Liquefied petroleum
LPS	Liquefied petroleum with safeguards to exhaust, fuel, and electrical systems

Source: Adapted from NFPA Standard 505 and UL Standard 558 and 29 CFR 1910.178.

than the balanced truck. A narrow-aisle truck with forks on extending arms is illustrated in Figure 18-9.

The narrow-aisle truck does not always have forks on extending arms and, therefore, requires floor space under the bottom shelf of the rack to accommodate the outrigger. The balanced truck does not require this space, nor does a truck with the extending arms.

Figure 18-9 demonstrates the need for good visibility and adequate clearances when loads are being maneuvered at these heights. It is also easy to understand why a guard over the operator is required on all lift trucks.

Another type of storage facility is becoming more common in large industrial plants and warehouses. It is referred to as the automatic storage and retrieval system (AS/RS), or high-rise storage. It permits storage of materials at even greater heights than is possible with a powered industrial truck. Stacker cranes are mounted overhead with a rigid mast extending down to floor level and are guided by a rail on the floor. The crane travels along a given aisle, and a fork

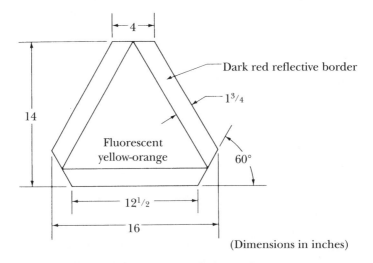

(Dimensions in inches)

Figure 18-8. Standard emblem for slow-moving vehicles. From ASAE S276.2/ANSI B114.1-1971, and OSHA 1910.145(d)(10).

on this mast places or retrieves unit loads on either side of the aisle. The aisle, however, is merely the space needed to move these unit loads. In some cases an operator rides on a platform close to the fork, but in most cases no people are allowed in the storage area except to repair and maintain the crane. The entire system is controlled by a computer. This provides a very fast and accurate storage system, but care has to be taken to ensure that no person enters the storage facility.

Storage bins and drawers are the common receptacles for small parts that are used frequently but in small quantities. There are two or three hazards associated with them. One is the height at which many of them are placed. If they are more than 1.5 meters (5 feet) high, users are inclined to stand on whatever is available, such as another bin, in order to see into them. If it seems necessary to use bins above this height, a substantial step device should be provided, but limiting the height of bins is a better solution.

Figure 18-9. Narrow-aisle forklift truck with a reach-fork® maneuvering a load at more than shoulder height. Courtesy The Raymond Corporation.

A fairly common injury occurs when someone pulls a bin or drawer out too far and the bin and all its contents spill onto the floor or onto the person. Some bins have a latch that permits the bin to be lifted out but prevents it from being pulled out inadvertently. However, many bins do not have this feature. If bins are taken out, or fall out, they quickly become battered. It is then difficult to maintain them in an operable condition, free of sharp edges.

Some companies have established a policy that limits the mass a worker is permitted to lift. Beyond that limit, some mechanized means of handling must be provided. Such limits vary from as little as 26 to 33 pounds up to 110 pounds, depending on the type of industry and the capabilities of the employees. As mentioned in Chapter 4, a major factor in the hazard of lifting is the position of the spinal column. One of the common errors is depicted in Figure 4-5; however, in both Figures 4-5 and 4-6 the box has no convenient handles.

Storage facilities are not limited to manufacturing and warehouse areas. Offices include many storage facilities, and they involve as many hazards as other storage areas. As in other areas, there is a tendency to pile things higher and higher, even on top of storage cabinets. The need for access to such stored material entices people to step up on anything nearby, including boxes, chairs, and shelves.

Many of the hazards involved in the storage of materials are strictly a matter of people not recognizing the risk, but some hazards could be avoided by the safer design of equipment. Filing cabinets have been involved in many incidents. They can easily tip over if they are not counterbalanced or secured. To make matters worse, if other drawers are not latched closed, they start to slide out as soon as the cabinet starts tipping.

Several things can be done to prevent this type of accident. For one thing, filing cabinets without latched drawers should not be used. The best solution

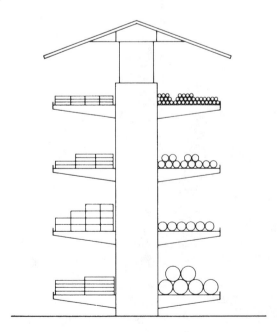

Figure 18-10. Double-arm rack for outdoor storage of materials such as lumber and pipe.

is to secure the cabinet to a firm object, such as a wall, the floor, or an adjacent cabinet. It is unlikely that two adjacent filing cabinets would be used in such a way that both would tip over together, but it would be even safer to fasten several cabinets together.

Latches can be made that will prevent someone from pulling out a drawer if another drawer is already out. Such latches have been designed, but many cabinets do not have this feature, particularly older units.

Durable materials are commonly stored outdoors. These materials include lumber, pipe, large castings, coal, and waste materials. In most cases they are handled with mechanical equipment, but occasionally they are handled manually. Frequently these materials are piled on the ground without the aid of storage racks or other facilities.

The major hazard in this type of storage is unstable piles that fall easily. Pipe, lumber, and waste materials often fall into the path of people walking by. On occasion, too, people walk over these materials and fall because of the instability of the pile. The use of racks not only reduces the danger of falling materials, but also prevents people from walking over them. Racks also provide some protection for the material. The racks shown in Figure 18-10 are called Christmas tree racks or arm racks. The type shown here holds materials on both sides, but frequently, one-sided arm racks are used. These racks permit access to one piece of pipe or lumber at a time or to a whole armful.

QUESTIONS

1. List some features that would make manually carrying a load safer.
2. List several safety features required on an overhead crane's pendent control box.
3. What types of fuel do powered industrial trucks typically use in an indoor facility and why?
4. A 6-inch diameter pipe, 10 feet long and weighing 250 pounds, is suspended by two vertical wire rope slings. How much weight is supported by each wire rope?
5. What type of powered industrial truck uses wheeled outriggers to balance the weight of the load?
6. A loaded lift truck, with a tall load, is travelling through an area with many pedestrians. How should the driver operate the truck, forward or reverse, and why?
7. Why should a facility not use a simple flat steel plate with a powered industrial truck to load a trailer?
8. What type of lift truck would you recommend for use in a chemical storage area with combustible dust present?
9. Why are pallets used in preference to skids in high storage areas?
10. Why would it be dangerous for a person to enter the storage area in an AS/RS system?

BIBLIOGRAPHY

American National Standards Institute. ANSI/ASME B20.1 -2009 Safety Standard for Conveyors and Related Equipment, ANSI/ITSDF B56.1-2010 Safety Standard for Low Lift and High Lift Trucks.

Ayoub, M. and A. Mital, 1989. *Manual Materials Handling.* New York: Taylor & Francis.

Eastman Kodak Company staff, 1984. *Ergonomic Design for People At Work*: The Design of Jobs, Vols. I and II. New York: Van Nostr and Reinhold.

Noro, K.,ed., 1987. *Occupational Health and Safety in Automation and Robotics.* New York: Taylor & Francis.

National Fire Protection Association Standard NFPA 505, *Powered Industrial Trucks*, 2011. Quincy, MA: NFPA.

National Safety Council. *Accident Prevention Manual for Business and Industry.* 11th ed. Chicago: NSC.

Rossnagel, W. E., L. R. Higgins, and J. A. MacDonald, 2009. *Handbook of Rigging.* 5th ed. Hightstown, NJ: McGraw-Hill.

U.S. Department of Labor, 2002. *Materials Handling and Storage*, OSHA 2236. Washington, D.C.: Government Printing Office.

*The manufacturers of industrial trucks offer excellent literature on the design of containers, storage facilities and the safe operation of their trucks.

19

Wood and Metalworking Operations

Michael Lowish, CSP

Introduction

There are a tremendous number and variety of wood and metal manufacturing industries. Many aspects of working with wood and metal may affect the safety and health of workers. Some of the more common ones will be discussed in this chapter. Obviously, it would be impossible to cover all such operations in any one book. This chapter will focus on the use of wood and metal as raw materials. There are many good sources of more detailed information on specific aspects of both industries. Some of them are listed in the bibliography.

The methods of machining and finishing both wood and metal involve operations, equipment, chemicals, and techniques that bear on the safety and health of workers performing the operations. The inherent characteristics and variability of the materials affect how they behave while being machined. The techniques for machining include sawing, milling, routing, abrasive planing, abrasive cutting, slicing, turning, stamping, bending, rolling, shearing, punching, forging, grinding, planing, and drilling. In essence, each technique comes down to a tool's performing a task on a material. The location of the operation to be performed is the *point of operation*, which is particularly important since it is a key location where guarding and various methods of protecting the operator must be employed. *Power transmission* components of machinery, such as gear drives, belts and pulley, chain drives, and shafts, among others, are related to the overall mechanics of making the machine perform a given task. These components may all cause injuries from in-running nip points or moving parts. These machine components are normally enclosed by guarding; however, maintenance activities in both industries may expose personnel to these components. *Other moving parts* may include any component of the equipment that could cause injury. Examples may include being struck by moving parts or being pushed against fixed objects. Proper lockout/tagout techniques must be followed to prevent serious accidents. Both woodworking and metalworking may also present toxic exposures to workers that must be controlled.

Material Considerations

Internal stresses are common to both wood and metal. Understanding the causes of the various stresses is important to successful manufacturing and to the safety of personnel working with the products. As trees grow, there are, naturally, parts of the tree that are in compression and parts that are under tension. Also, lumber and wood products must be dried as part of basic manufacturing. The migration of moisture out of the cells of the wood causes shrinkage, and this shrinkage may cause internal stresses. Wood has a very distinct, directional grain that is not found in metal. This imparts strength, function, and beauty in the finished product but also directly affects how the material will behave during machining. As wood is machined, the stresses in the wood will be relieved, causing movement that may cause binding against tooling or warpage. Understanding and control of these natural forces is important both to obtaining maximum yields and to safely working with the materials. Internal stresses may be present or created in metals by casting, forging, heat treating, welding, and some machining operations. These are normally relieved by various heat-treating techniques used to impart specific, desirable characteristics in the finished product. Heating and chemical treatments of metals may present significant health exposures to workers that must be controlled. Metals have some directional characteristics, but they are minor compared with those of wood. Machining techniques for wood create a variety of waste byproducts, from dust to chips. Hazards associated with them may include fire and explosion, engulfment, respiratory hazards, wood thrown out of machines, broken tooling, and others. Metalworking also creates waste byproducts in the form of chips, dust, waste chemicals used for cooling, lubricants for cutting tools, and various preparations to the metal surfaces. These may also present hazards from fire and explosion, respiratory and skin contact hazards, broken tooling, and debris thrown by equipment. Engineering controls are most commonly used to mitigate these potential hazards in both industries. Guarding, noise enclosures, exhaust systems, explosion suppression systems, personal protective equipment, and employee medical monitoring are all methods used to protect employees working in both industries.

Both wood and metal products may be physically large and heavy, requiring the use of material-handling equipment such as lift trucks, hoists, cranes, conveyors, and other aids. Machined parts made of wood seldom have edges sharp enough to cause cuts, but depending on the operation and species of wood, splinters may be a significant hazard, which is particularly dangerous for employees working with veneers. It is very common during metal manufacturing to create razor-sharp edges on both the product and the waste metal from machining. Like working with wood, splinters of metal are an ever-present hazard.

Machining

Machining techniques for wood and metal have a lot in common, and many machines used in industrial woodworking operations are based on designs originally developed by the metalworking industry. Primary differences are in the speeds of both cutting tools and feed rates. Woodworking tooling runs faster and cutting angles are more aggressive than with metalworking tools. Machining techniques unique to metalworking include the use of coolants and lubricants with cutting tools and the use of heat and electric current to perform machining operations. The forming of metal by stamping and bending presses and by roll formers is also unique to metalworking. Almost all metal machining is performed with the part held by mechanical means. In industrial woodworking, parts are also clamped by the machine or separate jig or fixture. In some woodworking operations, the operator may hold the part being machined. Handheld tools such as drills, sanders, grinders, chisels, and punches are common to both wood and metalworking activities, and while these tools have slight differences, the operations and techniques of use are very similar.

Modern tooling (drill bits, saw blades, cutters, mills, etc.) used in both wood and metalworking are very similar in how they are made and what they are made of. The use of carbide- and diamond-tipped cutting tools is very common and necessary due to the nature of the materials, feed rates, and production demands of modern manufacturing. Differences are primarily in the geometry of the actual cutting surfaces (see Figure 19-1).

It requires significantly more force for a dull tool to perform properly than a sharp tool. This applies equally to hand tools and machine tools. In addition, dull tools generate heat and will leave a substandard surface on the part being machined. The planning of jobs to consider the methods of machining available, the characteristics of the raw materials, feed rates, and other factors are critical to tool management. Dull tooling is far more likely to grab or bind, which can cause parts to be pulled out of holding devices and thrown. Modern abrasives are used more and more for machining operations. Like traditional cutting tools, abrasives function as miniature cutting tools. If the abrasive becomes loaded with the material being removed, it not only does not cut, but also generates heat from friction.

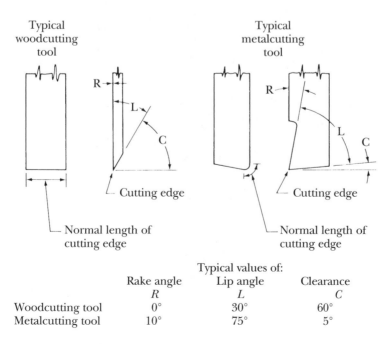

	Typical values of:		
	Rake angle	Lip angle	Clearance
	R	*L*	*C*
Woodcutting tool	0°	30°	60°
Metalcutting tool	10°	75°	5°

Figure 19-1. Comparison of geometry of woodcutting and metalcutting tools.

Metals such as aluminum, magnesium, titanium, and uranium, among others, can present significant fire hazards when small chips or dust are generated. Methods of controlling dust and chips must be engineered into the process when machining these materials. Reference sources such as NFPA provide specific guidance for designing systems that work safely with these materials.

Grinding equipment is common in the wood and metalworking industries—in the former for sharpening tooling and fabricating cutters. These uses, in addition to fabricating metal parts, are very common in modern metalworking. Grinding presents several serious potential hazards that must be controlled. Figure 19-2 provides key safety considerations when using a floor grinder. Grinding wheels and tooling are made in a wide variety of shapes, sizes, and compositions and are designed to be used in a specific manner. When used improperly or for other than its intended use, it is quite possible for the tool to shatter, exposing anyone in the surrounding area to serious injury. Due to the nature of vitrified grinding wheels and tools, they should *never* be used without proper enclosures and guarding in place. ANSI and OSHA standards provide extensive guidance on proper selection, use, and protection of personnel when using grinding equipment. On common bench and pedestal-mounted grinders it is common to mount wire brushes. As the wire brush is used, the bristles will break and are thrown out and away from the machine. For this reason it is essential that the wire brush be used with the same enclosure required

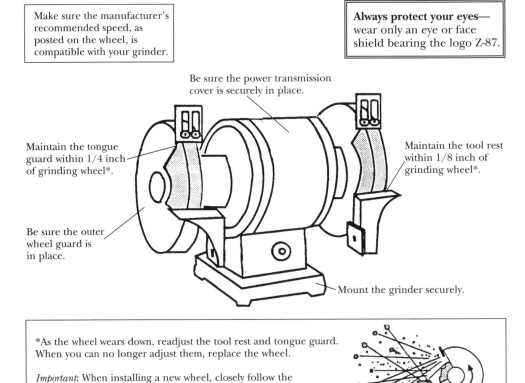

Make sure the manufacturer's recommended speed, as posted on the wheel, is compatible with your grinder.

Always protect your eyes— wear only an eye or face shield bearing the logo Z-87.

Be sure the power transmission cover is securely in place.

Maintain the tongue guard within 1/4 inch of grinding wheel*.

Maintain the tool rest within 1/8 inch of grinding wheel*.

Be sure the outer wheel guard is in place.

Mount the grinder securely.

*As the wheel wears down, readjust the tool rest and tongue guard. When you can no longer adjust them, replace the wheel.

Important: When installing a new wheel, closely follow the manufacturer's instructions. An improperly installed wheel can break (right) and cause injury.

Figure 19-2. How to use a bench or floor grinder safely.[1]

around grinding wheels and that safety glasses and face shields be worn by operators. Additional hazards are presented by the material being ground and by the waste material generated during grinding operations. There are a wide range of potential health exposures that must be addressed by ventilation, PPE, and proper techniques based on the material and equipment being used. Larger grinding wheels used on surface grinders are used with cooling/cutting fluids. Since these grinding wheels are porous, they will absorb some of these fluids. To prevent the wheel from collecting fluid on one side and becoming out of balance, it is very important that the wheel be running when the fluid feeds are turned on and that it be run without fluid to sling out any fluids after machining is completed.

Laser machining is now very common in both wood and metal manufacturing industries. Lasers offer advantages of being able to cut or etch without creating internal stresses on adjacent surfaces. Holding fixtures must simply position the parts on a bed that either moves under the laser or that allows the laser to travel over the parts. Wood and metal parts that would be difficult or impossible to fabricate using conventional tooling are easily cut. This equipment is always computer-controlled, so there is little direct operator intervention. Safety and health considerations include exhaust ventilation of fumes created as the laser burns into the part and stray light emissions.

It is very important to clamp parts being machined in woodworking and metalworking, since tooling may tend to grab the part. Unclamped parts may be either violently thrown or spun, causing serious injury.

Principles of Guarding

Guarding considerations include access to various parts of the equipment and material being machined and barriers to materials being ejected from the equipment. Hazards can be to personnel, both the machine operators and others who may be in the area. The ANSI B-11 Standards on machines can be found in Appendix E.

Tooling

Cutting tools may be divided into two categories: single-point tools and multiple-point tools. Single-point tools, such as chisels or lathe tools used in both industries, have only one cutting edge. Figures 19-3 and 19-4 show examples of shields on wood and metalworking lathes. Multiple-point tools have a series of cutting edges arranged either in a straight line, as on a bandsaw blade, or around a circle, as in a circular saw blade, drill bit, milling cutter, or cutterhead.

Tooling may be moved against a stationary workpiece or held stationary while the work moves by or into the tool. In modern equipment there is normally a combination of the two. In either case the tooling will present sharp edges if it is exposed, and if it breaks or grabs a part, flying debris can present a hazard to personnel in the area. Design considerations for guarding should both prevent access to the points of operation and contain any possible flying debris. Abrasive tools may not appear to present sharp edges, but the millions of tiny cutting surfaces placed in movement by powered equipment are very capable of quickly abrading flesh and bone. Fine particles created by abrasive machining may present fire and explosion hazards if not exhausted from the

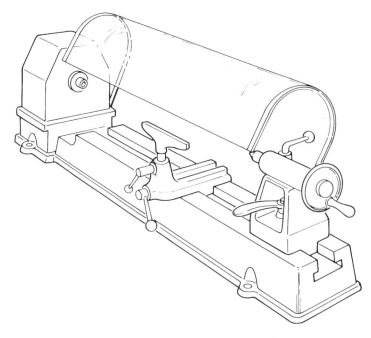

Figure 19-3. A hinged plastic shield covering the work area on a wood-turning lathe.

Figure 19-4. The hinged plastic shield travels with the tool on this metalcutting lathe.

equipment and may create toxic exposures to personnel. Collection points for exhaust systems must be designed to capture debris before it is allowed to accumulate inside equipment or escape into the atmosphere. In addition, to be effective, dust and fume collection ventilation systems must be designed so that they do not pull contaminated air into the worker's breathing zone.

Operator Considerations

With both wood and metal manufacturing operations, a strong cultural commitment to working safely by management, supervision, and employees is essential to avoid unnecessary accidents and serious injuries. Whether jobs are custom, short-run, or long-term production runs, it is essential to set up and use proper setups, holding fixtures, clamping, guarding, and safety devices designed to protect the workers. In woodworking operations, wood shapers and stationary routers are among the most dangerous machines since there is an infinite variety of setups and cutter shapes available to the operator. Proper setup will greatly reduce or prevent possible access to moving cutters; however, it is often easier and less time-consuming for operators not to use holding fixtures on short-run projects, thus exposing their hands to rotating tooling. If a workpiece is grabbed by the tooling, the worker's hands are frequently pulled into the path of the point of operation and parts are violently thrown from the machine.

Power presses and press brakes are among the most potentially hazardous metalworking machines. Pinch points and shearing actions are the major hazards. Tooling and parts are capable of being ejected violently from this equipment if the machine is not properly set up. Improperly maintained or modified presses and brakes may malfunction, causing the machines to double trip, catching an unsuspecting operator reaching into the point of operation. Operators of this equipment may be tempted to reach into the point of operation; if they notice that a part of the piece is misaligned, they might reach into the machine to straighten it. They may also be tempted to reach into the machine if they are working on a wage-incentive basis in an effort to find shortcuts to the press's cycle time. The only truly effective way to prevent serious injuries is to prohibit any operator hand-in-die operations. This can be accomplished by the proper feeding and ejecting of parts, guarding, and training. Various guarding methods that prevent any possible access to the point of operation during the downward stroke of a press can be very effective, but all require maintenance and management systems. Many of these machines may be activated in a number of ways. Two-hand controls that require simultaneous activation by both of the operator's hands are effective if they have not been modified or defeated. When more than one worker on a machine is present, duplicate controls must be in place. This does not, however, protect other workers who may be in the vicinity of the equipment. Protective measures that do not require both hands to be clear of the point of operation during a machine cycle may eventually allow the operator to have an accident. Holding parts with tools while feeding does not prevent the operator from feeding parts by hand, unless additional safeguards are in place. Parts positioned and held by operators on a press brake often do not provide adequate protection when the brake is activated by a foot pedal. Light curtains must be set up and maintained so that workers cannot reach over, under, or around the protective barrier. Allowing more than one worker to operate a press or brake with only single controls is an invitation to disaster since the "helper" is unprotected.

Chemicals and Finishing

A lot of chemical hazards may be present or created in both woodworking and metalworking operations. Awareness and proper management are essential to maintaining a safe working-environment.

Both wood and metal products are assembled with various types of adhesives that may present health exposures to employees. Finishes used for both materials may present significant fire and explosion hazards, as well as significant health risks if improperly handled.

Strong reactive bleach is sometimes used to match colors in early stages of some furniture finishes. A number of the adhesives used contain toxic chemicals that may be hazardous to the respiratory system and skin. Catalyzed wood finishes are very durable but contain chemicals that are toxic and that may sensitize employees so that they will have a reaction upon contact with only trace amounts. Most wood finishes and a lot of adhesives used in wood industries are flammable and must be stored properly and used in properly ventilated areas. Dip tanks are frequently used to apply stain to furniture. OSHA and NFPA both have specific standards detailing the proper storage and application of flammable and combustible materials. A selected list of NFPA codes and standards is shown in Table 19-1. Guidelines on the storage of flammable and combustible materials, spray finishing, dip tanks, personal protective equipment, and respiratory protection may all apply to modern woodworking operations.

In metalworking operations the range of possible chemical hazards is even larger. Degreasers, acid and caustic baths, salt baths, various plating solutions, coatings, and finishes may all present hazards that must be controlled. Many of these materials may be reactive with one another or may break down in the presence of other operations typically found in the metalworking industry. One example is the creation of phosgene gas when halogenated solvents (commonly found in parts cleaners) are exposed to the energy released from arc welding. Both parts cleaners and welding equipment are commonly located together in maintenance shops.

Practically every industrial facility has both gas and electric welding equipment. They may be used both for manufacturing and the maintenance of equipment throughout the facility. NFPA 51B, the OSHA welding standards, and manufacturer's literature are all excellent reference sources for the safe use of welding and cutting equipment. While a comprehensive examination of the welding and cutting processes is beyond the scope of this chapter, it is important to explore health and safety considerations of welding in both the woodworking and metalworking industries.

Table 19-1. National Fire Protection Association (NFPA) Codes

30	Flammable and Combustible Liquids Code, 1996
33	Spray Application Using Flammable or Combustible Materials, 1995
34	Dipping and Coating Processes Using Flammable and Combustible Liquids, 2007
51B	Fire Prevention During Welding, Cutting, and Other Hot Work, 2009
484	Combustible Metals, 2009
664	Prevention of Fires and Explosions in Wood Processing and Woodworking Facilities, 2009

Safety and health considerations for both wood and metalworking operations involve planning jobs, selecting the proper tools and equipment for the job, proper operator training, good maintenance, and effective management systems.

QUESTIONS

1. What is meant by the term "point of operation"?
2. What are common elements that would be part of the power transmission components of machinery?
3. What is the common objective for lockout/tagout with both wood and metalworking machinery?
4. Name common reasons that parts may bind against tooling.
5. How can internal stresses be relieved in wood? In metals?
6. What fire and explosion characteristics do both wood and metals have in common?
7. What are the primary differences in machining techniques between wood and metal?
8. Why is a dull tool more hazardous than a sharp tool?
9. What are four metals known for their hazardous properties from dust and small chips?
10. Why is it important to clamp parts being machined?
11. Name two design considerations for effective machine guarding.
12. Dust exhaust systems must perform what two functions?
13. What is the advantage of using two hand trip devices on machinery?
14. What are two hazards common to many wood and metal finishing techniques?

NOTES

1. *Tech Guide No. 2.* Published by the Environmental Science and Technology Laboratory of Georgia Tech Research Institute, 1988.

BIBLIOGRAPHY

Clark, Edward, John A. Ekwall, C. Thomas Culbreth and Rudolf Willard, 1987. Furniture Manufacturing Equipment. Raleigh: Department of Industrial Engineering, North Carolina State University.

National Safety Council, 2009. *Accident Prevention Manual for Business and Industry,* 13th Edition. Chicago: NSC.

Willard, Rudolph, 1970. *Production Woodworking Machinery,* 3rd Edition. Raleigh: North Carolina State University Press.

29 CFR 1910. *Occupational Safety and Health Standards—General Industry,* 2009.

20

Prevention through Design

John W. Mroszczyk, Ph.D., PE, CSP

Introduction

The world around us includes buildings, structures, vehicles, tools, facilities, equipment, products, hardware, machines, and materials. A failure in any one of these systems can result in injuries, fatalities, illnesses, or damage to property or the environment. The human element adds another dimension. Equipment operators, machine operators, workers, consumers, maintenance workers, construction workers, employees, managers, and members of the public interact with these systems in a number of ways. The possibility of a failure in the human-system interface adds to the overall risk of injury or damage.

All too often, and for various reasons, the risk is managed after the design has been released. There may be a rush to get the design off the drawing board, and:

- hazards that arise in actual use are often not taken into account;
- it is often taken for granted that the system operates the same throughout its life cycle as it was intended to do on the drawing board;
- human error is not considered;
- instructions are not followed or the system is not assembled properly;
- wear of system components may not have been properly tested under various maintenance schedules.

This approach to design results in expensive retrofits, increased maintenance, increased insurance costs, increased downtime, and lower productivity, not to mention injuries, fatalities, and property damage.

There is another approach: Safety through Design, Design for Safety, and Prevention through Design (PtD) are the same names for a design philosophy that anticipates hazards and designs hazards out or reduces the risk in the design phase. The idea of addressing hazards in the design phase, rather than relying on the operator's consistent obedience to safety rules, dates as far back in time as 1946 (NSC 1946). The concept of Prevention through Design has been defined by NIOSH as:

> Addressing occupational safety and health needs in the design process
> to prevent or minimize the work-related hazards and risks associated

with the construction, manufacture, use, maintenance, and disposal of facilities, materials, and equipment.

Eliminating hazards in the design phase has a number of benefits:

- Injuries, fatalities, and damage to property and environment are reduced or eliminated
- Retrofitting to correct design shortcomings is reduced or eliminated
- Productivity is increased
- Quality is improved
- Downtime is reduced
- Overall costs are lower

Prevention through Design is gaining national and global acceptance. NIOSH has launched a seven-year national PtD initiative. This approach can be applied to any system, including all work sectors, buildings, structures, consumer products, vehicles, equipment, and facilities. The OSHA Alliance Construction Roundtable has supported Design for Construction Safety, the application of the Design for Safety methodology to construction projects.

Hazard Identification

The first step in the Prevention through Design process is to identify the hazard(s). A *hazard* is the potential to do harm or damage. A list of hazards can more easily be identified using the physical categories listed in Table 20-1. For example, since gravity is always present, areas where a

Table 20-1. Physical Categories That Can Be Used for Hazard Analysis (Mroszczyk, 2010)

Category	Manifestation in Workplace
Gravity	Falls from elevation, falling objects
Slopes	Upsets, rollovers, unstable surfaces
Water	Drowning
Walking/working surfaces	Tripping, slipping
Mechanical hazards	Rotation, reciprocation, shearing, crushing, cutting, entanglement, drawing-in, pinch points
Stored energy	Springs, pneumatics, hydraulics, capacitors, liquids and gases under pressure
Electrical	Contact with live parts, contact with parts that may become live under faulty conditions, electrostatic, current, voltage, sparks, arcs
Thermal	Contact with hot or extremely cold objects or substances
Chemical	Corrosive, fire and explosion hazards, contact or inhalation of harmful fluids, gases, fumes, and dusts, toxic
Biological	Allergens, carcinogens
Radiant Energy	High noise levels, nuclear, X-rays, light, lasers
Ergonomic	Lifting, repetitive motion

person could fall from or where an object could fall on a person should be considered. If the facility is near water or if large open water tanks are accessible, then the chance a person could drown should be addressed. There will almost always be walking/working surfaces, so slip and trip hazards should be eliminated. Any system that has moving parts most likely will have mechanical hazards. If the facility contains chemicals, then corrosion, explosion, and toxicity should be analyzed. Biological materials should be checked for harmful effects.

Questions that should be raised during a hazard analysis might include:

- What happens if maintenance is not done?
- How does wear and tear affect the operation?
- What are the ways in which the device can be assembled incorrectly?
- What are the ways that the device can be misused?
- Does the device have a finite life?
- What are the intended use(s)?
- How will the system actually be used in the field?
- What are the foreseeable misuses?
- What are the ways in which operator error can affect the system?
- What is the potential chain of events that could lead to an injury?
- What are the health hazards?
- Are the controls properly designed and displayed so that an operator is not confused?
- What are the failure modes of the system?
- Does the end user need special training?
- How can the system be abused?
- Is redundancy provided for critical parts of the system?
- Is the system safety affected by the weather or a power outage?
- Are there ways in which stored energy could be released?

There are *apparent* hazards and *hidden* hazards. Apparent hazards, as the name implies, are either obvious or have been recognized in the past. These include unguarded belt drive systems, heavy lifting tasks, and open-sided work platforms. Standards can be used as a source to make a list of apparent hazards. Checklist analysis is another tool that is useful to account for apparent hazards. Checklist analysis consists of yes-or-no questions that are derived from published standards, codes, and industry practices.

A hidden hazard is an unapparent source of injury. Hidden hazards may be related to operator error, wear and tear, lack of maintenance, or may result from a chain of events. There are a number of analysis tools that can be used to uncover hidden hazards. These analytical methods are discussed briefly.

A what-if hazard analysis is a structured brainstorming session. The analysis starts with assembling an experienced, knowledgeable team. The individuals chosen for the team should be experienced in design, standards, regulations, previous maintenance issues, previous assembly problems, power failures, field problems, past history, potential failure modes, foreseeable use and misuse, and possible operator errors. Information such as video tapes of operation, design documents, and maintenance procedures should be gathered. The team then poses a series of what-if questions. Besides the design analysis questions listed above, a what-if analysis might also include the following:

> What if the operator fails to follow procedures?
> What if procedures are followed, but are incorrect?
> What if there is an equipment failure?
> What if there is a utility failure?
> What if there is a change in weather?
> What if the operator is not trained?

The answers to the what-if questions can help to uncover hidden hazards in the system.

Fault tree analysis (FTA) is a schematic flow diagram that identifies those areas in a system that are most critical to safe operation in a logical manner. FTA is a top-down analysis. It starts with the undesirable event at the top (head event), followed by primary and secondary events that could cause the head event. A primary event is a characteristic inherent in the system itself. A secondary event is due to an external source. For example, if the system is a light bulb, an undesirable event would be the failure of the light bulb. A primary event might be a worn filament. A secondary event might be a surge in voltage. Next, the relationship between the causal events and the head event is determined in terms of AND and OR Boolean operators. Once the fault tree is done, the logic tree will indicate those events that can be eliminated so that the undesirable event at the top cannot occur based on Boolean logic.

Job safety analysis (JSA) or job hazard analysis (JHA) is another analytical tool that can be used to uncover hidden hazards. A complete JSA requires a look at each production step or each machine function. In general this would include receiving raw materials, production, assembly, packaging, and shipping. First you identify each step, then consider every potential hazard you can think of, or anything that can go wrong. Next to each hazard suggest actions that can be taken to eliminate the hazard or the occurrence.

Failure mode and effects analysis (FMEA) is a systematic tabulation of the failure mode(s), the effects of each failure, the safeguards that exist, and the additional actions that can be taken. The FMEA process is done for each item or function to be analyzed. An example of how FMEA might apply is provided by an analysis of a chemical storage tank. The physical hazard category is chemical. The related human traits include inattention. Two failure modes are considered: the tank can rupture or it can be overfilled. Events that can cause a rupture of the tank include a weld failure and internal corrosion.

Risk Assessment

The next step after the hazard(s) have been identified is to assess the risk. Risk is a measure of the probability of a hazard-related incident occurring and the severity of harm or damage. There are many risk-assessment methods, both qualitative and quantitative. The following is a fairly simple, straightforward qualitative method for assessing risk. The severity of the harm is estimated as one of four levels:

Catastrophic—death, amputation, permanent disability

Serious—serious debilitating injury (able to return to work after a period)

Marginal—significant injury requiring more than first aid, able to return to work at same job

Negligible—no injury or injury requiring no more than first aid, little or no lost work time

Probability of occurrence is estimated as one of the following levels:

Frequent—Likely to occur often

Probable—Will occur several times

Occasional—Likely to occur sometime

Remote—Unlikely, but possible

Improbable—So unlikely as to be near zero

Once the severity of harm and the probability of occurrence have been estimated, the risk level can be determined from the matrix in Table 20-2. For example, a cut finger (negligible severity) that might occur occasionally would be a 4C risk (acceptable). A finger amputation (catastrophic severity) that could occur several times (probable) would be a 1B risk (unacceptable).

Table 20-2. The Matrix Below can be used to Assess Risk Once the Severity and Probability Have Been Estimated.

	Severity of Harm			
Probability of Harm	**Catastrophic**	**Serious**	**Marginal**	**Negligible**
Frequent	1A	2A	3A	4A
Probable	1B	2B	3B	4B
Occasional	1C	2C	3C	4C
Remote	1D	2D	3D	4D
Improbable	1E	2E	3E	4E

Hazard Risk Assessment	**Suggested Action**
1A, 1B, 1C, 2A, 2B, 3A	Unacceptable, risk reduction is necessary
1D, 2C, 2D, 3B, 3C	Unacceptable if personnel exposed to hazard
1E, 2E, 3D, 3E, 4A, 4B	Acceptable with management review
4C, 4D, 4E	Acceptable without management review

Hierarchy of Engineering Controls

Once the hazard(s) have been identified and the risk assessed, the next step is to apply the hierarchy of engineering controls (Table 20-3). The top priority in the hierarchy of engineering controls is to design out the hazard or reduce the risk to an acceptable level. An example of designing out a hazard would be to specify a staggered bolt hole pattern on a critical metal bracket so that the bracket cannot be installed upside down. The next course of action, if the hazard cannot be eliminated by an alternative design, is to apply safety devices. An example of a safety device would be an interlocked machine guard or a dead man's switch. When the hazard cannot be eliminated or the risk reduced by incorporating safety devices, then warning devices should be applied. An example of a warning device is a beacon light or horn. If these measures are impractical or ineffective, then warning signs and labels, training, operating procedures, and administrative procedures should be implemented. If the risk cannot be reduced with any of these interventions, then the design should be terminated.

Table 20-3. Hierarchy of Engineering Controls

1. Design out the hazard or reduce the risk to an acceptable level
2. Incorporate safety devices
3. Provide warning systems
4. Apply administrative controls
5. Recommend personal protective equipment
6. Terminate the design

For many situations, a combination of these measures might apply. It should be noted that the first two measures, design out the hazard or apply a safety device, take precedence over warnings, operating procedures, training, and administrative procedures. The latter design measures are not as reliable because they depend on humans to follow the warning or the procedure.

Coated Flight Bars for Mining Machines

Noise-induced hearing loss is a common health problem in mining. Continuous mining machines are used to cut, gather, and remove coal and non-coal materials from underground rooms and pillar mining operations. These machines rank the highest amongst equipment where operators exceed 100 percent noise dosage. There are three subsystems that generate sound: the cutting head drum, the conveying system, and the duct collection system.

The conveying system includes steel bars that are perpendicular to the conveyor chain, called flights. The flights span the width of the conveyor and move the aggregate to the rear of the machine. The principle noise source is from metal-on-metal contact between chain flights and the conveyor deck.

A PtD solution was proposed (see Figure 20-1). Coating the flights bars with urethane reduced the risk of hearing loss. The coated flight bars demonstrated a 35-percent reduction in total work shift noise dose. The PtD solution also had an added benefit: the coating extended the life of the chains.

Figure 20-1. This photo shows a PtD solution for reducing the noise levels in mining operations. Coating the chain conveyor and flight bars protects mine operators' hearing and extends the life of the chain. *(Photo courtesy of NIOSH.)*

Commercial Fishing Deck Winch

Commercial fishing is a very dangerous occupation. Fatal injuries, caused by unsafe machinery and equipment, remain a significant problem in the industry. Powerful, capstan-type winches used to reel in large fishing nets pose an entanglement hazard to deck workers on fishing boats.

The capstan winch is usually mounted in the center of the deck. The winch's drum rotates while the crew works on deck. Fishermen who lose their balance or are inattentive can become caught in the fishing line as it winds around the drum. The winch provides no entanglement protection and the controls are usually out of reach of the entrapped person.

A unique PtD solution—an emergency-stop (E-stop) system situated on the top of the winch—addresses the serious machinery hazard posed by a capstan deck winch. When pushed, the emergency switch arrests the drum's rotation in less than 180° rotation, sufficient to limit serious entanglement injury (see Figure 20-2).

Asphalt Paving Equipment

More than half a million workers in the United States face exposure to fumes from asphalt, a petroleum product used extensively in road paving, roofing, siding, and concrete work. Health effects from exposure to asphalt fumes include headache, skin rash, sensitization, fatigue, reduced appetite, throat and eye irritation, cough, and skin cancer.

A capstan-type winch with fishing lines wound around.

A fishing vessel captain demonstrating the use of an emergency-stop (e-stop) mounted on the winch.

Figure 20-2. The photo on the left shows the entanglement hazard posed by powerful, capstan-type winches used to reel in large fishing nets. The photo on the right shows the PtD solution: an emergency-stop (E-stop) system situated on the top of the winch. (*Photos courtesy of NIOSH.*)

The PtD solution incorporated ventilation controls and "warm mix" asphalt. Warm-mix asphalt is the generic term for a variety of technologies that allow the producers of hot-mix asphalt pavement material to lower the temperatures at which the material is mixed and placed on the road. Reductions of 50 to 100°F have been documented, resulting in less fumes. Such drastic reductions have the obvious benefits of cutting fuel consumption and decreasing the production of greenhouse gases. In addition, potential engineering benefits include better compaction on the road, the ability to haul paving mix for longer distances, and the ability to pave at lower temperatures. Figure 20-3 shows the before and after effects these design changes.

Figure 20-3. This photo shows the before-and-after effects on asphalt fume emissions in highway-class pavers following PtD design interventions. (*Photos courtesy of NIOSH.*)

QUESTIONS

1. Is it reasonable for the manufacturer of industrial equipment to rely on the end user to faithfully perform maintenance when the lack of such maintenance could create a hazard that could cause an injury?

2. Which of the following benefits can be realized by applying Prevention Through Design principles?
   a. Downtime is reduced
   b. Retrofits are reduced or eliminated
   c. Quality is improved
   d. Injuries and fatalities are reduced or eliminated
   e. All of the above

3. True or false: Applying warnings or relying on instructions take precedence over eliminating the hazard by engineering design or incorporating safety devices,

4. Questions that could be raised during a hazard analysis include all but the following:
   a. What are the failure modes of the system?
   b. How can the system be abused?
   c. Is there a market for the system?
   d. How does wear and tear effect the system?

5. You are designing a new product and have inquired into the field use or similar products. You discover that similar products are being misused in a certain way that leads to injuries. Should you consider this misuse when designing the new product?

6. All of the following members should be considered when forming a "what if" design analysis team except:
   a. Safety manager
   b. Design engineers
   c. Accountants
   d. Customer service personnel familiar with field problems
   e. All of the above

7. You are designing a large water storage tank that will be open at the top. There will be a catwalk within several feet of the top of the tank? Should you consider lifesaving equipment in your design?

8. You have determined that a particular design hazard can result in an amputation and that the occurrence is at least once during the ten-year life of the design. Should you design out the hazard?

9. The following resources can be used to identify apparent hazards.
   a. ANSI standards
   b. NFPA standards
   c. ASSE Safety Professional Handbook
   d. OSHA website
   e. All of the above

10. You have done a fault tree analysis and have determined that event A and event B must occur in order for the undesirable event to occur. Which one of the events, A or B or both, should be eliminated by engineering design?

REFERENCES

American National Standards Institute, 2000. B11.TR3: Risk Assessment and Risk Reduction-A Guide to Estimate, Evaluate and Reduce Risks Associated with Machine Tools. Washington, D.C.: ANSI.

American Society of Safety Engineers (ASSE), 2009. "Prevention Through Design: an ASSE Technical Report TR-Z790.001." Des Plaines, IL: ASSE.

Cervarich, M., 2009. "Warm-Mix Asphalt: Preventing Exposure at Its Source." *PtD in Motion.* Issue 5, July 2009. NIOSH.

Christensen, W. and F. Manuel, 1999. *Safety through Design.* Itasca, IL: NSC Press.

Hagan, P, et al., eds, 2009. *Accident Prevention Manual for Business and Industry, Engineering and Technology,* 13th Edition. Itasca, IL: National Safety Council.

Haight, J., ed., 2012. *The Safety Professionals Handbook.* Des Plaines, IL: American Society of Safety Engineers.

Hammer, W., 1972. *Handbook of System and Product Safety.* Englewood Cliffs: Prentice-Hall.

Heidel, D., J. Collins, and E. Stewart, 2009. "Prevention through Design in Health Care Settings. *The Synergist,* November 2009.

Kovalchik, P. G., R. J. Matetic, A. K. Smith, and S. B. Bealko, 2008. "Application of Prevention through Design for Hearing Loss in the Mining Industry." *Journal of Safety Research* 39(2): 251–254.

Lincoln, J., D. Lucas, R. McKibbin, C. Woodward, and J. Bevan, 2008. "Reducing Commercial Fishing Deck Hazards with Engineering Solutions for Winch Design." *Journal of Safety Research* 39: 231–235.

Manuele, F., 2005. "Risk Assessment & Hierarchies of Control." *Professional Safety.* May 2005.

Mroszczyk, John, 2010. "Take Your Pick: Error Proof Humans or Your Organization's Environment." Proceedings of the ASSE Rethink Safety: A New View of Human Error and Workplace Safety Symposium, San Antonio.

Mroszczyk, John, 2007. "Designing for Construction Safety: Opportunities for Design Professionals." *American Society of Safety Engineers Construction Practice Specialty Newsletter,* Vol. 7, No. 1.

National Institute of Occupational Safety and Health (NIOSH), 2010. Prevention through Design Plan for the National Initiative. Cincinnati, Ohio: CDC NIOSH.

National Safety Council (NSC), 1946. *Accident Prevention Manual for Industrial Operations,* Chicago: National Safety Council.

Planek, T., editor, 2008. Special Issue: Prevention through Design. *Journal of Safety Research* 39(2).

Stewart, E., 2009. "Prevention through Design: Three Examples from Health Care That Integrate Worker Safety with Sustainable Design." *PtD in Motion,* Issue 5, July 2009. NIOSH.

U.S. Dept. of Defense, 2000. MIL-STD-882D Standard Practice for System Safety.

Woodson, Tillman, and Tillman, 1992. *Human Factors Design Handbook,* 2nd Edition. New York: McGraw-Hill.

21

Case Studies

John W. Mroszczyk, Ph.D, PE, CSP

Introduction

The case studies presented in this chapter are real-life situations. Each case provides a context to illustrate a particular hazard, the lack of appropriate safety-engineering measures, and what are the proper safety-engineering measures. For each case, a brief depersonalized background will be given, the hazard will be identified, and the engineering intervention to eliminate or mitigate the hazard will be discussed. The proper application of safety-engineering principles and practices, as each case shows, would have prevented an injury or fatality.

Case #1 Concrete Block Machine

Description

Any machine that has moving parts should be examined for adequate safeguards. The following case study of a concrete block machine illustrates this principle. The machine is shown in Figure 21-1. The machine can be run in automatic or manual mode. In the automatic mode, a steel pallet is indexed into the machine and raised by a lift to the mold. The mold is filled from a hopper at the top and vibrated. The pallet is lowered, stripping the block from the mold. The pallet and block are then pushed onto a receiving conveyor. The machine has a number of limit switches that will cause it to pause if a pallet is not in the proper position. The machine will resume operation when the pallet is corrected without pressing a restart button unless the machine has been switched to manual mode.

The machine can also be run in manual mode, which is used when making machine adjustments. A switch on the control panel allows the operator to change the machine operation between automatic and manual. The control panel is located a distance away and behind the operator when facing the machine. A beacon light on the control panel flashes when the

Figure 21-1. Concrete block machine.

machine is in automatic mode. The MANUFACTURER did not design or equip the machine with barrier guards to protect operators from dangerous moving parts.

On a particular day, the machine stopped, due to a misaligned pallet. The OPERATOR reached into the machine to straighten the pallet, believing the machine had been turned to manual mode. However, this was not the case. Once the OPERATOR straightened the pallet, the machine resumed operation, pinching his forearm in the moving parts. The OPERATOR did not see the flashing beacon light because of its location relative to the machine.

Discussion

In this case the obvious hazard is the moving machine parts. Human behavior should always be considered when performing a risk analysis for a machine. Without proper safeguards, an operator may inadvertently come into contact with moving parts due to inattention, fatigue, forgetfulness, or by deliberate risk taking (Mroszczyk, 2010). In this case, the proper safety engineering safety devices would be an interlocked barrier guard. Such a guard would have prevented access to the moving parts. The interlocking feature would stop the machine if the guard was removed and prevent a restart until it was put back in place. Interlocked barrier guards have been a standard safeguarding method dating back to the 1940s (NSC, 1946).

In addition, the beacon light was not placed in a proper location. To be effective, a visual warning such as a beacon light should be placed where the OPERATOR is most likely to be looking (Woodson, 1992). In this case, the beacon light provides a visual reminder to the OPERATOR working on the machine that the machine is in automatic mode. The beacon light should have been located where the operator would most likely see it when working on the machine. This warning device is an additional safety measure that would have alerted the operator that the machine was still in automatic mode and would resume operation when the pallet was straightened.

Case #2 Cardboard Scrap System

Description

This case study involves a cardboard scrap system. The system consists of a shredder, a shredder hood, air-conveyance ductwork, blowers, a cyclone, and a baler. Figure 21-2 shows the shredder, shredder hood, air-conveyance ductwork, and blower. Cardboard scrap is placed on a conveyor at the front of the shredder. The material is then "pinched" between two rollers and shredded by an auger near the back of the machine. The auger is a rotating cylinder with hardened steel blades. The shredded materials leave the auger and are drawn by the air system into the shredder hood. The material is then conveyed to a cyclone located on the roof before dropping into a baler.

The system would clog in one of two areas: The shredder auger would clog from overfeeding. The conveyance duct would clog in the wide-to-narrow transition behind the shredder hood. To clear a clog in the conveyance duct, it was the operator's practice to remove an access cover in the duct and insert a metal rod to free up the material. The auger could be cleared by unbolting and removing the interlocked auger cover, clearing the machine, then bolting the cover back in place. Clearing the auger this way was a cumbersome process that encouraged operators to take shortcuts. The COMPANY procedure was to turn off the shredder and lock out the controls before attempting to clear any clog.

The machine would clog fairly regularly. The operators found it much easier to clear the auger by inserting a metal rod through the conveyance-duct access opening up to the auger. On this particular day, the OPERATOR attempted to clear the auger but either forgot to turn the shredder off and

Figure 21-2. Cardboard Scrap System.

lock out the controls or thought it had already been done. The noise levels in the plant made it difficult to know which equipment was operating. The OPERATOR removed the duct access cover and inserted the metal rod as far as the auger. The rod contacted the moving auger. The end of the rod that was outside the duct recoiled and struck the OPERATOR in the eye causing serious injury.

Discussion

It should have been foreseeable to the MANUFACTURER that the auger and the conveyance duct would clog up. Clogging can be minimized by designing an exhaust system that provides adequate air velocity to convey the material. The design duct velocity to convey various materials can be found in a number of reputable sources (ACGIH, 2010). In this case, the duct velocity was inadequate. The material was not transported properly, which clogged the duct and increased the number of times the operator would have to clear the machine.

Reliance on a lockout procedure in this case was not successful in preventing an injury. Lockout is an administrative procedure that is prone to human error. It should only be relied upon when there are no design alternatives. It is foreseeable from an understanding of human behavior that the OPERATOR may forget to turn the machine off and lock out the controls. It is also foreseeable that the OPERATOR may take a shortcut and attempt to clear a jam in the auger through the duct access opening rather than unbolting the auger cover. The proper safety-engineering intervention in this case would be to interlock all access covers and to require a manual restart of the equipment by the operator after the access covers were put back in place.

Case #3 Automatic Garage-Door Openers

Description

This case study illustrates that standards and government regulations can lag behind that state of the art and how complying with a voluntary standard does not mean a product is safe. Automatic garage-door openers are devices that open and close garage doors. The openers are controlled by switches mounted on the wall inside the garage and by remote control from outside the garage. A typical garage-door opener consists of a power unit, electric motor, track, trolley, and drive. The opener is mounted from the ceiling above the door. Figure 21-3 shows a typical residential garage-door opener. The door is attached to the trolley with a connecting arm. When the control switch is activated, the motor turns and moves the trolley along the track. Limit switches shut off the power to the unit when the door reaches the fully open or fully closed position.

The hazard associated with garage-door openers is entrapment between the leading edge of the door as it closes and the garage floor. A number of child-entrapment deaths have occurred over the years. The circumstances surrounding these incidents can be categorized below (FOIA, 1991). They constitute the foreseeable use and/or foreseeable misuse of the product and should have been have been well-known to the industry.

Figure 21-3. Automatic residential garage-door opener.

- Child found trapped underneath a garage door for unknown reasons
- Child trapped underneath a garage door while retrieving a toy
- Child tripped and fell under moving garage door and became trapped
- Child pushed button to close garage door and attempted to run or ride a bicycle out of garage when struck by door
- Children played a dangerous game called "beat the door." They activated the door switch then tried to run under the door as it closed.

For many years, door closer MANUFACTURERS provided mechanical contact-reversal systems when a more reliable infrared light beam was available. The mechanical-contact systems were intended to reverse the door when there was resistance to the door closing. These systems generally consisted of a clutch between the motor and the drive mechanism, a drive mechanism motion sensor, and a controller. When the forces on the drive mechanism exceed the force transmittal capability of the clutch, the clutch slips, the drive mechanism stops or slows down, the change is detected by the motion sensor, and the controller reverses the door.

Discussion

These mechanical contact-reversal systems were not successful in preventing entrapment fatalities and injuries for a number of reasons that include:

- A child's neck is pliable. Unconsciousness can result at forces well below the force it takes to reverse the door.
- Settlement in the garage floor or a misadjusted down-limit switch can leave an entrapment gap between the edge of the door and the garage floor even after the motor has shut off.

- When a child is trapped beneath the door, the door continues to compress the child as the door moves closer and closer to the point that the down-limit switch cuts off the power.
- The mechanical clutches required regular adjustment. The owner may decrease the sensitivity in order to get the door to close. The force at which the door reverses when contacting an object, such as a child's body, is heavily dependent on the clutch adjustment.

The 1981, 1986, and 1991 editions of UL 325 did not adequately address the entrapment hazard. Entrapment injuries and fatalities continued to occur even with UL 325 rated door operators. The 1981 edition of the standard required that a door not remain in contact with a 2-inch-high obstructing object for more than 2 seconds. This was not effective, because a child's chest or neck is not a rigid object. Even if the door is adjusted properly, a child's body can still be compressed during the 2-second time when the reversal mechanism is not required to work. The 1986 and 1991 editions of the standard replaced the 2-inch, solid object with one that was resilient in the first inch, capable of being compressed to 1 inch under 100 pounds. This resilient object did not replicate a child's neck, and like the 2-inch, solid object, could not prevent entrapment in a gap created by a settlement in the floor. Reversal of the door, if it did occur, would only take place after entrapment.

A memorandum obtained through the Freedom of Information Act indicates there were 132 entrapment incidents, including 65 deaths, from March 1982 through June 1991 (FOIA, 1991). Yet was not until the early 1990s that UL 325 and the US Consumer Product Safety Commission required MANUFACTURERS to have secondary entrapment protection.

Conclusions

This case study illustrates a number of issues:

- Voluntary standards only provide minimum requirements for safety. The early UL-rated door openers did not prevent entrapment injuries and fatalities.
- A product that meets a standard when new may not perform the same way after it has been used. In this case, the clutch-reversal mechanisms would go out of adjustment, or the garage floor could settle. Either of these foreseeable events would render the door-reversal mechanism ineffective in an emergency.
- The safety content of standard and the ability to protect the public varies widely. The UL standards did not adequately address the entrapment hazard until the early 1990s. The tests prescribed in the UL standards from 1981 through 1991 did not adequately address the entrapment hazard when it was known that injuries and deaths were occurring from entrapment.
- Standards and government regulations typically lag behind the state of the art. Electric-eye beam sensors were available in 1981. These devices were fail safe and did not rely on object contact and constant adjustment to reverse the door. It was not until the early 1990s that

there was a federal regulation and updated UL standard that required these devices when they had been available for nearly a decade.

Case #4 Unguarded Floor Opening on a Bridge-Construction Project

Description

A GENERAL CONTRACTOR was hired for a bridge-construction project. The GENERAL CONTRACTOR hired a SUBCONTRACTOR for the steel work. The SUBCONTRACTOR was installing cross beams between the girders. He had personal fall protection and was tied off to a horizontal lifeline. At the same time, the GENERAL CONTRACTOR was installing tongue-and-groove planks between the girders. They did not have fall protection and were walking back and forth on the planking as they were putting the planks down. There was a call for coffee. The GENERAL CONTRACTOR went for coffee, leaving a small section at the far end unfinished and did not place a guardrail, cover, or even warning tape or cones to guard the open hole. The SUBCONTRACTOR, standing about midway where the planking was finished, looked up and saw what appeared to be a complete, safe, working/walking surface with girders along both sides (see Figure 21-4). He unhooked and walked in the same direction as the GENERAL CONTRACTOR. It was nighttime, and there were lights shining toward him. When the SUBCONTRACTOR got near the far end, he instinctively looked up toward a mound of dirt. In doing so, he did not see an open hole where the planks had not been put down. He fell through the opening and 20 feet down to the ground.

Figure 21- 4. Bridge-construction project.

Discussion

OSHA 1926.501 requires that workers on a working/walking surface be protected from falling through holes more than 6 feet above lower levels by personal fall-arrest systems, covers, or guardrail systems. In addition to OSHA regulations, the GENERAL CONTRACTOR had their own written fall protection program that required holes to be protected by a cover labeled "Hole—Do not remove." If a hole could not be covered, then guardrails or personal fall-arrest systems should be used. Construction-safety best practices emphasize *fall prevention* over *fall protection*.

It is better to use guardrails and covers than safety nets or fall arrest systems (OSHA, 2010).

In this case, the GENERAL CONTRACTOR created a hazard when they left an unprotected hole in the planking. A cover or temporary guardrail would have prevented this incident. The GENERAL CONTRACTOR did not comply with OSHA regulations, construction-safety practices, and their own fall-protection program.

Case #5 Electric-Oven Door-Hinge Design

Description

This case study illustrates the need for instructions to be clear, well thought out, and correct. Oven doors are commonly designed with spring-loaded hinges. In this case, the MANUFACTURER designed a hinge that hooked onto two rods within a pocket. The hinge was held in the pocket by a trim plate that was secured with one screw. In order to remove the door, the MANUFACTURER'S instructions stated to remove the screw from the trim plate, open the door to the full position, place a pin in the pinhole in the hinge, close the door, remove the hinge trim, then lift up on the door until the door came out of the pocket.

An appliance-repair TECHNICIAN was called to a residence to replace a thermostat in such an oven. The doors were to be removed in order to facilitate removal of the oven removal from the wall cutout so that the thermostat could be replaced. The TECHNICIAN could not remove the screw from the trim plate without opening the door to the full position. Next, the screw from the trim plate was removed (see Figure 21-5). The trim plate fell off because it was held in place solely by the screw that had just been removed. But, before the TECHNICIAN went further, the hinge suddenly and unexpectedly snapped away from the oven and amputated the TECHNICIAN's thumb.

Discussion

There is a hidden hazard associated with the removal of the door. The door hinges are spring loaded. Once the hinge plate is removed, the hinges will snap back. If this hazard could not be eliminated by engineering design or a safety device, then instructions were appropriate. However, the MANUFACTURER'S instructions were defective because they did not warn that the hinges were spring loaded and could snap back when the trim-plate screw was removed.

Figure 21-5. Electric oven door hinge design.

The instructions also stated to remove trim-plate screw before placing a pin in the hinge pinhole. Placing the pin before removing the trim-plate screw would prevent the hinge from snapping back. The MANUFACTURER did not comply with UL 858 Underwriters Laboratories Standard for Safety for Electric Ranges, which requires the manufacturer to provide clear and proper instructions for safe maintenance and use.

Case #6 Pump-Jack Scaffold

Description

This case involves a pump-jack scaffold at a residential construction site. The pump-jack scaffold consisted of two 32-foot-long wood poles, two pump jacks, wooden planks, some metal braces, and 2 ½ × 5/8-inch wood strapping. The poles were constructed from 8 foot long 2 × 4s spliced together to form a 32-foot-long, 4 × 4 pole. The 2 × 4s were attached together with deck screws and nails. There were no mending plates at the splices. The poles were attached to the house with metal brackets and braced in a makeshift fashion using the wood strapping.

The planking spanned the two jacks that were set at a height of approximately 24 feet. Two workers were on the planking. As they moved to the right, toward the right pole, the pole buckled and failed. The entire scaffold collapsed. The workers fell approximately 24 feet to the ground and were seriously injured.

The pole failed at two locations along its length. Both locations were at a splice joint (see Figure 21-6). There were also large knots in the wood at the splices.

Figure 21-6. Failed pump-jack scaffold.

Discussion

Unfortunately, these scaffold poles are used all too often in construction, particularly residential construction. A 4×4 pole made up of 2 x 4s will only be as strong as a single 2×4 at the splice location, unless a mending plate is used. Knots in the wood will further degrade the strength of the pole.

Conclusions

ANSI A10.8 and OSHA 1926.452(j) provides the requirements for pump-jack scaffold poles. Wooden poles should only be used up to 30 feet. Pump-jack scaffolds should not carry a working load that exceeds 500 pounds and be capable of supporting 4 times the maximum intended load. When wood poles are used, the pole lumber should be straight grained and free of large knots or defects. When 2×4s are spliced together, a mending plate should be installed at the splices.

Case #7 AGV System

Description

This case study involves an Automated Guided Vehicle (AGV) system. AGVs are driverless vehicles with automatic guidance systems and follow prescribed or preprogrammed paths. These vehicles are used in material handling to move, pick up, and deposit pallets or containers in factories to move materials or parts from station to station or in flexible manufacturing systems. AGV systems are also used in office settings to deliver mail.

Figure 21-7. Automated guided vehicle (AGV) system.

The early AGV's followed electromagnetic wires embedded in the floor carrying a current with a specific frequency. An antenna mounted underneath the AGV detected the frequency and followed the wire. Traffic management was accomplished by in-zone blocking. The guide path would be divided up into zones. Transmitters embedded in the floor at each zone boundary would detect a vehicle as it passed. The AGV would interrogate the transmitter. The vehicle would only be released into a zone when it received a "clear" signal. Magnets embedded in the floor were also used to stop and release an AGV. Mechanical bumpers mounted on the front of the vehicles acted like giant emergency-stop buttons, stopping a vehicle if it contacted an object or person.

More modern AGV systems are laser guided. Reflectors are installed throughout the plant. Each AGV has a laser transmitter and receiver mounted on a rotating turret. The AGV calculates its position based on the distance and angle of the reflected laser light compared to stored target coordinates. A proximity laser scanner and laser scanner interfaces replaced mechanical bumpers to provide non-contact obstacle detection.

One of the most important safety issues with AGV systems is the avoidance of vehicle-vehicle contact, vehicle-fixed-object contact, and vehicle-person contact. This particular case study involves vehicle-vehicle contact with a wire-guided system in a manufacturing plant. An AGV with trailers (see Figure 21-7) would be released from a charging station, move forward in a slightly curved path, and stop at a loading station where manufactured parts were manually loaded on the right side of the trailers. The operator would then press a button, and the AGV would continue in a counterclockwise loop, cross the incoming path from the charging station, and into a second loading station where the left side of the trailers would be manually loaded. The AGV could be jogged forward in the loading station with a switch controlled by the operator. After being released from the second loading station, the AGV would activate a transmitter in the floor, releasing another AGV from the charging station into the loop.

The USER wanted to increase production. The system DESIGNER installed a second charging station with an additional path so that two AGVs could be in the loop at the same time. While the operator at the second station was loading trailers, an AGV released from the charging station and entered the loop. One of the trailers from the entering AGV contacted the end of the last trailer of the AGV parked at the second station at the same time the operator stepped off the stand. The trailer from the parked AGV was pushed toward the stand, crushing the worker's leg.

Discussion

The DESIGNER in this case failed to ensure there was adequate clearance between the vehicles when they introduced a second vehicle into the loop. The added path was curved. Towed trailers are prone to *off-tracking*, a phenomenon where the wheels of the trailer(s) can track outside the wheels of the lead vehicle when following a curved path. This increased the likelihood that a trailer on an AGV entering the loop could contact the end of a trailer parked at the second loading station. The use of a "jog" control instead of an "index" control allowed variability in the placement of the trailers at the second loading station. The transmitter in the floor should have been located further downstream so that an AGV would not release from the charging station until the AGV had cleared the second station.

Case #8 Tile Floor

Description

This case study involves a ground-floor entryway to a condominium complex. The floor was covered with smooth tiles (see Figure 21-8). The slip resistance of the tiles was measured using a variable incidence tribometer (VIT) English XL after the tile was installed. The testing was done in accordance with the test-instrument manufacturer's instructions. The instrument was checked using a calibration tile. The slip resistance on the entryway floor tile was measured as .75 dry and .20 wet.

Discussion

The building owner or property manager should anticipate that an entryway is likely to be wet during inclement weather. Historically, a slip resistance of .50 or higher is considered safe for pedestrians walking in a normal manner on a level surfaces. The .20 wet measurement is well below the .50 minimum. A recent study relating the slip resistance as measured with a VIT to the probability of a slip indicates a greater than 50 percent probability of a slip when the slip resistance is .20 (Burnfield, 2006).

Figure 21-8. Tile floor.

This tile was a poor choice for an entryway. The building owner or property manager should have anticipated that this area is likely to be wet and chosen a floor covering that has a slip resistance of .50 or higher under wet conditions.

Case #9 Catwalk at an Incinerator Plant

Description

This case study involves a catwalk at an incinerator plant. The plant processes solid waste, converting it into electrical energy. Employees were required to enter the incinerator for maintenance. One of the incinerators was located on the 4th level of the plant. The only access to the incinerator was by stepping from a catwalk over an I-beam and across a wooden plank that was placed between the catwalk and the incinerator. The plank spanned a 4-foot distance between the catwalk and the incinerator, open on both sides. There was a four-story drop from the plank, and the plank was not secured.

An EMPLOYEE stepped from the catwalk and over the I-beam. One foot hit the plank, and the other missed. He fell off the plank, four stories down to the lower level.

Discussion

OSHA General Industry regulations and other safety practices are very clear that open-sided platforms 4 feet or more above the floor be guarded with a railing,

Figure 21-9. Catwalk at an Incinerator Plant.

intermediate railing, and toeboards. Platforms should be secured to permanent structures. The failure of the EMPLOYER to comply with OSHA regulations created a hazardous and unreasonably dangerous situation. A secured catwalk with guardrails and toeboards was feasible and relatively inexpensive. Figure 21-9 shows the catwalk and the incinerator after the incident, when a platform with railings was installed between the catwalk and the incinerator. The photo shows the catwalk (metal grating) on the left, the I-beam, and the incinerator door on the right. The platform with yellow railings between the catwalk and the incinerator was installed after the incident.

Case #10 Sidewalk Defect

Description

This case involves a sidewalk defect. Figure 21-10 shows a change in elevation between two sections of a concrete sidewalk. One of the sections has settled, creating a 7/8 inch change in elevation. Studies have indicated gait dynamic differences for step-up heights of ¼ inches or more (ASTM, 1990).

Discussion

Safety standards require walking surfaces to be flat or planar (ASTM, 2009). When this is not possible, surface defects up to ¼ of an inch are allowed.

The 7/8 inch change in elevation shown in Figure 21-10 exceeds safety guidelines. The PEDESTRIAN in this case was walking along the sidewalk, tripped, and fell at the defect. The hazard could have been corrected several ways. One way is to break up and remove the concrete and pour another

Figure 21-10. Sidewalk tripping hazard.

section. Another method is to trowel concrete to transition the low to high area, or to grind down the higher area. Highlighting paint along the raised edge can be used in the interim until a permanent repair or reconstruction is done. Highlighting paint was added after the incident.

REFERENCES

American Conference of Governmenal Industrial Hygienists (ACGIH), 2010. *Industrial Ventilation*, 27th Editon. Cincinatti, OH: ACGIH.

American National Standards Institute (ANSI), 2006. "ANSI/ASSE A1264.2 Provision of Slip Resistance on Walking/Working Surfaces." Des Plaines, IL: American Society of Safety Engineers.

American National Standards Institute (ANSI), 2001. "ANSI/ASSE A10.8 Safety Requirements for Scaffolding." Des Plaines, IL: American Society of Safety Engineers.

American National Standards Institute (ANSI), 2007. "ANSI/ASSE A1264.1 Safety Requirements for Workplace Walking/Working Surfaces and Their Access." Des Plaines, IL: American Society of Safety Engineers.

American Society for Testing and Materials (ASTM), 2009. "ASTM F1637 Standard Practice for Safe Walking Surfaces." West Conshohocken, PA: ASTM International.

American Society for Testing and Materials (ASTM), 1990. "STP 1103 Slips, Stumbles, and Falls." Philadelphia, PA: ASTM.

Associated General Contractors (AGC), 2009. *Manual of Accident Prevention in Construction*, 9th Edition. Arlington, VA: Associated General Contractors.

Burnfield, J., and C. Powers, 2006. "Prediction of Slips: An Evaluation of Utilized Coefficient of Friction and Available Slip Resistance" *Ergonomics*. Vol. 49, No. 10.

Consumer Product Safety Commission, 1985. "News from CPSC: Deadly Game: Kids Try to Beat Closing Garage Door." Bethesda, MD: CPSC.

Code of Federal Regulations Title 16 Commercial Practices Part 1211 (CPSC), 1992. *Safety Standard for Automatic Residential Garage Door Operators.* Washington, GPO.

Grob, D., 1981. "UL Increase Safety Requirements on Residential Garage Door Openers." *Lab Data,* Winter/Spring.

Hammer, Willie, 1972. *Handbook of System and Product Safety.* Englewood Cliffs, NJ: Prentice-Hall.

Industrial Truck Standards Development Foundation (ITSDF), 2005. "ANSI/ITSDF B56.5 Safety Standard for Guided Industrial Vehicles and Automated Functions of Manned Industrial Vehicles." Washington ,D.C.: ITSDF.

Kolb, J. and S. Ross, 1980. *Product Safety and Liability.* New York: McGraw-Hill.

Miller, Richard, 1987. *Automated Guided Vehicles and Automated Manufacturing.* Dearborn, Michgan: Society of Manufacturing Engineers.

Mroszczyk, J., 2010. "Prevention of Fall Fatalities & Injuries," *Blueprints.* American Society of Safety Engineering Construction Practice Specialty, volume 10, Number 2.

Mroszczyk, J., 1994. "The Use of Standards in the Design Process," ASSE Engineering Division Newsletter, Fall.

Mroszczyk, J., 2010. "Take Your Pick: Error Proof Humans or Your Organization's Environment," *Proceedings of Rethink Safety: A New View of Human Error and Workplace Safety* Symposium (Nov 4-5) San Antonio, TX. Des Plaines, IL: American Society of Safety Engineers.

Mroszczyk, J., 2003. "Safety Practices for Automated Guided Vehicles (AGV's)" ASSE Council on Practices and Standards Technology Committee. Des Plaines, IL: American Society of Safety Engineers.

National Safety Council (NSC), 1946. *Accident Prevention Manual for Industrial Operations,* Chicago, IL: National Safety Council.

National Safety Council (NSC), 1993. *Safeguarding Concepts Illustrated,* 6th Edition. Chicago, IL: National Safety Council.

OSHA Construction eTool: Unprotected Sides, Wall Openings, and Floor Holes http://www.osha.goc/SLTC/etools/construction/falls/unprotected.html downloaded February 22, 2010.

Underwriter's Laboratories (UL). "UL325 Door, Drapery, Gate, Louver, and Window Operators and Systems." Northbrook, IL.; Underwriter's Laboratories, 1975, 1981, 1986, 1991, 2002.

Underwriter's Laboratories (UL), 2005. "UL 858 Underwriters Laboratories Standard for Safety for Electric Ranges." Northbrook, IL.; Underwriter's Laboratories.

US Occupational Safety and Health Administration (OSHA), 1980. *OSHA 3067, Concepts and Techniques of Machine Safeguarding.* Washington, D.C.: U.S. Department of Labor Occupational Safety and Health Administration.

US Consumer Product Safety Commission Memorandum dated August 6, 1991 obtained through Freedom of Information Act (FOIA).

Woodson, W., B. Tillman, and P. Tillman, 1992. *Human Factors Design Handbook.* New York: McGraw-Hill.

APPENDICES

Appendix A

Temperature:

Units: K, C – temperature

Metric:

degrees Kelvin, °K = degrees C + 273.15 (273)

English:

degrees Celsius, °C = 5/9 (degrees F – 32)

degrees Rankine, °R = degrees F + 459.69 (460)

degrees Fahrenheit, °F = 32 + (9/5) degrees C

Some common values used in safety:

degrees F	degrees C	degrees K
20	–6.7	266
32	0	273
40	4	277
68	20	293
70	21	294
77	25	298
100	38	311
135	57	330
150	66	339
165	74	347
175	79	352
190	88	361
200	93	366
212	100	373
286	141	414
300	149	422
500	260	533
600	315	588
1000	538	811
1100	593	866
1500	816	1089
2000	1093	1366
2500	1371	1644
3000	1649	1922

Appendix B

Most common English-metric conversions

1. Length:
$$m = .3048 \times ft.$$
$$cm = 2.54 \times in.$$
$$mm = 25.4 \times in.$$

2. Area:
$$m\ squared = .0929 \times sq.\ ft.$$
$$cm\ squared = 6.452 \times sq.\ in.$$
$$mm\ squared = 645.2 \times sq.\ in.$$

3. Volume:
$$m\ cubed = 3.785 \times 10^{-3} \times gal.$$
$$= .02832 \times cu.\ ft.$$
$$= 1.639 \times 10^{-5} \times cu.\ in.$$

$$cm\ cubed = 16.387 \times cu.\ in.$$

$$L = 3.785 \times gal.$$
$$= 28.32 \times cu.\ ft.$$
$$= 1.639 \times 10^{-2} \times cu.\ in.$$

4. Mass and weight:
$$kg = .4536 \times lbs.$$

5. Density:
$$kg/m\ cubed = 16.02 \times lb/cu.\ ft.$$

6. Weight-force:
$$N = 4.448 \times lbf.$$
$$kgf = .4536 \times lbf.$$

7. Pressure:
$$Pa = 2.491 \times 10^{-4} \times in.\ H_2O$$
$$= 6894.76 \times psi$$
$$= 10^5 \times bar$$

$$kPa = .0479 \times lbf/feet\ squared$$

$$kgf/cm\ squared = .0703 \times psi$$

8. Temperature:
$$K = degrees\ C + 273.15$$

9. Power:
$$W = .2931 \times Btu/hr.$$
$$= 745.7 \times hp$$

$$kw = 1.758 \times 10^{-2} \times Btu/min.$$
$$= 1.055 \times Btu/sec.$$

$$metric\ hp = .001843 \times ft.\text{-}lbf/sec.$$
$$= 1.0139 \times hp$$

10. Energy:

$$J = 1.356 \times \text{ft-lbf.}$$
$$= 1055.1 \times \text{Btu}$$

$$\text{kgf-m} = .1383 \times \text{ft-lbs.}$$

11. Velocity

$$\text{m/sec.} = 5.1 \times 10^{-3} \times \text{fpm}$$
$$= .3048 \times \text{fps}$$

$$\text{cm/sec.} = .51 \times \text{fpm}$$

12. Flow rates:
    (water)

$$\text{m cubed/sec.} = 6.31 \times 10^{-5} \times \text{gpm}$$
$$\text{m cubed/min.} = 3.785 \times 10^{-3} \times \text{gpm}$$
$$\text{L/sec.} = .0631 \times \text{gpm}$$
$$\text{L/min.} = 3.785 \times \text{gpm}$$

   (air)

$$\text{m cubed/sec.} = 4.7 \times 10^{-4} \times \text{cfm}$$
$$\text{m cubed/min.} = .0283 \times \text{cfm}$$
$$\text{L/sec.} = .47 \times \text{cfm}$$

13. Other:

$$\text{seconds} = \text{mins.} \times 60$$
$$1/\text{seconds} = 1/\text{mins.} \times 1.667 \times 10^{-2}$$
$$\text{minutes} = \text{hrs.} \times 60$$
$$1/\text{minutes} = 1/\text{hr.} \times 1.667 \times 10^{-2}$$
$$\text{candela} = \text{candles} = \text{lumens/steradians}$$
$$\text{radians} = \text{degrees} \times 1.745 \times 10^{-2} = \text{minutes} \times 2.909 \times 10^{-4}$$
$$= \text{RPM} \times 6.283$$
$$\text{coulombs} = \text{ampere-hrs.} \times 3600 = \text{faradays} \times 9.649 \times 10^{4}$$
$$= 1 \text{ ampere-second}$$
$$1 \text{ ft-c} = 10.76 \text{ lux} = 1 \text{ lumen/sq. ft.}$$
$$= 10.76 \text{ lumen/m squared}$$
$$1 \text{ lux} = .093 \text{ ft-c}$$
$$5 \text{ ft-c} = 54 \text{ lux}$$
$$1 \text{ footlambert} = 3.426 \text{ cd/m squared} = 3.1416 \times \text{cd/sq. ft.}$$
$$= .2919 \times \text{cd/m squared}$$
$$\text{RPM} = \text{degrees/sec.} \times .1667 = 6 \text{ degrees/sec.}$$
$$= 01667 \text{ rps}$$
$$\text{degrees} = \text{radians} \times 57.3, \text{ minutes} = \text{radians} \times 3437.7$$
$$\text{seconds} = 2.06 \times 10^{5} \times \text{radians}, \text{ Rev.} = .159 \times \text{radian}$$

Appendix C

Properties of common materials

Material	Yield Strength kN/m²	Tensile Strength kN/m²	Density kg/cm³	Other Characteristics of Interest
Aluminum				
2014	97	186	2.8	nonmagnetic
6061 T6	276	310	2.6	nonmagnetic, brittle
Magnesium	221	290	1.9	nonmagnetic
Copper	276	345	8.9	nonmagnetic
Yellow brass	414	510	8.7	nonmagnetic
Alum. bronze	276	565	7.7	nonmagnetic
Gray cast iron	207	207	7.1	odd shapes, brittle
Carbon steel				
1010 HR	179	324	7.8	(hot rolled)
1020 CR	352	421	7.8	
1040 CR	485	586	7.8	
Stainless steel				
301 annealed	276	758	7.8	nonmagnetic
304 annealed	241	586	7.8	nonmagnetic
410 annealed	270	483	7.8	magnetic
Wood				(strength perpendicular to grain)
pine		2.1	0.35	
birch		6.3	0.55	
oak		5.7	0.64	
Plastic				
polyethylene		28	0.95	max. temp. 95°C, translucent
polycarbonate		62	1.2	max. temp. 130°C, transparent
acrylic		55	1.19	max. temp. 95°C, transparent
ABS		21–48	1.06	max. temp, 130°C
PVC		41	1.4	max. temp. 140°C
phenolic		52	1.4	max. temp. 150°C

Appendix D

ELECTRICAL SAFETY CHECKLIST

* **Plugs and Connector -**
    - Check for stray strands and loose terminals.
    - Are grounding plugs used where required by code?
* **Extension Cords -**
    - Are they temporary?
    - Are they too long? proper type?
    - Check for worn or frayed spots, splices, etc.
    - Are grounded cords used where required by code?
* **Office Equipment and Appliances -**
    - Are grills, toasters, and heaters away from combustibles?
    - Are cords, plugs, and connectors in good condition?
    - Are there too many extension cords and multiple plugs?
* **Receptacle Outlets -**
    - Check for use of grounding-type receptacles.
    - Check for continuity of the grounding connection.
    - Check for missing faceplates and any signs of arcing, excess heating,
      or loose mountings.
    - Is there continuity between the frame and grounding?
* **Lighting Fixtures -**
    - Are all fixtures labeled and grounded?
    - Are any fixtures too close to combustibles?
    - Check for burned out bulbs or tubes.
    - Are any fixtures covered with dust, dirt, etc.?
* **Equipment Grounding -**
    - Check machinery grounds to be in good order.
* **Electrical Service -**
    - Are weatherheads in good condition?
    - Are lightning arresters, surge capacitors, grounding conductors,
      and grounds in good condition?
    - Are switches safely located and accessible?
* **Switchrooms and Motor Control Centers -**
    - Are they clean, with no storage?
    - Is ventilation adequate?
    - Anything unusual—odors, noises?
    - Is voltage normal?
    - Check battery circuit breaker trip system.
    - Are all switches properly identified?
    - Are fire extinguishers suitable, charged, and in place?
* **Wall-mounted Electrical Equipment -**
    - Is it protected from physical damage?
    - Check for missing or open covers?
    - Are any live parts exposed?
* **Electrical Equipment for Hazardous Locations -**
    - Is all equipment enclosed and proper?
    - Check for excessive dust on motors.
* **Emergency Equipment -**
    - Check/test all exit and emergency lights to be in good working order.
    - Check fuel levels for emergency generators.
    - Check batteries for proper electrolyte and charge.

Appendix E

AMERICAN NATIONAL STANDARDS INSTITUTE
ANSI B-11 Series of Standards on Machines

ANSI B11.1-2009	Safety Requirements for the Construction, Care, and Use of Mechanical Power Presses
ANSI B11.2-1995 (R2005)	Safety Requirements for the Construction, Care, and Use of Hydraulic Power Presses
ANSI B11.3-2002 (R2007)	Safety Requirements for the Construction, Care, and Use of Mechanical Power Press Brakes
ANSI B11.4-2003 (R2008)	Safety Requirements for the Construction, Care, and Use of Shears
ANSI B11.5-1988 (R2008)	Safety Requirements for the Construction, Care, and Use of Ironworkers
ANSI B11.6-2001 (R2007)	Safety Requirements for the Construction, Care, and Use of Lathes
ANSI B11.7-1995 (R2005)	Safety Requirements for the Construction, Care, and Use of Cold \ Headers and Cold Formers
ANSI B11.8-2001 (R2007)	Safety Requirements for the Construction, Care, and Use of Drilling, Milling, and Boring Machines
ANSI B11.9-2010	Safety Requirements for the Construction, Care, and Use of Grinding Machines
ANSI B11.10-2003 (R2009)	Safety Requirements for the Construction, Care, and Use of Metal Sawing Machines
ANSI B11.11-2001 (R2007)	Safety Requirements for Gear and Spline Cutting Machines
ANSI B11.12-2005 (R2010)	Safety Requirements for Roll-Forming and Roll-Bending Machines
ANSI B11.13-1992 (R2007)	Safety Requirements for the Construction, Care, and Use of Single-and Multiple-Spindle Automatic Screw/Bar and Chucking Machines
ANSI B11.15-2001 (R2007)	Safety Requirements for Pipe, Tube, and Shape Bending Machines
ANSI B11.16-2003 (R2009)	Safety Requirements for Powder/Metal Compacting Presses
ANSI B11.17-2004 (R2009)	Safety Requirements for Horizontal Hydraulic Extrusion Presses
ANSI B11.18-2006	Safety Requirements for Construction, Care, and Use of Machines and Machinery Systems for Processing Strip, Sheet, or Plate From Coiled Configuration
ANSI B11.19-2010	Performance Criteria for Safeguarding
ANSI B11.20-2004 (R2009)	Safety Requirements for Integrated Manufacturing Systems
ANSI B11.21-2006	Safety Requirements for Design, Construction, Care, and Use of Lasers for Processing Materials

ANSI B11.22-2002 (R2007)   Safety Requirements for Turning Centers and Automatic, Numerically Controlled Turning Machines

ANSI B11.23-2002 (R2007)   Safety Requirements for Machining Centers and Automatic, Numerically Controlled Milling, Drilling and Boring Machines

ANSI B11.24-2002 (R2007)   Safety Requirements for Transfer Machines

Index

A

A-weighted decibel (dBA), 94
  NIOSH recommended noise exposure limit, 96
  OSHA noise exposure limit and action level, 96
Accessibility Code A117.1, 165
Accident(s). *See also:* Incidence rates
  costof, 4-5
  defined, 3-4
  federal reporting requirements, 29
  involving workers and machines, 67-68
  stress and, 73-74
  three-phase injury sequence (Haddon), 296
Acoustical materials, 231-235, 240
ADAAG/ANSI A 117.1, 10, 161
AFL-CIO Workers' Institute for Safety and Health, 8
Agricultural machinery, and whole-body vibration, 62
Air cleaning devices, 123-24
  comparison of, 126-27
Air contaminants, 272-73
  OSHA standard 1910.1000, 273
Air handling units (AHUs)
  chemical filtration in, 125
Air sampling, 275-77
Airborne contaminant(s)
  EPA Clean Air Act, 278
  exposure limits, 115
  collectors and filters for, 122-23
  control of, 115-24
  OSHA standard 1910.1000, 115-16, 274
  particulates, hazardous, 91
Aisles
  dead-end, 169, 171
  mirrors at intersection of, 172
ALARA (as low as reasonably achievable), 99, 262
Alarms, personal, 107
Alveoli, 90-91
Ambulation, 132-33
American Chemical Society, 3
  recommendations on contact lens use, 87
American Conference of Governmental Industrial Hygienists (ACGIH)
  fume hood airflow rates (face velocities), 117
  heat exposure limits, 113
  "Threshold Limit Values for Chemical Substances and Physical Agents", 273
  TLVs for airborne contaminants, 115
American Society of Agricultural Engineers (ASAE), 14

American Society of Heating, Refrigerating, and Air-Conditioning Engineers (ASHRAE)
  ASHRAE 110 standard for testing and management practices for safe laboratory hood systems, 117
  ventilation standards, 113-14, 117
American Society of Mechanical Engineers (ASME)
  Pressure Vessel Code, 298
  "Rules for the Construction of Stationary Boilers and for Allowable Working Pressures", 14
American Welding Society or the American Boiler Manufacturer's Association, 14
Americans with Disabilities Act (ADA), 10, 161
  fire alarm systems, 198
American National Standards Institute (ANSI), 1-15
  A9.1, 161
  A10.8, 434
  A117.1-1992, "Accessible and Usable Buildings and Facilities", 135, 145
  A1264.1-2007, "Safety Requirements for Workplace Walking/Working Surfaces and Their Access", 143
  B11.0–2000, "Safety of Machinery—General Requirements and Risk Assessment", 359, 361-62, 368, 409
  B11.TR1, "Ergonomic Considerations for Machine Tools", 367
  B11.TR6, 386
  C.1, 161
  F1506 arc flash protective clothing, 314
  RIA R15.06 – 1999 (R2009), Industrial Robots and Robot Systems – Safety Requirements, 296
  S1.4(1983), specifications for sound-level meters, 223
  S12.6-1997 (ANSI 2008), 97
  Z9.5, 117
  Z10 standard, 6
  Z136.1, *Safe Use of Lasers*, 266
  Z358.1-2009, standard for emergency eyewash stations, 105
  Z359, Fall Protection Standard, 19, 21
  ASME 20.9-2009, "Safety Standards for Conveyors and Related Equipment", 392
  ASME B30.10-2009, 393
  ISEA Z89.1-2009, standard on hard hats, 89
  ITSDF B56.1-2010, requirements for powered industrial trucks, 395-96
  National Electrical Safety Code (ANSI C2), 307
    color codes, 66

Amperes, 303
   Anthropometric data, 70, 330-33
   hand/arm strength, 333
Arc flash, 315-16
   hazard analysis, 316
   PPE for, 100-102
   protection boundary, 316
Asbestos, 91, 279
ASTM International
   D-2047-74 "Test Method for Static
      Coefficient of Friction of Polish-Coated
      Surfaces as Measured by the James
      Machine", 140
   E-415 for determining the sound
      transmission class (STC) of materials,
      234-35
   F-13 "Pedestrian/Walkway Safety and
      Footwear", 138
   F-2412-2005 "Standard Test Methods for
      Foot Protection", 90
   F-2413-2005 "Standard Specification
      for Performance Requirements for
      Protective Footwear", 90
   F-2508 "Standard Practice for Validation
      and Calibration of Walkway Tribometers
      Using Reference Surfaces" (ASTM
      2011), 141
Attitude(s)
   and capacity to do work, 73
   towards safety, 6
Auditory
   capabilities, 48-50. See also: Hearing
   signals, 49
Automated Guided Vehicles (AGV)
   case study involving, 434-36

B
Back
   injuries, NIOSH, 57-59
   pain, from using computers, 76
Back belts, NIOSH Publications 81-122 and
   94-122, 57
Baghouse filters, 123
Barriers
   in buildings to prevent movement of
      smoke and gases, 166
Bayer, Dr. Charlene, 114
Board of Certified Safety Professionals
   (BCSP), 11
Boiler codes, American Society of Mechanical
   Engineers, 14
Boolean algebra, 3
British Petroleum (BP), 10
Brungraber, Dr. Robert, 140
Building codes
   Accessibility Code A117.1, 165
   IBC, 162
Building construction
   concrete strength (compressive) at elevated
      temperatures, 190

low-carbon steel, 189
Building design, 164-67, 189-95. See also: Life
   Safety Code (LSC)
   electrical systems, 305, 307-308, 312
   fire-resistive and noncombustible
      construction, 181, 189, 192
      occupant load, 171
   paths of egress, 165
   performance-based, 162, 188
   stairwells, 195
   structural steel yield strengths at elevated
      temperatures, 189
   for use of explosive materials, 250-55
   velocity of sound through materials, 219
   whole (WBD), 188
Building materials
   characteristics at elevated temperatures,
      189-92
   noise absorption of common, 232-235
Bump caps, 90
Bureau of Labor Statistics (BLS), 5, 29, 76
   analysis of injuries and illness cases
      involving days away from work, 2010, 54
   Census of Fatal Occupational Injuries
      (CFOI), 5
   eye injuries studied, 85
   report on elderly workers, 10

C
Carbon monoxide, 125, 172
   detectors, 198
Carcinogens, OSHA classification (29CFR
   1990), 22
Carpal tunnel syndrome, 54-55, 59
Catch points, 291, 293
CE mark, 15
Certification of products, 23-25
Chemical(s)
   absorption of hazardous, through skin, 88
   burns to eyes, 85, 86
   cartridges for respirators, 92
   hazardous, 111
   industrial explosions, 10
   used in finishing wood and metal
      products, 412
Chemical protective clothing and gloves,
   88-89
   degradation of, following exposure to
      hazardous chemicals, 88
Coal Mine Health and Safety Act, 16
Cochlea, 94, 219
Color codes (ANSI), 66
Common law, 1-2
Communication between employees and
   management, 11
Compressed air, OSHA Standard 1910.242(b),
   299
Compressed Gas Association, 93
Computer(s)
   ergonomic issues in using, 76

lighting for, 76
in the workplace, 76-80
Concrete, effect of temperature on strength of, 190
Confined Space Standard, OSHA, 18
Confined spaces, 175-76
Consumer Product Safety Commission requirement for garage doors, 430
Containers, pressurized, 297-300
Controls, 50-51. *See also:* Design of controls
U.S. Air Force, experiments with shape codes for, 50
Conveyors, 392-93
nip points on, 392
Corporate social responsibility (CSR), 10
Crane(s), 393-95
contact with power lines, 320-21
hooks and other lifting devices, 393-94
Cryogenic liquids, 99
CSP designation, 4
Culture, EH&S, 6-7, 10, 411
Cumulative trauma disorders, 59
Cyclone separator, 123

D

Decibel
A-weighted(dBA), 94,96
of intensity, 221-22
Defend-in-place philosophy, 160
Department of Defense (U.S. DOD)
antiterrorism codes such as UFC 4-010-01 and UFC 4-021-01, 254
building antiterrorism codes, 15
UFC 4-021001 (2008), 168
Department of Transportation (U.S. DOT)
classes of explosive materials and placards, 249-250
classes of hazardous materials, 282-83
HM-181 requirements for shipping, 284
Dermatitis, 29, 88, 272
Design, 70. *See also:* Anthropometric data; Building design; Lifting; Safeguards, *and* Worker-machine systems
of air ducts, 113-14
analysis, life safety, 161
building, 164-67
computer workstation, 77-78
controls, 340-43, 352-55
electric circuit, 307-14
eliminating hazards through, 131-32
emergency buttons, 346-47
establishing hazard levels, 362-63
foot switches, 347-48
guards, 369-80
hand tools, 333
handwheels, cranks and levers, 351-52
mat actuators, 348-49
military equipment, MIL-STD-1472, 330
power tools, 339-40
prevention through, 415

push buttons, 341-44
responsibility overview, 360-62
rotary switches, 351
safety interlock switches, 349
safety pull switches, 350-51
stairs, 146-49
stimulus-operator-response (SOR) model, 64
toggle switches, 349-50
two-hand trips and controls, 344-45
of work methods, 52-53, 57
worker satisfaction and job, 69
workstation and tool, 60
DIN 24980, 15
Door(s)
fire, 192
openings, 165-66
in paths of egress, 165-66
and ramps, 156
smoke, 192
stairwell, 195
Dust explosions, 255-57
Class II, Group G atmospheres, 257
grain-handling facilities, NFPA 61, 256
NFPA 68, Guide for Venting of Deflagrations, 256
sugar, 10, 255

E

Ear
cochlea, 94
cross section of, 221
Earmuffs, 96-97
Earplugs, 96-97
flat frequency response, 97
NIOSH hearing protector noise attenuation, 97
Egress, 150. *See also:* Building Codes; Building Design; Door(s), *and* Exit(s)
analysis, *SFPE Handbook*, 168
components, specified in Life Safety Code, 171
door locks, in paths of, 166
GSA study of flow rates on stairs, 168
means of, 160
planning, 167-75
time, calculation of, 162
Electrical
arcing, 314-316
circuits, 303-304, 307-14, 316-18, 322
codes and standards, 307-14
conductivity of common conducting and insulating materials, 304-305, 308-309
current, resistance of human body to, 306
failures, 314
fuses and circuit breakers, 311-12
grounding, 309-10, 324
hazards, PPE use for, 100-102
insulation, 305, 313
preventive maintenance, 312

products, testing laboratories, 308, 313
shock and electrocution, 311
switches, 319
systems, separated or isolated, 311-12
units, 303-304
Electrocutions
annual U.S., 311
during industrial maintenance, 322
Emergency
exits, NFPA symbol 170, 164
human behavior in an, 162-64
Planning and Community Right-to-Know
Act of 1986, 285
Environmental Protection Agency (EPA)
earplug noise-reduction ratings, 97
Standard 40 CFR, Part 68, risk management
plan, 278
Ergonomic(s). *See also:* Work area
ANSI B11.TR1, "Ergonomic Considerations
for Machine Tools", 367
arm motion, 51
defined, 43
and human motor capabilities, 50-53
OSHA proposed standard for, 18
use of computers, 76
Errors, types of, in worker-machine system, 64
European Community Standards and
Directives, 15
European Norm standards and directives, 15
Exit(s). *See also:* Egress
access to, 159, 169
discharge, 159
doors, 159, 162-63
ISO 7010 Safety Sign and Symbol, 163-64
NFPA 170 symbols for, 163
planning, 172-75
Explosion(s). *See also:* Explosive material(s)
characteristics of, 247-250
chemical, 10
defined, 247
dust, 255-57
NFPA Standard on Explosion Prevention
Systems, NFPA 69, 247
proof enclosures, 307
Explosive material(s), 248
DOT classifications and placards, 249-250
dynamite, 254
NFPA reactivity rating of, 248
Tovex, 254
Extension cords, 321
Eye(s)
and face protection, 85-87
injuries, 85
light exposure to, 86
washes and emergency showers, 105

F

Face shields, 86
Factory Mutual, 25

Failure Mode and Effects Analysis (FMEA),
35-39, 418
Fall arrest systems. *See:* Personal fall arrest
systems
Fall protection
barriers, 144
requirements, OSHA, 104
Fall(s), 55
through holes, 431-32
injuries from, 131
open sided platforms, 437-38
sequence of, 133
from slips and trips, 134-45
Family Safety and Health magazine, 9
Fans and blowers, 114, 122
Fatalities, 22
by industry, 21
fromunintentional injuries, 131
Fatigue, mental and physical, 73
Fault tree analysis, 38, 418
Federal Bureau of Mines, 16
Federal Metal and Nonmetal Mine Safety
Act, 16
Federal Safety Appliance Act, 16
Field of vision, 46
Filters, particulate, 91
Fire(s). *See also:* Building design
barriers, 165
and building design, 181
chemistry, 182-84
classes, 199
detection and suppression, 198-201, 207
doors, 192
emergency voice/alarm communication
system (ECS), 168
emergencies, 162-63
escapes, 167
extinguishers, labels, 201
extinguishing agents, 199-201
flame growth, 184-86
fuel-load analysis, 187-89
model, Japanese Building Research
Institute, 211
safety practices, 195-98
sprinkler systems, 201-208
venting of heat and smoke, 208-211
walls, 192, 257
Flammable and combustible liquids, 183, 196
labeling, 196
maximum allowable size of containers and
portable tanks for, 194
purging containers, pipes and hoses, 252
storage, OSHA 1910.106 and NFPA 30,
194, 196-97, 279
type II safety can for, 193, 196
Floor
holes and openings, 143-44, 431-32
slipping hazard case study, 436-37
tripping hazards, 134-36

Flying objects, 295
    force of, 300-301
Footwear, protective, 90
Fork lift. *See:* Industrial lift trucks
Formaldehyde, 125
Frequency and severity rates, 28-29
    probability of incidence versus severity, 33
Fuels. *See:* Flammable and combustible liquids

G

Gates. *See:* Guards, automatic (gates)
General Duty Clause, 18
General Services Administration (GSA)
    study of movement rates on stairs, 169
Georgia Tech Research Institute, ventilation
    study, 114
Glare. *See:* Lighting
Glasses
    ANSI/ISEA Z87.1-2010 standard for
        impact-resistant safety glasses, 86
    side shields for, 86
Globally Harmonized System of Classification
    and Labeling Chemicals (GHS)and the
    Hazard Communication Standard, 21,
    278-79
    OSHA categories, 280-80
Gloves
    chemical protective, 88
    nitrile, 88
    and protective clothing, 88-89
Good Housekeeping Seal of Approval, 25
Grasp(s), 52
    avoiding simultaneous, in work design, 52
    reach and, 60
Gratt, Lawrence B., 3
Ground fault circuit interrupter (GFCI), 311-13
Guards, 369-380. *See also:* Safeguards
    adjustable, 370-73
    automatic (gates), 376
    for conveyors, 392-93
    fixed, 369
    hinged, 371, 410
    interlocked, 373-75
    presence-sensing devices, 381-385
    pullbacks and holdouts (restrains), 377-79
    safety mats, 386-87
    shields and awareness barriers, 380-81
    sweep and pushout devices, 379-80
    trial-ring, 377

H

Haddon, William, Jr., three-phase injury
    sequence, 296
Hand(s), 51
    deviation from the cupped position, 51
    and information from fingers, 50
    tools, design, 333-39
    using left and right simultaneously, 53
    Hard hats, classes of, 89-90

Hartford Steam Boiler Inspection and
    Insurance Company, 13-14
Hazard(s). *See also:* Electrical hazards *and*
    Mechanical hazards
    arc flash and electrical, PPE, 100-102
    of attire and personal effects, 104-105
    chemical, 88
    cryogenic liquids, 99
    designing out, 131-32
    elimination, 366
    event, determining probability, 363-64
    falling or flying objects, 85-86
    grinders, 408
    hand tool, 333-39
    identification, 416-18
    ionizing radiation PPE, 99-100
    isolating, 366-67
    levels, establishing, 362-365
    nanoparticle, 98
    noise, 94-98
    particulate, 90-94
    power tools, 339-340
    radiant light, 86-88
    rating/severity of harm, 363
    safeguarding, 359, 365
    tripping, 134-36
Hazard analysis
    arc flash, 316
    failure mode and effect analysis, 35-38
    requirement for use of PPE (OSHA), 84
    techniques, 33-39, 418
Hazard Communication Standard (GHS), 21,
    196, 278. *See also:* Emergency Planning and
    Community Right-to-Know Act of 1986
    labeling requirements, 283
    Spanish language, 23
Hazard control. *See also:* Safeguarding of
    machinery
    EPA Standard 40 CFR, Part 68 risk
        management plan for, 278
    and PPE, 83
    priorities, 39
Hazardous locations, electrical power or
    equipment in, 307
Hazardous materials. *See also:* Nonhazardous
    materials
    collection, 284
    DOT classes of, 282-83
    Emergency Planning and Community
        Right-to-Know Act, 278
    LC_{50}, 272
    LD_{50}, 272
    NFPA standards, 271
    Resource Conservation and Recovery Act
        (RCRA) requirements, 278
Hazardous waste, 284-85
    DOT's HM-181 requirements for
        shipping, 284
Head protection, 89-90

Hearing, 48. *See also:* Auditory capabilities *and* Sound(s)
    human frequency range, 48, 95
    loss, noise induced (HIHL), 95-96, 220
    phenomenon of, 219-22
    presbycusis, 220
    problems and the aging workforce, 221
    protection, 94-98
Hearing protection, 94-98
    ANSI standard S12.6-1997 (ANSI 2008), 97
    NIOSH hearing protector noise attenuation, 97
Heat detectors, 198
Heat stressand cold stress equation, 113
Heating and cooling, circulating fans, 113
HEPA filters, 125
Hoods. *See:* Laboratory fume hoods
Housekeeping, 11
Human behavior in emergencies, 162-64
HVAC systems and Sick Building Syndrome, 125

I

IAQ. *See:* Indoor air quality
IBM, commitment to safety, 7
Illness, 4
Illuminating Engineers Society (IES), 47, 154
Illumination
    and color, 46-48
    level, 46
    recommended, of selected areas, 48
Immediately dangerous to life and health (IDLH), 93
    Mining Enforcement and Safety Administration (MESA), 274
Incidence rates, 29-33
Indoor air quality, 112, 115-16
    biological contaminants, 125
    guidelines, ASHRAE, 124
    humidity, 113-14
    temperature and humidity, 111-115
    ventilation, 116-25
Industrial lift trucks, 395-399
    ANSI and OSHA requirements for, 395-96
    NFPA Fire Safety Standard for, 396
    type designations, 400
    and whole body vibration, 62
Industrial Robots and Robot Systems: Safety Requirements, R15.06, 15, 296
Injuries, cost of disabling unintentional, 4
Injury Facts, 5
Inspections, OSHA, 19
Integrated Fire Alarm Mass Notification System (FA/MNS), 168
Integrated Public Alert and Warning System (IPAWS), 9
International Organization for Standardization (ISO), 15
Ionizing radiation, 99-100

ISO
    7010 Safety Sign and Symbol meaning "exit" and "egress" pictorial, 163-64
    31000 Standard, 3

J

Japanese Building Research Institute fire model, 211
Job-safety analysis, 62-63, 418
Job satisfaction and machines, 69

L

Labels, NFPA, 704, 279
Laboratory fume hoods, 117-22, 197
    ASHRAE 110 standard for testing and management practices for safe, 117
    quantity of air captured by various, 120-21
    recommended airflow rates (face velocities), 117
    types of, 117
Ladders, fixed and portable, 149-50
Laptops, and workstation guidelines, 78
Laser
    Food and Drug Administration's Center for Devices and Radiological Health (CDRH) classifications, 266
    machining, 409
    radiation, 265-67
    safety glasses, 86
    warning sign for Class IIIb and Class IV, 267
Leadership in Energy and Environmental Design (LEED) Certification (USGBC), 9
Liberty Mutual Insurance, 6
Life safety, 159
    and building design, 164-67
    design analysis, 161
Life Safety Code (LSC), 14, 160-61
    common path of travel, 169
    criteria for determining the actual required egress capacity for stairways and level components, 171
    dead-end aisles, 171
    maximum distance to exit, 169
    performance-based fire safety design option, 188
    specification of egress components,
Lifelines, rope materials for, 103
Lifting, 53, 55-59. *See also:* Back belts
    devices, 56
    "Work Practices Guide for Manual Lifting" (NIOSH publication 81-122), 57
Light
    intensity of, 45
    radiation exposure to eyes, 86
Lighting
    ceiling, 47
    fluorescent, 47
    glare, 76

high-intensity-discharge, 46
    optimal, for computers, 76
Lightning, 325-26
Lockout/tagout, 1910.147, 300, 418
    of de-energized electric circuits, 322-23
Lost-time frequency rate, 28-30
Low carbon steels, ASTM A7 and A36, 189
Lowrance, Dr. William W., 3
Luminance, 45-46
    background, 45
    and illumination level, 46-48

M

Machining of wood and metal, 406-409
    grinders, 408
    material considerations, 406
    tools, 407, 409
Maintenance, 11
    electrical, 312
Management commitment to safety, 6-7
Maslow, Abraham, hierarchy of human needs, 70-71
Mass Notification Systems (MNS), 15
Material handling
    automatic storage and retrieval systems, 400-401
    conveyors, 392-93
    cranes and hoists, 393
    manual, 59, 63, 80, 391-92
McGregor, Douglas, concepts of worker motivation, 70
Means of egress, 160
Meany, George, 8
Mechanical hazards
    classification and control methods, 289-97
    definitions, 291
Metatarsal guards, 90
Mine(s), 176
    Federal Bureau of, 16
    Sago, methane explosion, 255
Mining Enforcement and Safety Administration (MESA), IDLH gas concentrations, 274
Mirrors, use of at intersections, 170
Monte Carlo modeling, 34, 38
MORT (Management Oversight and Risk Tree), 38-39
Motivation, worker, 70
    Theory Z, 70
Multi-employer work sites, 22
Musculoskeletal disorders
    back injury, 57
    from computer use, 78-79
    task analysis tools, 62
    work related, 54-63
    wrist, 60

N

Nanoparticles, 125
    PPE for exposure to, 98

National Bureau of Standards, calculation of egress time, 162
The National Council on Radiation Protection and Measurements (NCRP), 262
National Electrical Code (NEC), 14. *See also:* National Electrical Safety Code (ANSI C2)
    classes of hazardous locations, 307
    maximum safe current of conductors, 305
National Fire Protection Association (NFPA), 14, 412
    codes pertaining to design of facilities to accommodate explosives or explosive material, 255
    definition of explosion, 247
    fire-resistive and noncombustible construction, 189
    Life Safety Code (NFPA 101), 160-62, 189
    NFPA 13, Installation of Sprinkler Systems, 205
    NFPA 30, 194
    NFPA 51B, 412
    NFPA 60, Standard on Explosion Prevention Systems, 247
    NFPA 68, Guide for Venting of Deflagrations, 256
    NFPA 72-2010, 15, 168, 198
    NFPA 70B, The Electrical Equipment Maintenance Code, 322
    NFPA 70E, 101, 307, 314, 322
    NFPA 80, "Standard for Fire Doors and Fire Windows", 166
    NFPA 92A, "Smoke Control Systems", 210
    NFPA 92B, "Smoke Management Systems", 162, 210-211
    NFPA 101A, "Alternative Approaches to Life Safety", 162
    NFPA 170, "Standard for Fire Safety Symbols,' 163
    NFPA 220, 189
    NFPA 484, the standard for combustible metals, 256
    NFPA 505, "Fire Safety Standard for Powered Industrial Trucks", 396
    NFPA 654, the standard for prevention of fire and dust explosions from manufacturing, processing, and handling of combustible particulate solids, 256
    NFPA 704, 279-81
    NFPA 780 for lighting protection, 336
    NFPA 2112, protective clothing, 314
    reactivity symbol for materials, 248
    shocks and electrocutions, U.S., 311
    symbols for identification of hazardous materials, 281
National Institute of Occupational Safety and Health (NIOSH)
    contactlensuse recommendations, 87
    eye injury estimate, 85

fume hood airflow rates (face velocities), 117
Guide to Industrial Respiratory Protection, 93
hearing protector noise attenuation, 97
lifting guide, 57-59
Nanotechnology Program, 125
particulate filter classifications, 91
Prevention through Design (PtD) Initiative, 25, 415-16, 420-422
product certification, 25
Publications on back belts, 81-122 and 94-122, 57
respirator certification, 99
noise exposure recommendations, 95-96, 217
study on safety practices, 6-7
National Institute of Standards and Technology (NIST), Fire Dynamic Simulator, 162
National Safety Council (NSC), 5, 142
data collection and reporting, 29
electrocutions in U.S., 311
Family Safety and Health magazine, 9
report on construction worker fatalities, 10
NFPA 101. *See:* Life Safety Code (LSC)
Nip points (or in-running), 291-94. *See also:* ANSI/ASME 20.9-2009, "Safety Standards for Conveyors and Related Equipment" *and* Guards
belt conveyor, 292
chain and sprocket, 392
guarding, 292-93
Noise, 94-98. *See also:* Acoustical materials
enclosures, 240-42
exposure standards, 224-28
induced hearing loss (NIHL), 94-95, 217
measurement, 222-24
PEL limits, 217
reduction coefficients (NRC), 233-34
reduction ratings (NRR) for earplugs, 97
surveys, 228-31
Noise control, 235-45
at the receiver, 243
at the source, 235-39
in the path of transmission, 239-43
Nonhazardous materials, 286-87

O

Occupant load, 169
Occupational Safety and Health Act, 17-22
General Duty Clause, 18
performance vs. specification standards, 17-18
Occupational Safety and Health Administration (OSHA)
1910.106, 194
1910.109 for explosive magazines, 254
1910.147, Control of Hazardous Energy (Lockout/Tagout) Standard, 300, 323

1910.178, requirements for powered industrial lift trucks, 395-96, 398
1910.217, 368
1910.242(b), standard on compressed air, 299
1910.1000, air contaminants standard, 273
1910.1096, radiation regulations, 99
1910.1200, HazCom Standard, 21,196, 278
1926.452(j), 434
Air Contaminants Standard, 1910.1000, 115, 273
Cancer Policy (Part 1990), 17, 285
citations, 20
fall protection requirements, 104
filter lenses for protection against radiant energy, 87
fume hood airflow rates (face velocities), 117
inspections, 19-21
noise standard (29 CFR 1910.95), 96, 217, 225-28
regulations on contact lens use, 87
requirement for workplace PPE assessment, 84
respiratory protection standard (29 CFR 1910.134), 94
safety data sheet format, 280
Standard for Confined Spaces (29 CFR 1910.146), 176
standards for fixed ladders, 149
Subpart S, electrical hazards, 322
VPP program, reduction in injuries, 6, 23
Ohm's Law, 307
Organic solvents, absorption though skin, 88
OSHA 18001 Standard, 3
Ouchi, W.G., Theory Z, 70

P

Particulate filters, NIOSH classifications, 91
PE registration, 4
Performance appraisal, 11
Permissible exposure limits PELs (OSHA), 90, 112, 115
calculation for multiple exposures, 275
Personal fall arrest systems, 102-104
Personal Protective Equipment (PPE). *See also:* Personal fall arrest systems
for arc flash and electrical hazards, 100-102, 316, 320
barrier creams, 89
chemical protective clothing and gloves, 88-89
for cryogenic liquids, 99
earmuffs, 97
earplugs, 96-97
employer responsibility to pay for, 85
eye and face protection, 85-87
hazardous materials, 286
head and foot protection, 89-90

heat and cold, 89
for ionizing radiation, 99-100
for nanoparticle exposure, 98
noise, 94-98
OSHA requirements, 84
personal alarms, 107
radiation suits, 100
respiratory protection, 90-94, 99
safety glasses, lasers, 86
sharp objects and punctures, 89
Phons, 221
Pinch points, 289-92, 411
Platforms, temporary and movable, 150-52
Polycarbonate lenses for safety glasses, 85
Positioning sequence, in designing work methods, 52
Posture, high-stress, 62
Powered industrial lift trucks. *See:* Industrial lift trucks
PPE. *See:* Personal Protective Equipment (PPE)
Pressurized containers, 297-300
Prevention through design, 416. *See also:* Design
  case studies, 420-22
Probability, 34
Process hazard analysis (PHA), 256
Product certification, 23-25
Purging of containers, pipes and hoses, 252, 294

R

Radiation, 88
alpha and beta particle, 100, 261
background, 262
common sources of, 260
dose and dose limits, 262
electromagnetic spectrum, 259-60
exposure limits and guidelines, 99
gamma and X-rays, 100
ionizing, 99, 261-64
low-frequency, 268
non-ionizing, 264-68
units, 261-62
UV, 85, 125, 248, 267-268
warning symbol for ionizing radiation, 263
Railings, 143-45
for robots, 296
Ramps, 145-46, 167
Raynaud's Syndrome, 59, 62
Recommended exposure limit (REL), NIOSH, 96, 115
Record keeping, 28-33
Repetitive motion disorders, 59
Requirements for Packaging Machinery and Packaging-Related Converting Machinery, 15
Resource Conservation and Recovery Act (RCRA), (40 CFR 261), 278, 284

Respirators. *See:* Respiratory protection devices
Respiratory protection devices, 90-94
air-purifying, 91-93
chemical cartridges, 92
fit test, 92
supplied air, 93
Risk, 3
assessment, 359-62, 365, 418-19
establishing a level of, 364-65
management, 27-28
quantitative limits of, 27
Robots
and human interfaces, 295
railings around, 296
welding, 356
Ropes, 103

S

Safe, 2
Safeguarding of machinery, 359. *See also:* Guards *and* Risk assessment
Safeguards. *See also:* Guards
case study, concrete block machine, 425-26
characteristics of, 368-69
control and system reliability, 386-88
interlocks, 427-428
two-hand, 411
Safety
attitudes towards, 6-11, 307-308
cans, 193, 196
communication between employees and management, 11
credentials, 11
culture, 6-7, 10, 411
data sheets, 290
engineering, defined, 4
engineering controls, 420
glasses, 85
and health standards, mandatory, 15-16
hierarchy, 365
and risk, 3
and security, 15
Safety glasses, 85-87
and chemical protection, 86
filter lenses against radiant energy (OSHA), 87
with lasers, 86
Safety standards
European, 15
product, 430-31
voluntary, 430
Scaffolds, 150-52
pump jack, 433-34
SCBA, 93. *See also:* Respiratory protection devices
Scheduled charges for traumatic or surgical loss of member and impairment, 30
Schott, C. Donald, 6-7
Scissor lift, 153, 296

Severity rate, 2

SFPE Handbook of Fire Protection Engineering, egress of occupants in, 168

Shear points, 291, 293, 411

Showers, emergency, 105-106

Sick Building Syndrome (SBS), 124-25

Signals, visual and auditory, 49
    Class IIIb and Class IV lasers, 267
    NFPA identification of hazardous materials, 281

Signs and markings, 132
    on guards, 371
    ionizing radiation, 263
    poison, 282
    slow-moving vehicle, 400

Silicosis, 91

Slip(s), 136-41
    human factors, 134, 136-138
    resistance, 136-37, 141. *See also:* Tribometry testers, 137-40
    tile floor case study, 436-37

Society of Fire Protection Engineers, "Introduction to Performance-Based Fire Safety", 188

Society of Risk Analysis, 3

Sound(s)
    energy absorption, 232-235
    level meters, 223-25
    loudness, 221
    masking, 49
    octave bands, 220
    physics of, 217-19

Spinal column, 54

Sprinkler systems, 201-208
    heads, 203
    NFPA Standards 13, 13D, and 13R, 201

Squeeze points, 289-91

Stair(s)
    design, 146-49
    discharge rates on, 169-70
    flow rates on, 169-70

Stairwells, 166-67, 195

Static electricity, 323-326
    grounding, 324

Steinbuch, Dr. K., 72

Stevenson, William J., 257

Stimulus-operator-response (SOR) model, 64
    accident analysis using, 67-68

Storage racks, 399-403

Strains and sprains, 54-55

Stress
    and accidents, 73
    heat and cold, 113
    from noise, 217
    physical, 72
    sensory overload and, 72

Superfund Amendments and Reauthorization Act (SARA) 1986, 285

Sustainability, 10

USGBC Leadership in Energy and Environmental Design (LEED) Certification, 9

Systems engineering, 34

Systems safety
    analysis, 63-70
    engineering, 33
    worker-machine, 64-68

T

Tactile capabilities, 50

Task analysis tools, 62

Temperature and humidity, controlling indoor, 112-15

Tendinitis, 54-55, 59

Terrorist attacks, effects on codes and standards, 15

Theory Z, 70

Threshold limit values (TLVs)
    ACGIH, 115
    defined, 273

Time-weighted average (TWA), 96, 273-74

Tinnitis, 217

Toeboards, 144

Total heat load, 113

Toxic substances. *See also:* Hazardous materials
    PEL calculation for multiple exposures, 275

Training, 11
    NFPA 70E, 322
    OSHA required PPE, 84
    safety, 74-75
    29CFR1910, 17-18

Tribometers, 138
    variable incidence, 140, 436

Tribometry, 136-37

Trips and falls, 134-36
    sidewalk case study, 438-39

Trivex® safety glass lenses, 85

U

Underwriters Laboratories, Inc., 25
    UL 858, Safety for Electric Ranges, 433

Uniform Federal Accessibility Standards, 16

U.S. Chemical Safety and Hazard Investigation Board (CSB), 255

U.S. Military Design Standards
    MIL-STD-1472D for design of ramps, stairs, and ladders, 145-46, 149
    MIL-STD-882C, 363

UV
    sunlight protection, 86
    radiation from welding, 85

V

Ventilation, 116-25. *See also:* Airborne contaminant(s) *and* Sick Building Syndrome
    desiccant-based system (EPD), 114-15
    ducts, 121

fans, 122
    hazardous gases, 283
    portable exhaust, 118
    pressure loss, 119-22
    standards for indoor air quality, 113
Vibration, 61-62
    and air filtration, 125
    ANSI guide, 18, 53, 61
    isolation mounting device, 243
    from lift trucks, 62
    PPE protection against, 89
    whole-body, 61-62
Vibration syndrome, 59
Video registration and analysis (VIRA), 62
Vinyl chloride, 285
Violations, OSHA, 20
Visionand reflectivity of colors, 45
Visual acuity, 44-45
Visual signals, 49
Volatile organic compounds, 125
    acceptable ASHRAE guidelines for, 1
Volts, 304
VPP program, 6, 23

W

Walking surfaces, 134, 141-42
    maintenance of standing and, 152-54

Walsh-Healey Act, 16
Warnings, 132
WBGT Index, 113
Welding, 412
    eye injuries, 85
    robots, 356
White fingers. *See:* Raynaud's Syndrome
Whole-body vibration, 62
Williams-Steiger Occupational Safety and Health Act of 1970. *See:* Occupational Safety and Health Act
Work area, normal and maximum, 51
Worker-machine systems, 63-70
    stereotyped reactions, 66
Workers' compensation laws, 4
Workers' Institute for Safety and Health, 8
Wrist flexing, 60

X

Z

Z244.1, Control of Hazardous Energy Lockout/Tagout and Alternative Methods, 15
Z359 Fall Protection Standard, 19, 21

NOTES

NOTES

NOTES

NOTES

NOTES

NOTES

NOTES